Contents

Chapter 3 Farming systems and husbandry practice
Niall Bromage, Jonathan Shepherd and John Roberts
(section 3.6)

Chapter 4 Propagation and stock improvement
Niall Bromage

INTENSIVE FISH FARMING

EDITED BY

C. Jonathan Shephe

BVSc, MSc, PhD, MRCVS

AND

Niall R. Bromage

BSc, PhD

b

**Blackwell
Science**

© C. J. Shepherd and
N. R. Bromage 1988, 1992

Blackwell Science Ltd
Editorial Offices:
Osney Mead, Oxford OX2 0EL
25 John Street, London WC1N 2BS
23 Ainslie Place, Edinburgh EH3 6AJ
350 Main Street, Malden
  MA 02148 5018, USA
54 University Street, Carlton
  Victoria 3053, Australia
10 rue Casimir Delavigne
  75006 Paris France

Other Editorial Offices:

Blackwell Wissenschafts-Verlag GmbH
Kurfürstendamm 57
10707 Berlin, Germany

Blackwell Science KK
MG Kodenmacho Building
7-10 Kodenmacho Nihombashi
Chuo-ku, Tokyo 104, Japan

Iowa State University Press
A Blackwell Science Company
2121 S. State Avenue
Ames, Iowa 50014-8300, USA

First published 1988
Reprinted 1989, 1990
Reissued in paperback 1992
Reprinted 1995, 1996, 2001

Set by Setrite Typesetters Ltd., Hong Kong
Printed and bound in Great Britain by
MPG Books Ltd, Bodmin, Cornwall

DISTRIBUTORS

Marston Book Services Ltd
PO Box 269
Abingdon Oxon OX14 4YN
(*Orders*: Tel: 01235 465500
        Fax: 01235 465555)

USA and Canada
Iowa State University Press
A Blackwell Science Company
2121 S. State Avenue
Ames, Iowa 50014-8300
(*Orders*: Tel: 800-862-6657
        Fax: 515-292-3348
        Web www.isupress.com
        email: orders@isupress.com

Australia
  Blackwell Science Pty Ltd
  54 University Street
  Carlton, Victoria 3053
  (*Orders*: Tel: 03 9347 0300
          Fax: 03 9347 5001)

A catalogue record for this title
is available from the British Library

ISBN 0-632-03467-X

Library of Congress
Cataloging-in-Publication Data

For further information on
Blackwell Science, visit our website:
www.blackwell-science.com

Chapter 5   Nutrition and growth
Takeshi Watanabe

Chapter 6   Fish health and disease
Jonathan Shepherd

Chapter 7 Intensive marine farming in Japan
Takeshi Watanabe

Chapter 8 Fish culture in the United States
Nick Parker

**Chapter 9   The development of polyculture in Israel:
a model of intensification**
Shmuel Sarig

Chapter 10   Commercial development and future prospects
Jonathan Shepherd

Foreword

Intensive aquaculture is a very new phenomenon. Nearly twenty years ago, when Dr Shepherd and I were first beginning our involvement in the new industry of intensive aquaculture here at Stirling, it consisted of little more than a relatively small intensive rainbow trout industry and a lot of hope.

Now, almost twenty years later, we have a remarkable growth in both technology and production. Intensive aquaculture, whether salmon, trout, catfish, eel or tilapia, is now a significant source of high quality protein.

The success of the intensification process in aquaculture was initially dependent principally on the achievement of the nutritionist in formulating complete diets which were adequate for growth in the context of economic cost and the engineer in terms of water management and control. Recently, however, the various disciplines of applied animal biology have begun to make a very significant contribution, so that strain selection, reproductive cycle control, vaccine prophylaxis, gender definition and genetic engineering have all started to make their contribution to the increased efficiency of the industry.

Edited by two aquaculture experts from the academic and commercial worlds, who have been very much part of this development process, this volume seeks to provide a basic introduction to the concept of intensive fish culture and to indicate not only what is possible now, but where the future may be leading. Four fifths of the surface of our planet is water. Culture of these waters for the benefit of mankind is only beginning. This book should provide a milestone in that development.

Professor Ronald J. Roberts BVMS, PhD, MRCVS, FRSE, FIBiol
Director, Institute of Aquaculture,
University of Stirling, Scotland

Contributors

Dr C. J. Shepherd, Promotora de Recursos Marítimos SA, 15126 Merexo, Muxía, La Coruña, Spain.

Dr N. R. Bromage, Institute of Aquaculture, University of Stirling, Stirling, FK9 4LA, Scotland.

Dr N. C. Parker, US Fish and Wildlife Service, Southeastern Fish Cultural Laboratory, Marion, Alabama 36756, USA

Dr J. K. Roberts, BP Nutrition (UK) Ltd, Shay Lane, Longridge, Preston, Lancashire, PR3 3BT, UK

Mr S. Sarig, c/o Laboratory for Research of Fish Diseases, Nir David, DN 19150, Israel

Professor T. Watanabe, Laboratory of Fish Nutrition, Tokyo University of Fisheries, Konan 4−5−7, Minato-Ku, Tokyo 108, Japan

Acknowledgements

Firstly the editors are immensely grateful to the authors of the contributed chapters for their time and forbearance during the long gestation of this volume. It is also a pleasure to acknowledge the kindness of various individuals who have read and commented on particular sections. These have included Professor C. Y. Cho, Dr V. O. Crampton, Dr J. A. K. Elliott, Dr B. J. Hill, Dr D. J. Jones, Dr R. G. Kirk, G. W. Mace, Dr C. E. Nash, Dr D. W. B. Sainsbury and the postgraduate students at Aston University, as well as Professor R. J. Roberts who wrote the foreword.

The following graciously allowed us to use photographs and other material additional to that supplied by the authors: Agra-Europe (London) Ltd, *Aquaculture Magazine*, Dr R. Bootsma, BP Nutrition Ltd, British Trout Association, Bridgestone Corporation, Elsevier Publishers, Fischtechnik, Dr D. Ford of Aquarian Ltd, Golden Sea Produce Ltd, Mr R. W. Harrison, Mr R. Hughes, Institute of Freshwater Research, Drottningholm, Azienda Agricola Mandelli, Mr A. Purves, Mr C. G. Saunders-Davies, The Scottish Salmon Information Service, Mr R. R. Watret, Mr E. J. P. Wood and Mr C. Woods.

The cover illustration is by Graham Smith based on an idea from N. R. Bromage.

Last but not least we wish to thank Jean Macready, Julia Virley and Muriel Warwick for typing assistance and Graham Smith, Nigel Haverly and Terry Weir for help with the graphics, together with our families who supported us throughout.

C. J. Shepherd
N. R. Bromage
1988

Note

Exchange rate March 1988:
£ Sterling 1 = US$ 1.85 = ¥ 236

(see chapters 5, 7, 8, 9, and 10)

---

1 What is fish farming?

1.1 INTRODUCTION

This book is about the intensive cultivation of finfish and has been written in order to give an up-to-date picture of how this branch of the 'aquaculture' industry is developing worldwide. The cultivation of aquatic organisms under controlled conditions has now reached the stage where the annual world harvest from aquaculture is over 10 million tonnes compared with a total fish harvest of about 85 million tonnes. This 1986 estimate means aquaculture has grown at a compound rate of 10% per year since the 1973 estimate of about 3 million tonnes. Although we shall concentrate on finfish farming, which annually produces well over 4 million tonnes, it is important to remember that over half of the world's aquaculture production is shellfish (especially mussels and oysters) and edible seaweed. Finfish are farmed for various purposes, including the restocking of angling waters and as ornamental or bait fish, but it is the farming of fish for human food which is of prime importance.

Like conventional livestock production, fish farming is not new, but both have benefited from man's technical progress during the last hundred years or so, which has turned the art of animal husbandry into a science. Like the intensive poultry farmer, the modern fish farmer is increasingly moving towards purpose-built rearing systems with high stocking rates in order to achieve greater efficiency. Although meaningful comparisons of productivity are complicated, fish have certain advantages over land animals in their suitability for farming. Being cold-blooded, fish do not have to expend energy in maintaining body temperature. Also they do not have to support their weight, unlike land animals, and should therefore be inherently more efficient at converting food into flesh, while a fish farm has the added advantage of being a three-dimensional rearing area.

In general fish have a lower proportion of inedible bones and offal and there is, of course, growing interest in their health advantages as a food source compared with poultry and red meat. Given these various factors, it is no surprise that fish farming is now making such a large contribution to world fish supplies.

1.2 HISTORY

Fish farming was probably first practised as long ago as 2000 BC in China and in 475 BC Fan Lai produced his Chinese treatise on carp culture. The Bible refers to fish ponds and sluices (Isaiah chapter 19, verse 10), while ornamental fish ponds appear in paintings from Ancient Egypt. The Romans started a type of coastal fish farming which still exists in Italy. Later on during mediaeval times carp were stocked in castle moats and monastery stew ponds offering a ready supply of fresh fish throughout the year.

If the Chinese first started fish farming, mainland China remains the largest producer of farmed fish to this day, accounting for almost half of the world output with approximately two million tonnes of various carp species farmed each year. Carp farming gradually spread throughout much of Asia and westwards into Europe. During the 1930s European settlers introduced carp farming into Palestine while Brazilian biologists discovered how to bring fish into spawning condition using pituitary hormones. Around the same time trout farming became established in Denmark and elsewhere leading more recently to salmon farming.

Marine fish farming probably started in Indonesia around 1400 AD when young milkfish were first trapped in coastal ponds at high tide. Milkfish farming is still a major industry, particularly in Taiwan. However, much of the recent expansion in marine fish farming has occurred in Japan, starting with the upsurge in yellowtail farming during the 1960s.

In assessing the current status of fish farming, it is helpful to recall the recent expansion in world fisheries overall. After the second world war, the global fish catch increased rapidly from less than 20 million tonnes to about 65 million tonnes, before coming to a virtual halt in the early 1970s. A combination of escalating fuel costs and declining wild fish stocks has resulted in a current growth rate of less than 8% per year for the fishing industry worldwide and an annual global catch of around 85 million tonnes. This compares with a current growth in world population approaching 2% per year. Although it varies according to geographical region, at the present time fish represent about 6% of world protein supplies on average and about 24% of animal protein supplies, including that proportion − roughly a third − of the fish catch made into fishmeal for animal feed.

1.3 DIVERSITY

Compared with land animal farming, fish farming is a much more varied activity mainly because there are many more species of fish farmed, each with different characteristics. Whereas some fish are marine and others need freshwater, many fish (e.g. mullet) can be farmed just as readily in brackish waters. Certain migratory fish, such as salmon, breed in rivers where the eggs hatch and the fry live until ready to go off to the sea before eventually returning to spawn as mature adults; so the salmon farmer needs a freshwater hatchery as well as a seawater farm to rear the stock up to marketable size. Different species are adapted to live at different temperatures and tropical fish usually cannot be farmed in temperate climates unless the water is warmed artificially. Fish also vary greatly in their food requirements ranging from carnivores like eels or salmon, which require a high proportion of animal protein in their diet, via omnivores with less selective tastes, to herbivorous fish, such as grass carp. A further complication is the early part of the life cycle which may be straightforward or involve a variety of different larval stages, each with different food requirements.

It was largely to exploit the different food preferences of different species that the practice of 'polyculture' has come about, consisting of a combination of different fish species within the same production unit. This contrasts with 'monoculture' of a single species. In addition both systems may be integrated into other aspects of agriculture; for example, fish may be reared within flooded rice paddy fields or ducks may be reared in association with fish ponds.

A fish farmer may undertake the entire production cycle from spawning of mature adults through early-rearing and on-growing to production of marketable fish. Alternatively the cycle may be short-circuited in various ways, for example by using supplies of wild juvenile fish if they cannot be bred in captivity (Fig. 1.1). In

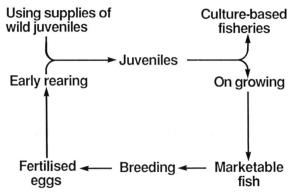

Fig. 1.1 Diagram of aquaculture operations.

the case of migratory fish it may even be possible to release hatchery-reared juveniles into the wild and rely on their homing instinct for recapture. This is sometimes called 'ranching' and is a variation of 'culture-based fisheries' in which wild fish populations are boosted by hatchery-reared juveniles.

The diversity of fish farming is also shown in the widely varying levels of intensification adopted, each of which may be entirely appropriate to prevailing local conditions. Thus at one extreme, very extensive fish farming systems may offer only marginal improvements on natural levels of fish productivity in the wild. For instance coastal lagoons may be stocked with wild-caught fry which are managed as they grow by removing predators, whereas feed requirements are met by the natural supplies of animal and plant life in the lagoon with no supplementary feeding by the farmer. At the other extreme highly intensive systems are designed to allow close control over the environment, feed requirements are met entirely by specially formulated diets and high yields of fish are harvested in a predictable fashion. Between these two extremes lie the majority of finfish-rearing systems, for example standing-water ponds to which fertiliser is added, together with some supplementary feeding, in order to enhance natural productivity within the ponds resulting in high yields of species such as carp, either alone or more commonly with other species in polyculture systems. These are often described as semi-intensive (or semi-extensive) systems.

As much as half of global finfish farming may be accounted for by various types of carp, in particular Chinese carp grown by polyculture. Much of this extensive and semi-intensive farming is not traded and represents subsistence production, but yields of up to 5 tonnes per hectare are commonly achieved by managing the ponds to make maximum use of the available ecosystem. This entails using organically fertilised ponds and varying the combination of species and stocking rates, with supplementary feeding undertaken only if the resulting increase more than offsets the cost of feed. Chinese carp polyculture is a highly specialised art of long tradition and perhaps the only significant improvement in recent times has been the adoption of induced spawning methods. The resulting industry is nonetheless still the most successful means of finfish farming in the world. Similar farming systems have consequently spread to India and Bangladesh which rear Indian major carp within polyculture ponds. Russia and Eastern Europe also have well established finfish rearing enterprises based on common carp and related species, but climatic and other factors have resulted in their having generally disappointing levels of productivity. To underline the global importance of farmed carp, nearly 40% of the world's annual freshwater fish harvest comprises farmed fish, mainly carp species.

Although nearly 75% of global finfish farming uses freshwater, brackish water and marine finfish farms now contribute approximately 1 million tonnes per annum. For instance, milkfish farming in Indonesia, the Philippines and Taiwan is traditionally centred on milkfish captured as fry in coastal ponds at high tide. Milkfish feed on algae which are therefore encouraged to grow by fertilisation and various techniques of pond management. Although it is not yet possible to induce spawning in milkfish, they probably represent the most commonly farmed coastal species in the world.

Notwithstanding the major proportion of global freshwater and coastal fish farming being conducted along the lines described above for carp and milkfish, recent development in countries such as Israel, Japan and the USA has been somewhat different. The evolution of finfish farming in the more industrialized market economies has become generally characterized by some or all of the following factors:

1. Use of purpose-built rearing units for each stage in the production cycle.
2. High stocking rates to achieve maximum yields of marketable fish in relation to available rearing space.
3. Use of scientifically formulated diets, usually in the form of a compounded pellet.

4. A high degree of automation applied to husbandry operations such as feeding, grading and harvesting.

5. Close management control over the production cycle, wherever possible stretching from egg via larval rearing to marketable fish and broodstock.

As a result, Japan and Taiwan probably have the world's most advanced and varied fish farming industries with fish such as yellowtail, red seabream and sea bass reared in cages and eels, milkfish, tilapia and mullet reared in brackish water ponds. Taiwan has experienced a more than three-fold increase in farmed output over the last decade due largely to intensive farming systems. Similar improvements are now being adopted in other parts of South East Asia (e.g. Thailand and Malaysia) with the farming of groupers, sea bass, etc. Parallel developments are also being seen in freshwater fish farming. Thus whereas Chinese polyculture skilfully exploits various ecological niches within fertilised ponds without any supplementary feeding, Israeli fish farms utilise the same principle (with a combination of common carp, tilapia, grey mullet and silver carp), but achieve significantly higher yields by using pelleted diets and pond aeration systems.

To sum up, the diversity of fish farming is a reflection of the very different types of fish farmed and the extent to which the farmer undertakes the entire cycle of operations within his farm. However, it also reflects the varying levels of intensification which may be adopted and it is the more intensive methods which will be described here in greater detail.

1.4   SELECTION OF SPECIES FOR FARMING

Over 20 000 different species of fish have been described but less than 100 are farmed commercially. In selecting those species of fish suitable for farming, various biological and economic factors are of particular relevance as follows:

1. Market price and demand: the farmer must be able to achieve a selling price which more than compensates for the various costs of production, while the volume of demand must be able to take up farm output without excessive price erosion. This may have to take account of potential competition from other farms or indeed from wild fish supplies.

2. Ability to feed for fast growth with efficient conversion: it is important to understand the nutritional requirements of the species in question. This enables the farmer to formulate diets in order to achieve fast rates of growth and an efficient conversion of feed into flesh. Although many fish farmers rely on fertilising the pond to encourage the growth of plankton and other organisms as fish feed, truly intensive farming increasingly relies on the sole use of artificial pelleted diets.

3. Hardiness and adaptability to heavy stocking: farmed fish must respond well to the stress of confinement and crowding. If high stocking densities lead to fighting, cannibalism or outbreaks of disease, it is unlikely that farming will be viable.

4. Simple larval development: the combination of a large egg and a single larval stage makes it comparatively easy to hatch and then bring onto feed such fish as trout. By contrast many marine fish have microscopic eggs and a series of different larval stages each needing different types of microscopic food. It is for this reason that progress in farming certain marine fish has been so slow.

5. Ability to reproduce in captivity: having to depend on catching wild fry in order to stock fish farms is a serious disadvantage. Discovering the hormonal basis of fish reproduction has made it possible to spawn certain species in captivity. However, it is still not possible to spawn eels or milkfish artificially and fluctating supplies of young stock can be a problem.

Table 1.1 lists the main types of finfish currently farmed, classifying each according to whether this takes place under freshwater, brackish water, or truly marine conditions. Certain families appear under more than one category because so-called 'anadromous' species of wild salmon, trout, ayu, striped bass or sturgeon normally breed in freshwater but migrate to live at sea for a major part of their life cycle, whereas 'catadromous' fish, such as eels, do exactly the opposite. Other natural migratory patterns involve the adult fish breeding at sea before the young move into brackish estuarine conditions to grow, as happens with mullet and milkfish. A striking feature of certain fish species is their ability to live and thrive under different conditions. For instance mullet can withstand widely different salinity levels with the result that they can be farmed successfully in freshwater ponds as well as marine enclosures. This characteristic of adaptability, be it to crowding or changing water

Table 1.1 Principal farmed finfish

Mariculture	Brackish water culture	Freshwater culture
1. Eels *Anguilla* species	1. Cichlids *Sarotherodon* species, *Tilapia* species	1. Ayu *Plecoglossus altivelis*
2. Flatfish e.g. turbot (*Scophthalmus maximus*)	2. Eels *Anguilla* species	2. Catfish e.g. channel catfish (*Ictalurus punctatus*)
3. Groupers *Epinephelus* species	3. Flatfish *Pleuronectes platessa*	3. Cichlids *Sarotherodon* species, *Tilapia* species
4. Mullet *Mugil* species	4. Groupers *Epinephelus* species	4. Cyprinids e.g. common carp (*Cyprinus carpio*)
5. Pompano *Trachinotus carolinus*	5. Milkfish *Chanos chanos*	5. Eels *Anguilla* species
6. Porgy e.g. black porgy (*Mylio macrocephalus*)	6. Mullet *Mugil* species	6. *Heterotis niloticus*
7. Puffer *Fugu* species	7. Pompano *Trachinotus carolinus*	7. Labyrinth fish Anabantidae
8. Salmonids *Salmo* species, *Oncorhynchus* species	8. Porgy e.g. black porgy (*Mylio macrocephalus*)	8. Milkfish *Chanos chanos*
9. Sea bass *Dicentrarchus labrax*	9. Puffer *Fugu* species	9. Mullet *Mugil* species
10. Seabream *Sparus* species	10. Seabass *Dicentrarchus labrax*	10. Salmonids *Salmo* species, *Oncorhynchus* species, etc.
11. Tuna *Thunnus* species	11. Seabream *Sparus* species	11. Sea perch e.g. Nile perch (*Lates niloticus*)
12. Yellowtail *Seriola quinqueradiata*	12. Sea perch e.g. cock-up (*Lates calcarifer*)	12. Snakeheads *Ophiocephalus* species
	13. Yellowtail *Seriola quinqueradiata*	13. Striped bass *Morone saxatilis* 14. Sturgeons *Acipenser* species 15. Whitefish *Coregonus* species

quality, is perhaps the most important biological factor determining the suitability of a particular fish species for farming. It is no coincidence that the most commonly farmed group of fish, namely the cyprinid or carp family, is highly adaptable to fluctuating temperature, oxygen level, poor water quality, etc., and can be heavily stocked without apparent stress. If a particular species of fish becomes stressed when crowded under farm conditions, it is likely to stop feeding and consequently growth and survival rates will be poor.

In surveying the range of principal intensively farmed finfish it is notable that economic factors of high price and demand are common to many established fish farming industries, outweighing in importance any single biological feature such as artificial spawning or simple larval development. Eels cannot be reproduced in captivity and few marine fish have a simple larval development, but the incentive of a high market price has led to their successful farming. The corollary is that relatively lower priced fish, such as plaice, have been readily farmed on an experimental basis, but the potential profit has been insufficient to encourage commercial plaice farming to become established. This is because production costs, such as the cost of feed, are relatively high and so flatfish farming has focused instead on species like turbot which command a higher price, more than compensating for the various costs incurred by the farmer.

1.5 STATISTICS OF GLOBAL PRODUCTION

Statistical assessment of fish farming has been fraught with controversy, but current world production of farmed finfish is well over 4 million tonnes per annum (or at least double this figure if molluscs, crustacea and edible seaweeds are included to give overall aquaculture production). Farmed finfish thus represent around 5% of the annual global fish and shellfish catch at about 85 million tonnes (aquaculture represents around 12%, see section 1.1) and this proportion will undoubtedly increase due to the effect on growth in conventional fishing of escalating costs and reducing wild fish stocks. Figure 1.2 estimated the overall geographical split in farmed finfish for 1984 and the dominance of Asia in world production. The problem of accurately quantifying fish farming output is that certain governments are not able to collect accurate statistics and also that different countries have made different assumptions; for example, whether or not to include the fish catches from reservoirs and other waters stocked by hatcheries with young fry as farmed fish production. Unfortunately this can result in large statistical discrepancies, as is well illustrated by India, which the United Nations FAO estimated as producing *c.* 720 000 tonnes of

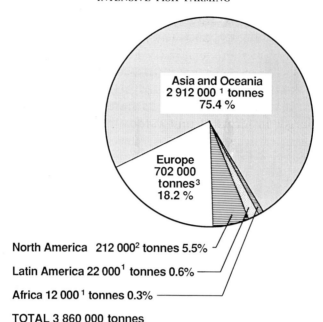

Asia and Oceania
2 912 000 [1] tonnes
75.4 %

Europe
702 000
tonnes[3]
18.2 %

North America   212 000[2] tonnes 5.5%

Latin America 22 000[1] tonnes 0.6%

Africa 12 000 [1] tonnes 0.3%

TOTAL 3 860 000 tonnes

Fig. 1.2 Estimated world production of farmed finfish by region for 1984 (after Huisman and Machiels, 1985; N. C. Parker, personal communication; FAO, 1985).

farmed fish in 1983. This compares with a far lower estimate of 131 000 tonnes by another analyst (T.−E. Chua) who re-examined Chinese and Indian aquaculture statistics to arrive at more conservative country estimates of finfish production, some of which are used in Table 1.2. Nonetheless, Chua concluded that world aquaculture increased at a rate of 10.5% per annum between 1972 and 1983 so that production had more than doubled over this period. It seems likely that global finfish farming has grown roughly in line with the growth in total aquaculture although different countries have emphasised different sectors. For example in Japan the growth of freshwater farming has stabilised and even declined at the same time that marine farming has been increasing steadily, as shown in Table 1.3. By contrast, freshwater finfish production in mainland China is currently expanding at an accelerating rate with little emphasis on high priced marine finfish.

Of particular interest is the variety of truly intensive fish farming industries which have already become established in various parts of the world, notably channel catfish, rainbow trout and yellowtail

(Table 1.4). Such farms rely entirely on artificial feeding, usually of higher-priced species, and whole new industries are currently emerging (e.g. grouper, seabream and various flatfish species). Although the volume of intensively farmed finfish at over 600000 tonnes per annum may currently represent only 15—20% of global farmed finfish on a weight basis, its significance is therefore considerably larger in terms of unit value.

Table 1.2 Main producers of farmed finfish

Country	Estimated production in 1983 in millions of tonnes	Principal species farmed
1. Republic of China	1.5*	Chinese carp; tilapia
2. USSR	0.3†	Carp; sturgeon; whitefish
3. Japan	0.3*	Yellowtail; seabream; salmonids; eels
4. Philippines	0.3*	Milkfish; tilapia; mullet
5. Indonesia	0.2*	Milkfish; tilapia; mullet
6. USA	0.2†	Channel catfish; salmonids
7. Taiwan	0.2*	Eels; milkfish; tilapia; mullet
8. India	0.1*	Indian carp
9. Bangladesh	0.1*	Indian carp
10. Romania	0.1†	Carp

*Chua (1986)
†Food and Agriculture Organisation (1985)

Table 1.3 Japanese production of farmed finfish from 1978—83 (tonnes)

Species	1978	1979	1980	1981	1982	1983
Yellowtail	121956	155053	149449	150907	146489	156170
Seabream	11315	12492	14973	18243	20648	25304
Horse mackerel	809	1461	2272	3195	3613	4266
Other marine fishes	928	1555	3023	2558	4258	5191
Trout	17166	16714	17698	17819	18230	17817
Ayu	7185	8455	7989	9492	10222	10318
Carp	29160	27452	25045	23784	24093	22397
Eel	32106	36781	36618	33984	36642	34489
Total	220625	259963	257067	259982	264195	275952

Table 1.4 Principal intensive finfish farming industries

Species	Estimated annual production in tonnes	Main countries involved
Atlantic salmon	40 000 (1985)	Norway, Scotland, Ireland
Channel catfish	134 000 (1982)	USA
Eels	30 000 (1985)	Taiwan, Japan
Polyculture (carp/tilapia/mullet)	11 000 (1985)	Israel
Rainbow trout	200 000 (1985)	Western Europe, USA, Japan
Red seabream	30 000 (1984)	Japan
Yellowtail	160 000 (1982)	Japan
Total	605 000	

1.6   EFFICIENCY, ECONOMIC DEVELOPMENT AND INTENSIFICATION

Herbivores are lower on the food energy chain than carnivores and so in biological terms they offer a more efficient means of producing animal protein. Plant protein is cheaper than animal protein and it should therefore be cheaper to farm herbivorous fish than carnivorous fish. But biological and economic concepts of efficiency do not necessarily coincide. If the fish farmer can get a much better price by farming carnivorous fish, then it may be worth his while spending more money on feed than would otherwise be necessary. In other words, any assessment of economic efficiency has to consider all the other costs of fish farming alongside feed costs. These may well include buying or hatching young stock, wages, water rates, pond construction and maintenance costs as well as the income generated by the resulting harvest. Clearly the economic viability of farming a particular species of fish hinges on a multiplicity of local factors, from the market demand for fish to the influence of site conditions on growth rate and construction costs, and also on the availability of suitably skilled personnel.

In comparing the economic efficiency of farming fish and land animals, it is tempting to study relative yields in relation to labour and land required. For instance, very intensive trout farms can produce over 100 tonnes of fish per man each year which compares favourably with conventional livestock units, whereas fish yields per unit area of 30 kg per square metre are higher than some of the most intensive pig or poultry units. However, such comparisons can be misleading and assume that either labour or land is the scarce resource on which maximum return must be achieved. In many arid parts of the world, it may be more appropriate to compare yield in terms of water requirements and fish may then represent a poorer return than other livestock. The situation is often complicated further by the possibility that swampy land or brackish water unsuitable for conventional livestock production may have potential for fish farming. An objective comparison of economic efficiency is best undertaken instead by costing all inputs (feed, stock, labour, etc.) and outputs (i.e. fish) and comparing returns on capital employed. This can be very complicated, particularly if market prices do not accurately reflect true costs and benefits, as is the case when evaluating many small-scale subsistence farms in the poorer regions of the world.

In order to compete effectively, farmed finfish must be offered to the market at similar or lower prices than the wild product. Also market size must be sufficient to absorb this increased supply without prices falling below farm production costs. Unfortunately much of the world's population has inadequate purchasing power for even low priced fish, whether farmed or wild. In practice this usually means that fish farmers in less developed regions either have to export to wealthier markets or use subsistence methods, thus reducing their efficiency of production. Outside Asia, fish farming has so far made a disappointing contribution to food production in such areas as Africa and Latin America, probably due as much to the lack of necessary husbandry skill as scarcity of capital and inadequate purchasing power. Various examples of successful subsistence fish farming in South East Asia show that small-scale and low intensity operations are not necessarily inefficient, particularly if they can be integrated with other agricultural activities. Indeed, more extensive fish farming methods represent a more efficient use of resources in less developed regions and intensive methods would be highly inappropriate under such circumstances. As the economies of such countries develop, however, so too can their fish farming industries be developed along more intensive lines.

The link between intensification of fish farm production and general economic development is well illustrated by comparing milkfish farming in Taiwan and the Philippines. Milkfish feed on algae and can be produced at relatively low cost in brackish water ponds under sub-tropical conditions. Taiwanese farmers achieve yields of over 2 tonnes per hectare, compared with about 700 kg per hectare in the Philippines due to intensive stocking with greater use of feed and fertilisers, etc. Although their production costs are higher, milkfish farmers in Taiwan receive higher prices and higher profits. The market price of milkfish in the Philippines is about 80% lower than in Taiwan, but it is still relatively expensive considering that the per capita income is about 110% lower than in Taiwan. Using per capita income as a measure of the level of the economy, milkfish is also more expensive to produce in the Philippines. Even for a particular species, fish farming can therefore evolve from an activity of subsistence level in poorer countries to become an efficient and highly profitable business in more developed countries. The latter have populations with the necessary purchasing power and hence market demand (as distinct from need) for fish. Their fish farmers in turn have the capital and buying power to invest in intensive farming methods. It follows that economic underdevelopment is a fundamental constraint on intensive fish farming. In theory this can be overcome by producing high price fish for export to wealthier markets elsewhere, but in practice this rarely happens without an inward flow of foreign capital. For example, Japanese companies operating in Taiwan helped to establish intensive eel farming which rapidly expanded in step with local economic development to the stage where 30 000 tonnes of farmed eels are now exported annually from Taiwan to Japan.

However, we are not so concerned here with extensive and semi-intensive (i.e. low-input) fish farming, despite its undoubted scope for producing high protein food and even generating foreign exchange in areas of real poverty and need − rather with commercial scale farming relying on high inputs of fish and feed to give correspondingly higher yields at harvest. In general it seems that the economic viability of fish farming is closely linked to management skill and to the intensity of operation. If the farmer has sufficient knowhow to raise and market his crop, other things being equal, he is more likely to succeed commercially the greater the degree of intensification. As stocking density increases, the farmer can exert more precise control over his stock. Closer management scrutiny allows him to optimise feeding rates giving more predictable growth rates, together with scope for planned harvesting regimes. As a result there is less feed wasted, unaccountable stock losses are fewer and the farmer can monitor and respond more readily to problems such as disease, predation, etc. Overall this tends towards a more efficient use of the capital tied up in fish, feed and the farm itself.

There are obvious limiting factors to intensification starting with the need to increase the level of artificial feeding. As the stocking density increases, oxygen levels decline and waste products accumulate from the fish. It may become necessary to increase flow rates and water exchange through the farm in order to maintain satisfactory water quality. If water is scarce it may become necessary to install recirculation and aeration systems to an extent where the cost becomes prohibitive. As the fish are crowded together, the risk of mass mortality due to system failures, such as blocked inlet screens, has to be taken into account, but this has to be weighed up against the likelihood of continuous unobserved losses from more extensive farms. An intensively managed fish farm therefore approximates to the controlled environment of an intensive poultry or pig farm. As far as possible the variables are brought within the farmer's control and the farming system is geared to achieve maximum yields under closely controlled conditions.

1.7 STRUCTURE OF THE BOOK

As shown above, farmed finfish represent less than half of total aquaculture production and intensive finfish farming is still a minor proportion of total finfish farming, particularly when measured by weight as opposed to value. This underlines the global dominance of more extensive methods of fish farming and their economic rationale throughout the less developed regions of the world. However, the technology is now available for farming an increasing range of finfish species by intensive methods akin to broiler chicken production. As with broilers, production economics favour intensification and those developed economies which can afford the necessary capital and technical back-up are now adopting intensive fish farming systems. The purpose of this book is to provide a detailed survey of the art and science of such intensive systems.

The approach chosen has been to examine firstly the biological requirements of fish as they relate to the feasibility of farming and the appropriate husbandry practice. In this connection the salmonid group of fish (salmon and trout) offers a useful model with exacting requirements covering both freshwater and seawater farming. Hence initial emphasis is on salmonid husbandry, with reference being made to other intensively farmed species where this is more appropriate (e.g. recirculation in eel farming).

Agribusiness development is sometimes said to revolve around the ability to 'breed, feed and weed'. So it is with intensive fish farming and current progress in the key technical areas of reproduction and stock improvement, nutrition, and disease control is therefore described in three separate chapters. These are followed by detailed descriptions of the development and current status of the intensive fish farming industries in Japan, the United States and Israel written by experts working in those countries. The Japanese chapter highlights the growth of marine finfish farming, whereas the US chapter concentrates on freshwater finfish, particularly channel catfish farming. The history of fish farming in Israel traces the development of intensive polyculture and brings out the underlying technical and economic factors.

To survive and prosper it is not enough that intensive fish farming is technically feasible. It must represent an economically efficient means of food production. The final chapter therefore focuses particularly on the economics and marketing of intensively farmed fish. This leads on to an exploration of the various constraints on further development and the book concludes by assessing the likely trends and prospects ahead.

2 Fish, their requirements and site evaluation

2.1 INTRODUCTION

In essence, fish farming is concerned with the transformation of inputs, such as eggs (or juvenile stock), feed ingredients and oxygen (via the water) into valuable outputs, i.e. marketable fish. This is summarised in Figure 2.1, which also draws attention to the need for economic inputs, such as capital and human labour. It is clearly impossible to gain any real understanding of practical fish husbandry without some appreciation of fish biology. The aim of this chapter is therefore to describe the structure and function of farmed fish with particular emphasis on their environmental requirements.

With intensively farmed fish the aquatic environment does not contribute to fish nutrition, as in other systems, but is nevertheless the vital medium in which the fish are stocked, fed, raised and into

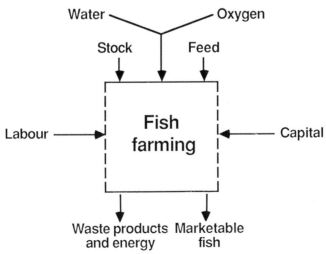

Fig. 2.1 The main inputs and outputs involved in fish farming.

which their waste products are voided. Of the various types of finfish intensively farmed, salmon and trout have probably the most exacting requirements for water quality and therefore the chapter has been written with special reference to salmonids. Other things being equal, the quantity of water available is usually the limiting contraint on how much fish can be produced at a particular site.

Hence any consideration of water sources for fish farming logically leads on to the problem of site evaluation. In addition to water, this must involve a brief introduction to such factors as land owner-ship, access, and suitability of the terrain for farm construction. The more detailed aspects of farm design and operation are ex-plored in subsequent chapters about specific fish farming industries.

2.2 FISH BIOLOGY

2.2.1 Preamble

Fish are cold-blooded or poikilothermic animals, their body tem-perature varying passively in accordance with the temperature of the surrounding water. Although fish as a group are tolerant of a wide range of temperatures, from just below 0°C up to 45°C, individual species generally have a preferred or optimum, as well as a more restricted, temperature range. Thus salmonids inhabit waters from 0−20°C although the optimum temperature for most species is 10−16°C. Changes in temperature within these ranges, however, significantly influence the biology of fish with the rates of all chemical reactions and processes within their bodies showing 50% increases with each 5°C rise in temperature.

The adoption of the aquatic habit has many other implications for the structure and physiology of fish. The 800−fold higher density of water, when compared with air, makes the streamlining and shaping of the body an important prerequisite for successful aquatic life, particularly if (like the salmonids) the animal is an active predator. Hence, most predatory fish, including salmonids, are ovoid in cross-section and torpedo-like or fusiform in shape.

The respiratory system assumes a greater significance for fish compared to terrestrial animals because water contains about one twentieth of the oxygen available in air, a proportion which is reduced still further by any increases in water temperature and/or ionic concentration. Fish are also exposed to much greater ionic and osmoregulatory challenges than terrestrial animals because their bodies are permanently immersed in a medium (i.e. water) which is not only the universal solvent but also the fluid in which gases must diffuse during respiratory exchange. Microbial infection and multiplication are also more likely to occur from the aqueous medium.

Some of the structural and physiological implications of an aquatic existence are considered further in the following description of the more important organ systems of salmonid fish. Figure 2.2 shows the position of these systems in trout and carp.

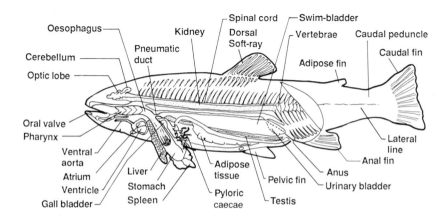

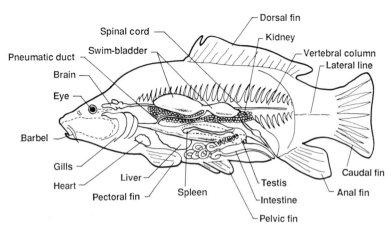

Fig. 2.2 Organ systems of trout (upper) and carp (lower).

2.2.2   The skin, musculature and locomotion

The external surface or skin of fish is composed of two layers, an outer epidermis and an internal dermis. It is from the underlying dermal layer that the characteristic scaly covering of fish is produced. The epidermis is a thin, fragile layer which is constantly sloughed off and renewed. It contains mucous cells which secrete the slimy outer covering of fish. This mucus protects the epidermal layer against abrasion and by lubrication makes the fish more streamlined and also difficult to hold. By rendering the skin less permeable, mucus also prevents entry of pollutant materials and microorganisms. The lipoprotein properties of the slime also tend to trap or bind heavy metals and bacteria, which are then lost as the mucus is washed off the surface of the fish.

The scales, which are composed of bone and connective tissue, grow in size in accord with other tissues of the body. Each scale overlaps the one behind producing a relatively impermeable but living cover. Growth rings or annuli are visible on the scales rather like those found in the main trunks of trees. Variations in thickness of these rings are produced by seasonal variations in diet or temperature or by spawning, and can be used to age fish. Unfortunately growth rings are not clearly seen in cultured fish because they are maintained under relatively stable conditions and fed at constant rates.

Much of the musculature of fish is composed of a series of muscle blocks arranged along the length of the body. Alternate contraction of the muscle blocks on each side of the fish flexes the body into a wave or S-shaped form. It is this wave form amplified by the tail which provides the major propulsive force for the fish. During this movement in most salmonids the head and anterior third of the body remain almost straight with the major flexion being restricted to the hind flanks and tail. This movement produces a normal swimming speed for salmonids of 5 mph although velocities three to four times this figure have been reported.

Aside from the tail or caudal fin, the remaining median and paired fins are controlled by muscles which are independent of the trunk musculature. Changes in position of the paired pectoral and pelvic fins allow braking and changes in direction and also provide lift. The median fins function primarily to prevent roll. Salmonids characteristically also have a further small median fin, known as the adipose fin, just anterior to the tail. In locomotory terms the function of this fin remains unclear. However, it is useful to fish farmers and fishery scientists wishing permanently to mark or identify individual fish or a group of fish because the adipose fin, once it is removed (easily accomplished with scissors or a razor without undue bleeding), is incapable of regeneration.

The only other structure which is of importance to locomotion is the swim- or air-bladder; this is an air-filled sac found immediately ventral to the spinal column (Fig. 2.2) which provides buoyancy for the fish. Occasionally, swim-bladders are incompletely filled and under such circumstances fish incline at an angle and swim vigorously just to maintain their vertical position in the water.

2.2.3   The sensory and nervous systems

As in higher vertebrates like ourselves, the nervous system of fish consists of a brain, spinal cord and nerves. The brain rarely fills the skull cavity and is small by human standards, only contributing 0.1–0.2% of the total weight of the fish as opposed to 2.0% for man. The brain itself is divided into three parts: fore, mid- and hind-brains which are concerned primarily with olfaction or smell, sight and coordination, and touch respectively. As might be expected, different parts of the brain and associated sensory systems are accentuated in different fish groups depending on the nature of their life style, in particular their modes of nutrition and movement.

Salmonids are active predators relying primarily on sight and secondarily on smell to seek out their prey. This sensory dependence is clearly evident from the characteristically large eyes and nasal apparatus of most salmonids. Their sight is particularly acute allowing simultaneous focusing of near and far objects, using what is effectively a lens with two separate focal points.

The sense of smell is also important for guiding anadromous species during their migration into fresh water to spawn. Often salmonids return to precisely the river or tributary of a river in which they themselves were spawned several years earlier. This unerring ability to return to the site of their ancestral origins is totally dependent on the discriminatory abilities of the nose and the olfactory epithelium lining and paired nasal pouches. Blinded fish are able to migrate successfully to spawn but not if their nasal pouches are plugged. Some fish are said to perceive odours when as few as five molecules of a chemical enter the nasal pouch; this represents perception in the range of one part per trillion!

Examination of the brain of salmonids shows large optic, olfactory and cerebellar lobes, i.e. those parts of the brain concerned with sight, smell and the coordination of movement and balance respectively. In contrast the cerebrum, which is the important centre of integration and innovative behaviour in higher vertebrates, is poorly developed in fish. Much of the behaviour of fish is inflexible and automative in nature, rather like the human 'knee-jerk reflex'.

Fish also have a paired auditory apparatus or 'ears', although these are used primarily for spatial orientation consisting only of the semi-circular canals of higher vertebrates and lacking the outer and middle-ear components of higher forms. Further information about position and movement is also provided by the lateral line sensory system which runs as a paired canal along the flanks of the fish.

2.2.4 The respiratory and circulatory systems

The uptake of oxygen from the water into the blood system of fish and the associated loss of carbon dioxide, respectively one of the raw materials and the major waste product of respiration or energy production in fish, are carried out by the gills. There are four pairs of gills in most bony fish and in salmonids, each gill consisting of a single gill arch on which are borne two rows of several hundred gill filaments. Each series of pairs of opposed filaments is known as a holobranch; individual filament rows are sometimes termed hemibranches (Fig. 2.3). This provides a total of approximately 1200 gill filaments in the complete gill assemblage. Each of the gill filaments carries along its length rows of semi-circular secondary lamellae mounted at right angles to the long axis of the filament. The secondary lamellae are the areas of respiratory exchange and their number varies according to the life style and hence the respiratory demands of the fish. Sluggish bottom-living species have as few as ten lamellae for every millimetre of filament length whereas the fast-swimming predatory salmonids have thirty to forty per millimetre. This provides large surface areas for exchanging oxygen and carbon dioxide, approximately $500-1000 \text{ mm}^2$ of gill or lamellar surface per gram body weight for active fish like the salmonids.

Each secondary lamella contains a complex of capillary channels

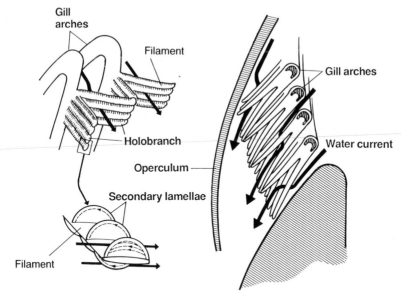

Fig. 2.3 Arrangements of gills, gill arches and secondary lamellae.

through which the blood passes. The capillary and lamellar membranes have very thin walls and hence respiratory exchange occurs readily between the blood and the surrounding water. Blood flow is so arranged that its direction is opposite to the water flow across the gills; this counter-current system improves the efficiency of the respiratory exchange. Approximately 60% of the oxygen dissolved in the water is removed during its passage over the gills, although this figure may fall to 10% under conditions of low dissolved oxygen, reduced gill function or higher temperatures. High levels of effluent or the presence of pollutants, poisons or microbial infections in particular may all damage the integrity of the lamellar membranes and hence reduce the surface area available for respiratory exchange. Even at these levels, however, the gills are considerably more efficient than our lungs.

During ventilation or 'aquatic breathing', water is passed in through the mouth, over the respiratory surfaces of the gills and out through the operculum. The uni-directional flow of water is produced by two mechanisms: in the first, water is forced across the gills by up and down movements of the floor of the mouth or by the speed of the water entering the mouth during active swimming of the fish; in the second, water is drawn or sucked across the gills by lateral movements of the operculi or gill covers. Valves at the mouth and edges of the operculi assist in maintaining the one-way flow of water across the gills. Occasionally, the direction of flow is reversed, invariably as a response to the presence of sediment or toxins; this cough reflex tends to flush out the offending particles. Food materials are prevented from entering the gill chamber by the presence of forwardly-pointing gill rakers on each of the gill arches. These are particularly well-developed in plankton- or filter-feeding fish.

The blood of fish, like that of humans and most other vertebrates, consists of a fluid, the plasma, in which are carried red and white blood cells or corpuscles. The red blood cells contain the iron-based pigment haemoglobin which, because of its strong affinity for oxygen, considerably enhances the oxygen-carrying capacity of the blood. Over 99% of the total oxygen carried is transported in the red blood cells and any disease which results in anaemia will profoundly affect the efficiency of this transport.

The blood circulatory system of fish consists of a single loop or circle of vessels involving the heart−artery−gills−arteries−organs and tissues−veins. This differs from the system in man in not having a separate circulation for oxygenation of the blood in the lungs (or gills). As a consequence the fish heart has only a single auricle and ventricle (Fig. 2.2). Auxiliary pumping, particularly to the gill surface, is achieved as a result of the elasticity of the conus arteriosus and the main arteries. Flexure of the body during locomotion also assists in the movement of blood. Blood pressures during contraction of the heart at 45−50 mm Hg are generally much lower in fish than in man, although salmonids tend to have arterial pressures in the order of 75−90 mm Hg. After the blood has been passed to the various organs and tissues of the body in the arteries, it is returned to the heart in the veins. The major veins themselves communicate with the two large Cuverian sinuses which pass along the back of the gill chambers to the auricle. The spleen and kidney are important to the blood system, as both are involved in the formation of the blood cells.

2.2.5   The osmoregulatory and excretory systems

Although the skin and scales together provide a relatively impermeable barrier to the passage of water and salts, uncontrolled movements of these materials still occur. Much more significant, however, are the gains and losses which occur through the gill surfaces which have to be permeable and large in surface area to allow for the efficient exchange of respiratory gases.

Different movements of salts and water occur in fresh and marine or salt waters. The body fluids of fish have salt concentrations of 300−400 mOsm/kg equivalent to 11‰ (parts per thousand) salinity or a freezing point depression of approximately −0.55°C. In freshwater, where the concentration of the surrounding water is generally less than 5 mOsm/kg, the fish tend to lose salts and gain water (Fig. 2.4). By contrast in seawater the fish have a lower concentration than the surrounding waters (1100 mOsm/kg, 35‰ or a freezing point depression of −2.03°C) and consequently they gain salts and lose water (Fig. 2.4). Successful maintenance of salmonids in either fresh or saltwater requires that the water and ionic composition of the body fluids is maintained at constant levels. This

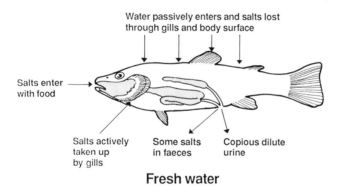

Fresh water

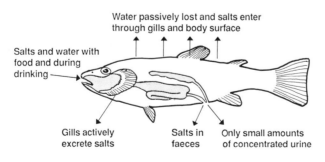

Sea water

Fig. 2.4 Osmoregulatory and ionic factors and their regulation in the freshwater (upper) and seawater (lower) environments.

means that regulatory processes must be available to reverse the uncontrolled gains and losses of water and salts through the skin and gills. Both the gills and kidneys are involved in this regulation. The position and structure of the gills have already been described. The kidneys are found immediately under the vertebral column above the swim-bladder. They are dark-red in colour and stretch the whole length of the body cavity.

In freshwater, fish drink very little but produce copious amounts of dilute urine; few salts appear in the urine because they are reabsorbed by the kidneys. Salts are also gained from the surrounding water by the active (energy requiring) uptake through the gills by special chloride cells found at the base of the secondary lamellae. Salts also enter the fish in the diet. This uptake of salts by the gills and kidneys, together with the excretion of water, equalises the salt loss and water gain experienced by fish in freshwater; in seawater, fish drink a lot (up to 15% of bodyweight per day), selectively excrete monovalent ions like excess sodium and chloride through the gills and produce small volumes of concentrated urine (Fig. 2.4). Again it is the chloride cells in the gills which are responsible for removing the excess salts present in the blood and passing them out into the external water. Many more chloride cells are found in the gills of marine fish, and in anadromous species of salmonids they increase and decrease in number at smoltification and during the spawning run into freshwater respectively. The net effect of these processes is that water is extracted from the seawater to make good the osmotic losses which occur passively through the gill and skin surfaces and the excess salts are excreted. These controls allow the three-fold differences in ionic concentration between the fish and the surrounding seawater to be maintained.

Aside from their osmoregulatory function the gills are also the site of excretion of ammonia, which is the major waste product of protein digestion and metabolism. Ammonia is extremely toxic to fish and hence it is essential that it is continually removed. Clearly the processes of salt and water regulation and excretion require that the gills and kidney are in a healthy condition. Damage to the gills by pollutants and/or microorganisms is just as important to the maintenance of salt balance as it is to respiration. Infections of the kidney of salmonids by BKD or PKD (see chapter 6) will severely reduce the ability of the fish to respond to osmotic change.

For this reason stocks of trout apparently showing no ill effects from BKD in freshwater may suffer high mortalities when transferred to seawater cage sites. Similar losses may occur during smoltification or spawning migrations into freshwater.

2.2.6   The digestive system and nutrition

The gut or alimentary tract of salmonids consists of a simple tube which stretches from the mouth through to the posterior external opening, vent or anus. As in other carnivorous fish the intestine of salmonids is relatively short, being up to three times as long in herbivorous fish. In salmonids the food which is eaten is rapidly swallowed. In the stomach the food particles are broken up as a result of the churning motion of the muscular stomach and the digestive action of the gastric acid and enzymes. After release of the partially-digested food from the stomach through a valve known as the pylorus, it is subject to further digestion and then absorption by the small intestine.

On dissection, one of the most prominent features of the salmonid gut is the set of pyloric caecae (Fig. 2.2). These are blind-ending tubes, originating from the pyloric area of the stomach, numbering 100 or more in salmonids but absent in the carp and many other cyprinids. Their exact function is unknown, although they are thought to be concerned with fat digestion. Like mammals, the salmonid small intestine receives digestive secretions from two macroscopic glands or organs, the liver and pancreas. The liver produces bile which is stored in the gall bladder and subsequently released onto the food after it has passed from the stomach. Bile helps with the solubilisation and absorption of fats. The liver is also concerned with the processing and assimilation of the digested food materials. For the salmonids the principal dietary nutrients are proteins and fats and consequently it is with the metabolism of fatty and amino acids that the liver is mainly concerned, after these end products of protein and fat digestion have been absorbed by the small intestine and transported to the liver in the hepatic portal vein.

The pancreas produces several digestive enzymes concerned with fat, protein and carbohydrate digestion, in addition to the hormones insulin, glucagon and somatostatin. In fish these hormones, particularly insulin, are concerned with the control of protein rather than carbohydrate metabolism, which is their more usual action in other vertebrates.

Damage to the liver by parasitic and microbial infections, in particular IPN virus disease of the pancreas (see Chapter 6), significantly affects the digestive functions of the glands resulting in reductions in the effectiveness of the conversion of the dietary nutrients into fish flesh. Many toxins, hormones, enzymes and other metabolites are neutralised by the liver. High fat levels in the diet can also lead to fatty degeneration of the liver.

2.2.7   The reproductive and endocrine systems

In both sexes of salmonid fish the gonads are paired organs lying just below the kidneys (Fig. 2.2). In immature fish both the ovary and testis are rudimentary thread-like structures anteriorly and dorsally placed in the body cavity. During maturation there are considerable enlargements in size of both the ovary and testis and at spawning they often constitute up to 30% and 10% of the bodyweight of the fish respectively. The testes are creamy-white in colour and smooth in outline, tapering backwards until they lead into the paired sperm ducts which at spawning release the spermatozoa or milt through the common urinogenital opening. The ovaries of salmonids are only partially enclosed by a membrane and at maturity they release the ripe eggs (ovulation) into the abdominal cavity. Here the eggs remain until, in wild fish, oviposition is induced by courtship behaviour or, in farmed stocks, they are artificially removed or stripped from the fish. Like the sperm the eggs leave the body through the urinogenital opening.

For most salmonids spawning is characteristically a seasonal event usually timed by the seasonally-changing cycle of day length. Growth of the ovary and testis and the enclosed developing gametes or sex products is controlled by hormonal secretions from the hypothalamus (part of the brain), the pituitary, thyroid and pineal glands and the gonads.

In conjunction with the nervous system, the endocrine glands and the hormones they secrete directly into the blood act to control most of the physiological and metabolic processes in the body (Table 2.1). Although the endocrine system is a heterogeneous collection of organs with no common origins, the major glands

Table 2.1 Major endocrine glands and hormones in fish

Gland	Hormone	Nature	Function
Hypothalamus	Releasing hormones	Peptides	Controls the pituitary gland
Pituitary Gland	Growth hormone (STH)	Protein	Controls growth
	Prolactin (LTH)	Protein	Controls ion balance
	Adrenocorticotropic hormone (ACTH)	Peptide	Controls interrenal gland
	Melanocyte-stimulating hormone (MSH)	Peptide	Controls colour change
	Thyroid stimulating hormone (TSH)	Protein	Controls thyroid gland
	Gonadotropic hormone (GTH)	Protein	Controls reproduction and gonads
Thyroid gland	Thyroxine (T4) Triiodothyronine (T3)	} Amino acids	Controls growth, reproduction, metabolism and nutrient assimilation
Interrenal (adrenal gland)	Adrenaline	Amino acid	Counteracts stress
	Cortisol	Steroid	Controls ion balance
Testis	{ Testosterone { 11-ketotestosterone		Metabolic effects. Control secondary sex characters.
Ovary	17β-oestradiol Oestrone Progestagens	} Steroids	Sperm production. Metabolic effects. Yolk and egg production
Pancreas	Insulin Glucagon Somatostatin	} Peptides	Protein metabolism and control of endocrine pancreas
Pineal gland	Melatonin	Peptide	Provides information about day/night and seasonal time
Ultimobranchial gland	Calcitonin	Peptide	Controls calcium levels
Stannius corpuscles	Hypocalcin	Protein	Controls calcium balance

found in fish and the hormones they produce are often the same or similar to those of other vertebrates (Fig. 2.5). This has advantages in that some of the hormone preparations and drugs developed primarily for medicinal or agricultural purposes can also be used for fish. Recently, there have been important developments with growth hormone which make its use as a growth promoter in fish a real prospect for the future.

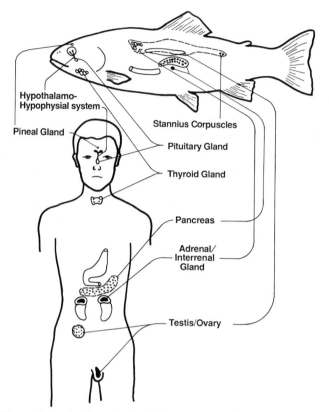

Hypothalamo-
Hypophysial system

Pineal Gland

Stannius Corpuscles

Pituitary Gland

Thyroid Gland

Pancreas

Adrenal/
Interrenal
Gland

Testis/Ovary

Fig. 2.5 The main endocrine glands of fish and their counterparts in man.

2.3 THE AQUATIC ENVIRONMENT

2.3.1 Water quality

The most crucial factor affecting the success or failure of a fish farming venture is the water supply. The quantity of water available determines the quantity of fish which can be produced, but if the water quality is unsuitable, then clearly the fish farm is bound to fail anyway. Each species has preferred ranges for the various parameters of water quality, such as temperature, dissolved oxygen and salinity, and ideally the fish farm should operate at the optimum levels of each parameter to achieve fast growth and efficient performance. However, in considering water quality, it is necessary to bear in mind that the various parameters interact with each other and also that a fish's preferred range and optimum requirement for, say, water temperature or dissolved oxygen will vary considerably over its life cycle.

In the following section the physical and chemical factors of water quality will be described with special reference to intensive farming of salmon and trout in temperate freshwater and seawater. Salmonids are generally reared under conditions of 'running water culture' in which a continuous flow of water maintains satisfactory water quality within the rearing unit. It is important to distinguish this arrangement from 'standing water culture' in which fish are reared in static water. It should be stressed that much intensive fish farming is undertaken at relatively higher water temperatures, often in standing water ponds with complicated daily cycles of water quality due to environmental factors and pond biology. Figure 2.6 illustrates the typical pattern of such changes as the pond warms up during the day and cools off at night; algae present in the water produce oxygen by photosynthesis during daylight hours. Carp and other species reared in such ponds are generally able to thrive over a wider range of environmental conditions compared with salmonids. Although the limits of tolerance for different water quality factors therefore vary considerably between different species, the same principles are common to all farmed fish.

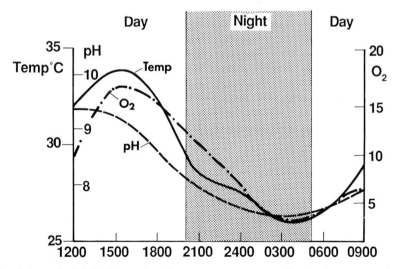

Fig. 2.6 Typical daily fluctuations in water temperature, dissolved oxygen and pH within a tropical fish pond.

Preferred spawning temperatures and development times for egg incubation and on to first feeding vary considerably between different species (see Table 4.5). For instance, the preferred temperature range for egg incubation in salmonids is 5−12°C and in carp is 17−25°C, although some species will withstand temperatures down to just over freezing (e.g. Atlantic salmon). By contrast salmonid growers have a preferred temperature range of 0−20°C. For instance rainbow trout will survive under ice cover at approximately 0°C and can survive temperatures of over 25°C for short periods. However, they do not thrive for prolonged periods above 21°C and become distressed at 25°C. The optimum temperature for salmonid growth is approximately 16°C, but varies somewhat with species, age and type of diet. As with other water quality parameters, fish are vulnerable to sudden fluctuations in temperature and can in general accommodate such changes more readily with increasing age.

Fish rely on the oxygen dissolved in water to support their metabolism and water's ability to take oxygen into solution depends on temperature, pressure and its dissolved salts. As shown in Table 2.2, the higher the water temperature, the less oxygen can be retained and seawater contains less oxygen than freshwater at the same temperature and atmospheric pressure. The solubility values given refer to water that is fully saturated with oxygen and under practical fish farming conditions this is frequently not the case. Also, they assume the normal atmospheric pressure of sea level (760 mm Hg), whereas at higher altitudes these solubilities will

Table 2.2 Solubility* of oxygen in freshwater and seawater

Temperature in °C	Solubility in freshwater in mg/l	Solubility in 35‰ salinity seawater in mg/l
0	14.6	11.3
5	12.8	10.0
10	11.3	9.0
15	10.2	8.1
20	9.2	7.4
25	8.4	6.7
30	7.6	6.1
35	7.1	5.7
40	6.6	5.3

*Assuming full saturation at normal atmospheric pressure

be further reduced. For example, at 11°C, the oxygen solubilities of fully saturated freshwater at sea level and 2000 m altitude are 11.0 mg/l and 8.9 mg/l respectively. The solubility of oxygen should be viewed against the minimum levels for salmonids of 5.0−5.5 mg/l for fish and 7 mg/l for eggs. By comparison, carp, catfish and tilapia can withstand dissolved oxygen levels of below 2 mg/l, provided it is for short periods (e.g. overnight). Clearly there is little oxygen available for salmonids towards the upper limit of their preferred temperature range and in any event the fish farmer needs to ensure the water is fully saturated with oxygen by means of splashboards, aerators, etc. Since fish will feed more readily as the temperature increases, and therefore use up the available oxygen at a faster rate, great care must be taken to avoid overstocking ponds and salmonids should never be fed above 20°C to minimise oxygen consumption. In this connection it should be remembered that oxygen uptake by fish is increased by a rise in water temperature and by a greater activity level and that small fish consume proportionally more oxygen per unit weight compared with larger fish. A detailed understanding of oxygen requirements can be used to calculate the water requirements of salmonids under different conditions (see section 2.3.2).

Acidity, alkalinity and pH have important implications for fish health. The pH value describes the acidity or alkalinity of water on a logarithmic scale from 0 (very acid) to 14 (very alkaline) with a neutral point of 7. The preferred range of pH for salmonids is approximately 6.4−8.4 and fluctuations in pH are particularly stressful. Low pH water is usually due to natural leaching of mineral acids from soils, swamps and forests. But the ability of water to resist changes in pH is perhaps more important than the pH value alone in terms of fish health. This 'buffering capacity' is mostly due to the presence of carbonates, bicarbonates and hydroxides which are characteristic of 'hard' waters associated with limestone areas. Hardness is expressed in terms of dissolved calcium carbonate and water sources can vary from very soft (e.g. 0−10 mg/l $CaCO_3$) to very hard (over 200 mg/l $CaCO_3$). Hard water is alkaline and although soft water is not necessarily acidic, it has a far lower ability to resist any increase in acidity due to a build-up of hydrogen ions. Hence spate conditions in rivers passing through calcium-deficient soils associated with igneous rock strata will result in a sudden drop in pH due to the organic acids running off the land. Such acidic water entering a fish farm can cause severe damage to fish gills and large losses.

INTENSIVE FISH FARMING

The main excretory products of fish are ammonia and carbon dioxide. In water the ammonia molecule reacts to form ammonium hydroxide which readily dissociates to the ammonium and hydroxyl ions as follows:

$$NH_3 + H_2O \rightleftharpoons NH_4OH \rightleftharpoons NH_4^+ + OH^-$$

Toxic	Non-toxic
form	Form

The ammonium ion (NH_4^+) is fairly innocuous to fish, whereas free ammonia (NH_3) is highly toxic and a level of 0.02 mg/l is generally regarded as the maximum acceptable limit for salmonids. The dissociation of ammonia in water is heavily dependent on pH and on temperature as illustrated in Table 2.3. Thus at a water temperature of 15°C, to achieve the 0.02 mg/l NH_3 maximum level would require a total ammonia level of 22.2 mg/l at pH 6.5, whereas at pH 8.5 a total ammonia level of only 0.3 mg/l would yield the same quantity of toxic free ammonia. In practice this means that below pH 7 ammonia is unlikely to represent a health problem in fish farms, but relatively low levels of total ammonia at higher pH levels are dangerous. Although seawater and alkaline freshwater are well buffered and do not fluctuate in pH, they therefore have the disadvantage of predisposing to ammonia toxicity under conditions of heavy stocking inadequate water exchange rates, polluted water and high temperature.

Table 2.3 Percentage of total dissolved ammonia which is unionised (toxic form) as a function of pH and water temperature

pH	Water temperature				
	5°C	10°C	15°C	20°C	25°C
6.5	0.04	0.06	0.09	0.13	0.18
6.7	0.06	0.09	0.14	0.20	0.28
7.0	0.12	0.19	0.27	0.40	0.55
7.3	0.25	0.37	0.54	0.79	1.10
7.5	0.39	0.59	0.85	1.25	1.73
7.7	0.62	0.92	1.35	1.96	2.72
8.0	1.22	1.82	2.65	3.83	5.28
8.3	2.41	3.58	5.16	7.36	10.00
8.5	3.77	5.55	7.98	11.18	14.97

Source: Trussell (1972)

In the wild many fish are 'euryhaline', which means that they can tolerate differing salinities and will move readily into and out of brackish water. Of the various farmed species with this ability, mullet are probably the best example, but carp, eels, seabass and tilapia can all be reared under conditions of varying salinity. Since those salmonids cultured for food consumption are anadromous in their native range (i.e. live for part of their life cycle in the sea, but need to return to freshwater to breed) it is not surprising that their adaptability to salt water has been a matter of study since the last century. It has been shown that salinity tolerance is first evident to a minor degree with salmonid eggs. Thereafter for juvenile rainbow trout important factors include the relation between salinity and age (or possibly weight) at acclimatisation, the rate of acclimatisation, water temperature, etc. For example, rainbow trout with a unit weight above about 100 g can be moved from freshwater directly into full strength seawater (32–34‰ salinity) without prior acclimatisation provided the water temperature exceeds about 6°C. Such direct transfer of Atlantic salmon smolts is only possible after smoltification has occurred, as evidenced by appearance and behaviour, and this process is partly triggered by changing day length or photoperiod. However, some Atlantic salmon farmers introduce salmon parr to approximately 20% seawater (about 7‰ salinity) six months before smoltification is due and slowly increase the salinity level using pumped seawater over the intervening period to take advantage of relatively warmer seawater temperatures. Seawater is generally a very stable alkaline medium except where it is diluted by freshwater entering via rivers, etc. Although the calcium content is only 0.4 mg/l, its total salts comprise 343 mg/l (at a salinity of 34‰) with a pH range of 8.0–8.4 and averaging 8.1. Seawater has therefore considerable buffering capacity and, combined with stability and abundance, this would usually offset the disadvantages of reduced oxygen solubility and a predisposition to higher free ammonia levels due to high pH, when compared with often limited supplies of freshwater for fish farming purposes. Brackish water will tend to have quality parameters intermediate between the particular freshwater and seawater profiles depending on the degree of mixing, which will itself be influenced by the greater specific density of seawater tending towards stratification with the lighter freshwater often lying on the surface.

Other physical and chemical parameters which can be important include various dissolved gases, suspended solids and the presence of certain minerals in otherwise unpolluted ground waters. For instance, carbon dioxide is absorbed preferentially to oxygen and low oxygen content is often associated with high carbon dioxide levels. When derived from certain rock strata by wells or artesian bores, water may have high levels of dissolved nitrogen and hydrogen sulphide. Like carbon dioxide, nitrogen may displace oxygen from solution, whereas hydrogen sulphide is highly toxic to fish in itself. The presence of these gases and the additional problem of super-saturation with dissolved gases, such as nitrogen, leading to Gas Bubble Disease among fish (see chapter 3; section 3.3.1 and chapter 6; section 6.4.3) are remedied by aerating the water supply to blow off the unwanted gas. Ground waters may also contain significant quantities of dissolved minerals such as ferrous iron and manganese. Under conditions of oxygen shortage in particular, these may be dissolved as bicarbonates and can form a colloidal precipitate of iron and manganese hydroxides on the outside of salmonid eggs or on the gills of fry. High levels of suspended solids and particulate matter can also damage fish, particularly on the gill surfaces. Suspended solids can build up due to inadequate removal of faecal wastes from ponds as well as from silt entering the farm, and may cause mechanical irritation as well as leading to increased ammonia levels and removing oxygen from the water by decomposition. The particular type of suspended solid greatly influences the safe upper limit, but as a general rule the maximum level of total suspended solids should not be over 30 mg/l. This and recommended upper limits for salmonids of other water quality criteria are given in Table 2.4.

2.3.2 Pollution, algal blooms and fish kills

From the foregoing it is clear that unless care is taken to ensure adequate water exchange, farmed salmonids will rapidly succumb to the effects of their own metabolism on water quality in terms of reduced oxygen, increased carbon dioxide, increased ammonia and suspended solids, etc. The same is true in principle for intensively farmed catfish or carp held in standing water ponds and artificial aeration may be necessary at higher temperatures, although the pond ecosystem may enable some detoxification of metabolic wastes. However, fish farms are particularly vulnerable to the effects of pollutants entering the water usually from agricultural or

Table 2.4 Suggested water quality criteria for optimum health of salmonid fishes*

Chemical	Upper limits for continuous exposure
Ammonia (NH_3)	0.0125 ppm (unionized form)
Cadmium[†]	0.0004 ppm (in soft water < 100 ppm alkalinity)
Cadmium**	0.003 ppm (in hard water > 100 ppm alkalinity)
Chlorine	0.03 ppm
Copper	0.006 ppm in soft water
Hydrogen sulfide	0.002 ppm
Lead	0.03 ppm
Mercury (organic or inorganic)	0.002 ppm maximum, 0.00005 ppm average
Nitrogen	Maximum total gas pressure 110% of saturation
Nitrite (NO_2^-)	0.1 ppm in soft water, 0.2 ppm in hard water (0.03 and 0.06 ppm nitrite-nitrogen)
Ozone	0.005 ppm
Polychlorinated biphenyls (PCBs)	0.002 ppm
Total suspended and settleable solids	30 ppm or less
Zinc	0.03 ppm

*Source: Wedemeyer, 1977
[†]To protect salmonid eggs and fry; for non-salmonids, 0.004 ppm is acceptable
**To protect salmonid eggs and fry; for non-salmonids, 0.03 ppm is acceptable

industrial processes. These can cause reduced performance or sudden mass mortalities of fish known as 'fish kills' and are described in relation to the various causes.

Many pollutants remove oxygen from the water, in particular organic material such as sewage or silage liquor. The resulting 'biochemical oxygen demand' (BOD) may leave insufficient dissolved oxygen to sustain the biomass of fish present in a farm, causing suffocation and an immediate fish kill. If pipelines are closed down for maintenance work, the aquatic organisms attached to the inside will die and when used again the pipe will discharge a slug of contaminated water with a high BOD. Riverbank maintenance and cutting aquatic vegetation upstream of a fish farm is another common cause of BOD-related fish kills.

Certain heavy metals are particularly toxic to fish and usually enter the water via industrial effluents. The main metals implicated are copper, lead, mercury and zinc, while cadmium, chromium, iron, manganese and nickel are also toxic. Defining maximum safety limits is difficult; acutely toxic levels usually lie within the range 0.1–1.0 mg/l active metal depending on the exposure time to fish and the water chemistry. Metals such as copper and zinc are especially toxic in soft water since the presence of calcium carbonate can precipitate the carbonate salts affording some protection. For example, approximate safe upper limits for copper are 0.006 mg/l in soft water and 0.03 mg/l in hard water. The various metals also interact; thus copper increases the toxicity of zinc under soft water conditions, whereas the additional presence of cadmium will exert a synergistic toxic effect in the presence of the other two metals. A particular feature of heavy metal poisoning is that the toxic effects may not become apparent until one or two days after exposure, when sudden mortalities may occur for no obvious reason.

Many non-metals are toxic to fish, ranging from simple molecules like ammonia and chlorine to complex organic compounds and hydrocarbons. Chlorine solutions are sometimes used to disinfect equipment on fish farms, as are iodophors (e.g. egg disinfection), and both should therefore be handled with great care to avoid accidental spillage. Insecticides and weed-killers are often a problem since they usually contain organophosphorous and organochloride compounds which are highly toxic. Sheep dips containing these products frequently enter hill streams causing fish kills. Poachers sometimes use cyanide preparations in order to steal fish which die without any external symptoms. Marine farms are sometimes threatened by crude oil pollution due to spillage from tankers: in practice the solvents used to disperse oil slicks are more toxic than the crude oil itself, although the latter may pose an additional threat of tainting the flesh with an objectionable taste. Fish taint can also occur due to various industrial effluents and a muddy off-flavour sometimes results from the action of soil bacteria or algae in earthen ponds.

Algal blooms have long been recognised as a considerable threat to warm water pondfish for several reasons. Firstly, the bloom can consume all available oxygen, either during nocturnal respiration or by increasing the BOD if the bloom suddenly dies, with resulting suffocation of the fish stock. Secondly, under certain conditions some algae are able to produce specific toxins which will kill fish unable to escape within a pond. Also the sheer density and sticky nature of the bloom may occasionally coat the gills and physically suffocate the fish. The species of algae best known as toxin producers in freshwater ordinarily belong either to the group of blue green algae (e.g. *Microcystis*) or to the phytoflagellates (e.g. *Prymnesium parvum*). Pondfish dying from oxygen depletion due to algal blooms are typically affected during the night and early morning, whereas fish suffering from a toxic algal bloom show nervous signs with losses peaking during the hours of bright sunlight. Although *Prymnesium* initially threatened the future of pondfish farming in Israel, the predisposing conditions and chemical means of control are now well understood. However, the more recent expansion in marine fish farming has highlighted the threat posed by toxic marine algae. These are dinoflagellates which cause massive kills of wild fish and are also involved in paralytic shellfish poisoning of man. Such algal blooms can discolour entire areas of sea, hence the name 'Red tide' and marine fish farms located in such areas stand to lose the entire crop. Aeration appears to mitigate some of the effects but is more feasible for a shore-based marine farm. In the case of a small local bloom, it may be possible to tow floating cages out of the path of the bloom if sufficient warning can be given. But certain coastal regions which regularly suffer from toxic marine algal blooms must be avoided by fish farmers given the present state of knowledge.

A fish kill is a sudden mass mortality occurring over a period of hours among fish which had previously been behaving and feeding normally. It is important to recognise that fish kills may have legal consequences in terms of liability, hence the need to act swiftly and prudently to mitigate the stock loss, if this is possible, and to collect all relevant information and samples.

The commonest cause of a fish kill is oxygen lack resulting from a systems failure. This can be simply due to inadequate water flow rates for the biomass of fish and the prevailing water temperature. More commonly it is due to blocked screens or failed pumps or aerators. Affected fish will be gasping and swimming at the surface, especially around the water inlet, and dead fish may show haemorrhages in the skin and gills and inside the abdominal cavity. Such problems are clearly the responsibility of the fish farmer and his staff and prompt remedial action may save a proportion of the stock. Should the fish have recently received an external chemical treatment, the possibility of overdosage should be considered. If, however, there is the possibility that the fish kill may be due to pollution entering the farm or to malicious poisoning, then it is advisable to call in the police and to take appropriate water samples and fish samples in front of independent witnesses. Clearly the occurrence of mortalities among any wild fish and aquatic animals outside the farm is important. Fish killed with cyanide appear to have died quietly and often have massive internal blood clots with the faint burnt almonds smell of cyanide. Toxic events due to algae or pesticides usually kill smaller fish first and survivors show convulsive, erratic swimming with lethargy. Dissolved oxygen levels will usually be within the normal range for pesticides and most industrial pollutants and may be above normal in the case of toxic algal blooms. But in the case of oxygen depletion due to a systems failure or a high BOD (e.g. from sewage or silage), dissolved oxygen levels will be very low if they are monitored in sufficient time, hence the urgency of comprehensive water sampling. Acute botulism should not be overlooked as a cause of sudden heavy mortalities in fish within hours of feeding moist diets which have started to putrefy, allowing *Clostridium botulinum* organisms present to manufacture the botulinum toxin. In a similar way sporadic botulism can occur if cannibal fish are allowed to scavenge on dead carcases which have not been removed from the bottom of earthen ponds. Fish kills can therefore be caused by a wide variety of different agents, but they are never caused by infectious diseases.

2.4 SITE EVALUATION

In attempting to assess the feasibility of farming a given species at a particular site, it is necessary to match the environmental requirements of the fish with the physical, chemical and biological characteristics of the site. It follows from the foregoing that the most critical factor in any quantitative evaluation of site potential is the water resource.

In order to determine the production capacity of a site, information on water quality must be supplemented by detailed knowledge of the quality of water available throughout the year. Provided the water requirements are known, the weight of fish which may be held at the site can be calculated. This will vary with water temperature and knowledge of the temperature profile of the site is necessary to predict growth rate. The likely duration of the production cycle from egg incubation to marketable fish can then be superimposed on the carrying capacity of the site in order to estimate the annual production capacity in terms of fish weight. Figure 2.7 illustrates the interaction of these various factors in evaluating site capacity for salmonid farming. The choice of husbandry system in order to exploit this potential will need to

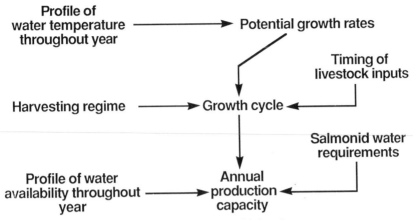

Fig. 2.7 Evaluation of site capacity for salmonid farming.

take account of local site conditions (e.g. terrain, access, flood risk) and it may not be feasible to utilise the full potential of the site in practice.

2.4.1 Temperature and growth cycle

For a given genetic strain of fish fed to appetite with an appropriate good quality diet, water temperature is the key factor controlling rate of growth. Figure 2.8 shows the exponential growth pattern in immature rainbow trout held at a constant water temperature. In this case an average growth cycle of 55 weeks from 'eyed' egg incubation to harvesting at 227 g unit weight was obtained at 14°C. Within limits, the duration of the growth cycle will vary in direct proportion to the average water temperature and Figure 2.9 illustrates that this is likely to range from an average cycle of approximately 9 months at 18°C to approximately 18 months at 9°C.

Under practical conditions of fluctuating water temperature, the fish farmer should endeavour to plan his operation so that peak temperatures coincide with the later (more exponential) part of the main growing cycle. In this way the bulk of his stock are exposed to summer temperatures when they have the potential for greatest weight gain. Grading and harvesting regimes also influence duration of the growth cyle as market size fish are separated from slower

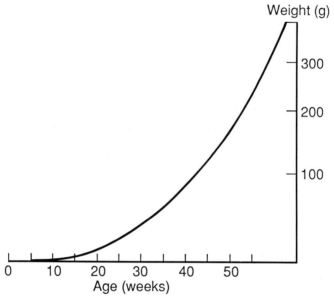

Fig. 2.8 Growth of rainbow trout at a constant water temperature of 14°C.

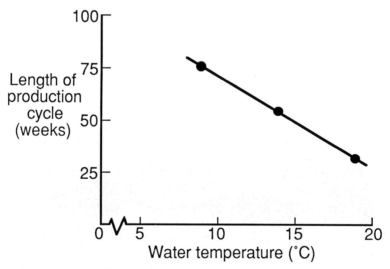

Fig. 2.9 Effects of varying average annual water temperature on the length of the production cycle to market size for rainbow trout.

growing fish within the same group and the overall farm stocks are manipulated to meet market demand. Although seasonal temperature changes may predispose to peak production during the summer, this will usually mean unsatisfied market demand during the colder periods of the year. It is then in the fish farmer's interest to produce fish during the period of reduced supply by use of early or late spawning strains or by altering the usual growth cycle. One means of achieving the latter is to accelerate the hatchery phase by use of heated water (chapter 3; section 3.3.2). Thus an increase in water temperature from 4°C to 10°C will mean a reduction of about 30 days in the incubation time for rainbow trout eggs. Unfortunately the large flow rates needed to ongrow salmonids mean that artificial heating of water is not normally a commercial option outside the hatchery unless waste heat is available (e.g. power station effluent, Plate 2.1). By contrast, eel farming requires lower flow rates and higher temperatures which can be achieved by means of recirculation (see section 3.6). It is worth noting that the relatively warmer seawater available in winter compared with ambient freshwater in certain regions (e.g. north-west Europe) has encouraged marine cultivation of rainbow trout to enable continuous feeding throughout the year and a consequently increased growth rate when compared with freshwater trout farms.

Plate 2.1 Using warm water from a coastal power station effluent to increase growth rates at an adjacent marine fish farm.

2.4.2 Water requirements of salmonids

Under intensive farming conditions salmonids require continuous water exchange in order to maintain appropriate water quality. This is in marked contrast to intensive farming of carp, catfish, etc., within standing water ponds. Not only do salmonids have a higher oxygen requirement, but they appear more sensitive to ammonia, carbon dioxide and suspended solids compared with these other species. Demand for oxygen, and hence water, increases with increasing water temperature. Also it is inversely proportional to fish size, because small fish need more water per kilogram of body weight than large fish. The activity level of the fish has an important influence and feeding, swimming, etc., will increase their oxygen demand. Even within the salmonid group, there are significant differences between fish species with rainbow trout having a greater oxygen requirement than salmon.

Detailed knowledge of how oxygen demand changes with water

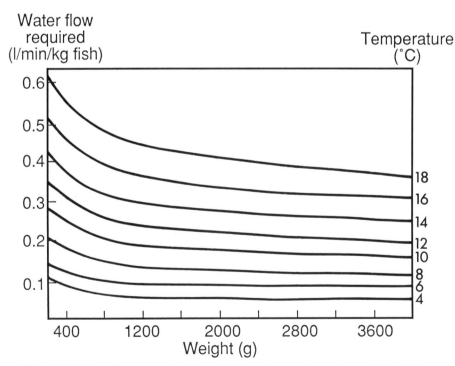

Fig. 2.10 Water flows required for Pacific salmon of varying weight held at different average temperatures (after Liao, 1971).

temperature and fish weight enables calculations to be made of water flow requirements. Clearly this is a critical factor in defining the production capacity of a particular water source. The dissolved oxygen uptake rates of different sizes and species of salmonids have been used to derive the water flow required under different conditions. Figure 2.10 illustrates the effect of different fish weights and water temperatures on the water requirements of salmon growers held in seawater 95% saturated with oxygen. By comparison it has been estimated that a production capacity of 100 000 salmon smolts requires 80–160 l/s of freshwater depending on circumstances. If one assumes a minimum dissolved oxygen level of 5.5 mg/l, then it is possible to tabulate oxygen consumption and hence water requirements of rainbow trout (Table 2.5). In this case the water requirements for 1 tonne of 200 g trout increase from 4.3 l/s at 6°C to 17.7 l/s at 16°C. Thus there is often a four-fold increase or more in the amount of water needed to carry the same stock at a particular site when comparing summer with winter, whereas in practice the trout farmer may have a far greater weight of fish present in the summer. For freshwater farms the situation is, of course, further exacerbated by the usual problem of reduced flow rates during the warmer months when the water is most needed.

For this reason the worst case production capacity of a running water site is usually calculated in relation to the minimum drought flow. For example, a freshwater stream with an absolute minimum recorded flow of 1000 l/s at a constant water temperature of, say, 12°C could carry approximately 90 tonnes of 200 g trout under

Table 2.5 Water requirements for 1 tonne of 200 g rainbow trout

Water temperature in °C	Oxygen consumption in kg/day	Water requirement in l/s	Water requirement in Imperial gallons/minute	Water requirement in cubic feet/second
6	2.6	4.3	56.8	0.15
8	3.4	6.2	81.8	0.22
10	4.3	8.6	113.5	0.30
12	5.1	11.2	147.8	0.40
14	6.0	14.3	188.8	0.51
16	6.8	17.7	233.6	0.63
18	7.7	20.9	275.9	0.74

Water requirement: assume 100% saturation of freshwater with oxygen and dissolved oxygen of effluent is 5.5 mg/l
Note 1: Imperial gallon = approx. 1.2 US gallons

extreme drought conditions (see Table 2.5). In practice this carrying capacity is somewhat less than annual production capacity as the farmer is likely to crop his stock through a succession of harvests and hence spread out the water requirements of his stock rather than harvesting all at one time with a consequent single peak harvest requirement for water.

If it were possible to have a continuous input of young fish and a continuous harvesting regime for market size fish, then a steady state production could be achieved representing the most efficient use of water. However, this possible strategy is usually limited by variations in water temperature coupled with seasonality of inputs and outputs. The resulting ratio of carrying capacity to production capacity will therefore vary from one site to another, but in various US and northern European trout farms with a production cycle of about 15 months, it is commonly in the region of 1:1.6. Under these circumstances in the foregoing example a site with a carrying capacity of 90 tonnes at minimum drought flow could produce about 144 tonnes of trout per annum assuming a water temperature of 12°C. The carrying capacity would increase still further at lower water temperatures and typically under ambient conditions in northern Europe annual production of 1 tonne of 200 g trout requires in the region of 4.5−7.5 l/s. However, growth rates are correspondingly slower and hence there is a trade-off between sites with low water temperature offering higher carrying capacity and slow growth as against higher temperatures offering reduced carrying capacity but faster growth.

Although the practical measurement of water temperature is cheap and straightforward, recording water flow rates is more difficult. High and low water flow meters are available but expensive. However, a simple weir device with a notch can be used instead on the basis that water height in the notch relates to flow rate. For cage sites and enclosures, water flow and exchange rates are influenced by such factors as wind and (in the sea) tidal currents, whereas a stock of fish swimming around a cage will in itself promote water circulation (and also inhibit ice formation in the winter); hence measuring water flow is less practicable or meaningful than at running water sites.

2.4.3   Other site constraints

Given suitable supplies of clean water in sufficient quantity offering commercially acceptable growth rates, it is obviously important that local site conditions enable practical exploitation in terms of fish farming development. To some extent these secondary constraints will be influenced by the type of husbandry adopted, notably in the choice between running water systems, which rely on continuous water exchange, and standing water systems (e.g. pondfish farms; section 2.4.4). However, it is important that legal rights to ownership or access be established initially. Also communications must be such as to enable regular supplies of stock and feed to be brought in and for fish to be transported off to market without difficulty. Daily road or boat access must therefore be maintained throughout the year. Services such as electricity and telephones are important and standby generators are vital if any of the water supply is pumped. Wherever possible the farm manager should live on the site in order to ensure round-the-clock surveillance. Otherwise stock theft will probably occur and problems such as a failure in the water supply will go undetected (chapter 3; section 3.5.1).

In constructing running water systems using gravity fed or pumped water, the terrain must be suitable for installing ponds and tanks without excessive expenditure on initial preparation works. Ideally land should be easily excavated with a suitable fall between the levels of the proposed inlet and outfall to the farm. It may be necessary to construct a dam upstream, either to act as a reservoir in times of drought flow or to reduce the risk of spate conditions causing flooding and consequent loss of fish. However, if wild fish stocks are present in the water course, it may be necessary to construct a fish ladder to enable them to circumvent the dam.

Certain fish farming systems rely on circulation through pens, cages or enclosures to give adequate exchange of water. It is therefore important that they be sited where currents are adequate to support high stocking densities without allowing accumulation of faeces beneath the fish. In the case of marine cages, the danger period of slack tide is therefore equivalent to minimum drought flow for running water systems in determining the upper limit of stocking. Too strong a current may distort the netting and damage fish and by their nature net pens are vulnerable to damage and the risk of fish escaping, hence the need for additional anti-predator nets above and around the fish in order to protect them from seals, fish-eating birds, etc. Perhaps the greatest risk to such systems is that of gales damaging the rearing units due to wind action on the exposed superstructure combined with the effect of waves. The greater the distance between the farm and the shore in the direction of the prevailing wind (i.e. the 'fetch'), the greater will be the consequent exposure. It is therefore necessary to select carefully for a sheltered location in the lee of the prevailing wind, while offering satisfactory water exchange. Whereas sites should also be well away from shipping lanes, marine farms in particular may on occasion come into conflict with yachts and other vessels seeking anchorage.

2.4.4 Selection of pondfish farming sites

Standing water culture of channel catfish, eels, etc., obviously requires much less water than is the case for the running water systems usually adopted for salmonid farming, but the underlying principles are similar. Thus the key factors are water supply, site topography and water retention. Continuous water loss occurs from standing water ponds due to evaporation, seepage and through draining down the ponds during husbandry routines, and these losses are offset by direct rainfall. For example, the Mississippi catfish growers expect a net loss of over 1 m annually by evaporation and estimate that a 20 hectare site will require a continuous inflow of about 100 l/s to compensate. As with salmonids, water quality is important and where a source of well water is available, it may be preferred to surface water (see section 3.3.1).

The topography of the land should facilitate low-cost excavation without the risk of flooding, as in flat, workable terrain with a fairly restricted catchment area. Although there is an increasing trend towards continuous stocking without regularly emptying the pond of water, it is preferable to construct levée ponds which can be completely drained without pumping. Typically the pond bottom has a slope of about 0.2% towards the harvest basin where a flexible standpipe is installed for drainage purposes. An important feature of the levée is that it should be wide enough on top to permit vehicular access for handling feed and fish (Plate 8.5).

The presence of clay soil clearly helps in water retention, whereas chemical sealants or physical membranes (e.g. butyl rubber) are commercially available. It is important during construction to ensure adequate compaction of the soil, both on the pond bottom and in levées to aid water retention. Although construction cost per unit area is lower for large ponds, smaller ponds lend themselves to more intensive management and are less vulnerable to the effects of wave erosion. Further information about pond layout and design is given in subsequent, more detailed, descriptions of particular industries (e.g. catfish: chapter 8; Israeli pondfish: chapter 9). The common overall objective must be to exploit the particular advantages of a given site in order to achieve the most cost-effective fish rearing conditions, taking into account the requirements of both fish stock and farmer alike.

3 Farming systems and husbandry practice

3.1 INTRODUCTION

The life cycle of fish embraces several disinct phases, such as spawning and egg fertilisation, hatching and larval development, and the subsequent growth of juvenile fish up to adulthood. The production cycle on a fish farm needs to take account of the differing biological requirements of fish as they grow though these various stages. Figure 3.1 depicts the functional components involved in trout farming, whether the aim is production of live fish or (more commonly) for slaughter as fresh, frozen or processed product.

A diverse array of farming systems is available to hold and rear fish intensively, both at the hachery stage and for on-growing to market size. Salmon and trout farming systems encompass the entire gamut of running water techniques, from excavated ponds, through tanks and raceways, to floating cages and enclosures. After describing the farming cycle on freshwater and marine salmonid

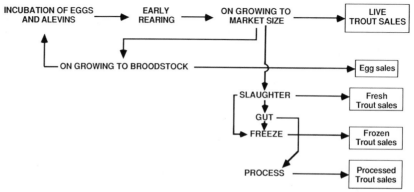

Fig. 3.1 Flow chart of the farming processes leading to different trout products.

farms, this chapter details the hatchery and on-growing systems commonly used in terms of farm design and also husbandry procedures. The latter entails feeding, grading, transport and harvesting methods as well as pond maintenance and effluent treatment.

Much of the information given specifically for salmonids has more general relevance to other intensively farmed finfish. However, there is growing interest in recirculation systems which have limited application to salmonids, hence the inclusion of a separate section on water reuse, written with special reference to intensive eel culture.

3.2   THE SALMONID FARMING CYCLE

As described in chapter 2 (section 2.3.1), growth rate is related to water temperature, level of nutrition and the genetic composition of the particular strain. Season and photoperiod may also influence growth rate and certainly affect the onset of sexual maturity, and hence play a crucial role in the growth cycle, dependent as it is on the availability of salmonid eggs or fry. For instance, mature rainbow trout held at ambient temperature in the northern hemisphere are artificially stripped of their eggs during the October−January period; the seasonality of spawning should therefore result in a seasonal pattern of harvesting. In practice, however, fluctuating water temperatures and uneven growth rates among the various farm stocks will tend to lessen this overall effect on the growth cycle. In rainbow trout the total farming cycle from production of eyed eggs (Plate 3.1) to harvest of 200 g fish typically varies from 10 to 20 months, depending on water temperature. However, many trout farmers purchase young stock from specialist fry or fingerling producers and concentrate their efforts in 'on-growing' for the table market.

In the case of anadromous salmonids, such as Atlantic salmon, the seasonal influence is mediated through both spawning and smoltification. Salmon fry during their first summer already show signs of a 'bimodal' distribution of unit weight with two distinct weight bands of fish emerging. The larger fry will become 'S1' smolts the following May or June, whereas the smaller fish will take a further year to smoltify as 'S2' smolts or even longer (e.g. as 'S3' or 'S4' smolts) before becoming adapted to seawater life (Plate 3.2). A similar pattern exists during the subsequent seawater

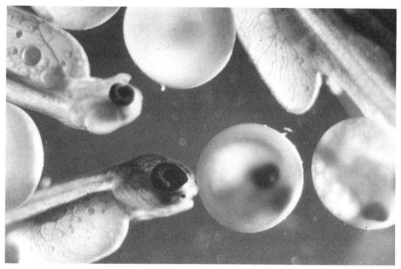

Plate 3.1 Close-up view of rainbow trout eggs and alevins (note embryonic eye visible within the egg).

Plate 3.2 Comparison of Atlantic salmon smolt (upper) and parr (lower).

phase, with a proportion of fish maturing as 'grilse' after approximately 12 months in the sea. The balance of the stock is usually harvested at a larger size the following summer after a further year at sea before they become mature, deteriorate in quality and start to lose value (Fig. 3.2). There is great variation in harvest

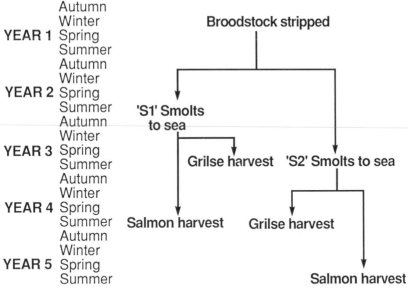

Autumn
Winter   **Broodstock stripped**
YEAR 1 Spring
Summer
Autumn
Winter
YEAR 2 Spring
Summer   **'S1' Smolts**
Autumn   **to sea**
Winter
YEAR 3 Spring
Summer   **Grilse harvest   'S2' Smolts to sea**
Autumn
Winter
YEAR 4 Spring
Summer   **Salmon harvest   Grilse harvest**
Autumn
Winter
YEAR 5 Spring
Summer   **Salmon harvest**

Fig. 3.2 Farming cycle for Atlantic salmon in Europe.

weights of Atlantic salmon, but they are typically about 1.5 kg for grilse and 3−4 kg for salmon (Plate 3.3). The total growth cycle from the egg to marketable fish ranges from about 2½ years for S1 grilse to about 4½ years for S2 salmon. Thus seasonal factors affect the availability of livestock inputs to the farm and hence the growth cycle.

Plate 3.3 Comparison of Atlantic salmon (upper) and grilse (lower).

3.3   THE HATCHERY

3.3.1   General procedures and water supply

A hatchery is that part of a farm or a specialised type of farm where the early incubation and hatching of eggs takes place. Hatcheries are also usually responsible for the swim-up or first feeding stage of development and the early growth of the fry. There is considerable variation amongst farms in the time that fry are moved out of the hatchery. Some maintain their fry in incubators and/or troughs within the hatchery until the fish are several weeks or months old, whilst others transfer their fry outside into larger, 1–2 m, diameter circular tanks or 1 m wide raceways of varying length, as soon as they reach 1000 fry/kg in size (i.e. an individual fry weight of 1 g or 450 fry/lb). Generally, the timing of the transfer of the fry from the hatchery depends on the availability of water and tank space; often it is carried out at the time of the first grading or sizing (section 2.6.3).

A characteristic feature of a good hatchery is the quality of its water supply. Salmonid eggs and fry require waters of the highest quality and although some hatcheries make do with surface water supplies, ground water sources are to be preferred because of their all-year-round constancy of temperature and their freedom from disease or pollution hazards. Ground waters are also usually free from silt, a contaminant whose presence in surface waters, particularly after heavy rains, can result in respiratory problems for both eggs and fry. Simple mechanical filters are now available, however, which can ameliorate this problem (e.g. sand filters). Spring and artesian well supplies are the best sources of ground-water, pumped systems being costly to run and subject to mechanical and electrical breakdown.

Ground water supplies, however, are not without their disadvantages. They may be poorly oxygenated and/or supersaturated with nitrogen, carbon dioxide and oxygen as a result of the greater solubility of all gases in water if it is subjected to increased pressure (e.g. when spring supplies are drawn from many metres below the surface). Supersaturation, particularly with nitrogen, can cause 'popeye' due to gas bubble disease when the high levels of gas come out of solution within the tissues of the fish (chapter 6; section 6.4.3). Ground waters may also suffer from high levels of mineral contamination (e.g. iron salts) as a result of the water being drawn from different soil strata. Again simple sand filters may be used to remove this contamination. Although constant, the temperature of ground waters is also quite low, particularly if compared with the temperatures of most surface waters during the spring and summer months. However, these minor difficulties are generally far outweighed by the advantages of having a disease- and pollutant-free water supply. Many of these problems can also be remedied with relatively simple and cheap methods. Thus the oxygen tensions can easily be improved by aeration. Aeration also serves to blow off any other dissolved gases present in the ground water. Splashing water over boards or through screens and/or the use of open channels or wide bore pipes to carry the water from its source to the hatchery also help in the removal of the supersaturated levels of gases. Supersaturation can also occur if there are air leaks to any pumps and associated pipework concerned with drawing up ground water supplies, or if any natural spring supply includes a high waterfall.

3.3.2 Egg incubation

After stripping from the brood female or henfish, salmonid eggs are soft and somewhat sticky to touch. At this stage and before fertilisation occurs, they are referred to as 'green' eggs. Green eggs can be moved from farm to farm for 24 hours following their stripping provided they are kept cool and do not come into contact with water. Usually, most hatcheries fertilise their eggs immediately, or at most after four to five hours, such as at the end of a day's stripping of a batch of broodstock.

  Following fertilisation and water hardening (see section 4.3), the eggs are transferred to a suitable incubator where they remain at least until they become 'eyed'. This is the stage of development when the eye of the fish embryo within the egg becomes darkly pigmented (Plate 3.1). The total egg volume of each batch of eggs, together with their size, should be measured within 24 hours of fertilisation. Some farms prefer to wait until eyeing to perform these measurements.

  After the first 24 hours has passed, under no circumstance should eggs be disturbed in any way for the next 10–15 days of incubation, as this will result in very high mortalities. At one time dead eggs used to be removed manually each day but this can disturb and damage adjacent healthy eggs. Now most hatcheries treat all their eggs every day or on alternate days with a 3−5 mg/l solution of malachite green. This minimises the fungal colonisation of dead eggs and prevents the transmission of infection to neighbouring healthy ones. Eggs should also be shielded from light by black covers over the incubators to avoid any damage to the eggs by ultra-violet radiation. Hatcheries should only be equipped with subdued lighting and on no account should sunlight be allowed to fall directly on eggs.

As soon as water hardening and any necessary measurements of the size and numbers of the fertilised eggs have been completed, they should be laid down in an appropriate incubator. Broadly speaking there are two major types of system for incubating eggs. The first is the hatchery tray and its many variants and the second is the hatchery jar or vertical incubator (Fig. 3.3).

Hatchery trays, sometimes known as Californian egg baskets, vary in size depending on the manufacturer, but are usually 40−50 cm square by 15 cm deep. Their base is perforated with elongated or rectangular holes whose aperture is such that it retains eggs but allows the free passage of water. Each tray is usually filled with a

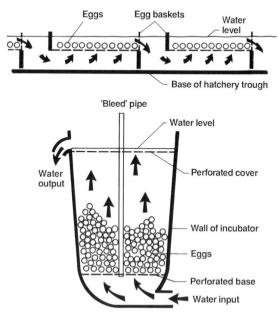

Fig. 3.3 Sectional views of conventional hatchery trays in a trough (upper) and a vertical incubator (lower).

single or double layer of eggs. Generally, most trays take about a litre of eggs, i.e. 6000−8000 salmon or 10000−20000 rainbow trout eggs, the exact number depending on the physical size of the eggs. The outside dimensions of each tray are such that they can be accommodated in series along the length of a fry trough or raceway. Water enters at one end of the trough and leaves from the other after passing in turn over the eggs in each of the serially arranged hatchery trays. Troughs can be made from concrete, aluminium or wood, but most commonly from fibre-glass. Each trough usually takes five or six hatchery trays, although some farms have troughs which operate with ten or more. The trays and troughs are usually built so that the perforated base of the tray is supported several centimetres above the floor of the trough. This encourages the flow of water through the perforations in the tray and up through the eggs.

Hatchery trays of modern design closely fit the troughs and only allow water to pass under the front or leading edges of the trays. The back edge of the tray projects down to the bottom of the trough making a water-tight seal (Fig. 3.3). Consequently, the only way that water can pass is up through the perforations in the bottom of the hatchery tray. This flow is further directed because water can only pass out through perforations in the back edge of the tray. Similar flow patterns also occur through the remaining trays in each trough. Earlier design promoted the upward passage of water through the bottom of the hatchery trays by having separate baffles before each tray to divert the water flow down-wards. The angling of the mesh of the tray base also encourages this movement. Although a range of flow rates is used by different hatcheries, in general each litre of eggs requires 3−5 l/min inflow. Thus conventional hatchery troughs with five or six trays usually run with 15−20 l/min total inflow to the trough. Water supplies to the troughs in the hatchery can be provided by pipes and valve controls. However, many hatcheries prefer an open duct system with adjustable elbow-bends to the individual troughs. The open channels are easier to clean and inspect, and the elbows less likely to block than valves. The open channels also help alleviate any problem with supersaturation.

There are many forms of egg incubation system which are basic modifications of the hatchery tray. Egg boxes, also known as Atkins or Montana boxes, are three to four times deeper than trays and can therefore accommodate up to 5 l of eggs per box, the eggs being stacked in multiple layers rather than in the one or two layers of the egg trays described above.

Hatchery trays can also be vertically stacked with up to 20 trays per unit. The water enters at the top of the stack where it is first forced up through the perforated base and eggs of the uppermost tray as described above. The outflow from this tray is then collected and splashed down to the inflow of the tray beneath and so on to the bottom tray. The splashing at each level maintains the oxygen levels for the successive trays of eggs. Such trays are often circular (Plate 3.4) and each tray can be slid out from the stack for inspection and cleaning. Of course this inspection should not be carried out during the early stages of incubation unless absolutely necessary because of the risk of damage to the eggs. The vertical stacking of trays in addition to saving floor space is also economical of water, using only 15−20 l/min for each 16-tray incubator.

The second type of incubator is the hatchery jar or vertical incubator (Plate 3.5) which holds 5−100 l of eggs. At its simplest this is an upturned jar or bottle with the bottom cut out. Water enters from the bottom of the jar through its upturned neck and then overflows over the top having passed through a solid column of eggs (Fig. 3.3). Large incubators are often made from refuse or dustbin-like containers. In all cases vertical incubators for salmonid eggs need to have a flat perforated screen at their base to retain

Plate 3.4 Close-up view of eyed trout eggs in a circular egg tray.

Plate 3.5 Vertical egg incubator and associated egg trays.

the eggs and allow the inflow of water. Often the perforated base is covered with a layer of gravel, the purpose being to spread out the water inflow so that all parts of the egg column are supplied. Some units have a pipe or tube of wider dimension than the water inflow tube, running from below the perforated base to a position above the top of the incubator. This is used to bleed off any air bubbles entering with the water, which would otherwise interrupt or short-circuit water flows. The bleed pipe can also be used to administer fungicides.

Vertical incubators are used by most hatcheries which have to handle large numbers of eggs because, like the stacks of hatchery trays, they save floor space and conserve water. Some vertical incubators, holding up to 100 l of eggs, run on as little as 20 l/min. However, most hatcheries operate with 3 l/min for each 4−5 l of eggs, less than a quarter of that required for the conventional hatchery tray.

The small volumes of water required for vertical incubators and the vertical stacks of hatchery trays also open up the possibility of artificially controlling the temperature of the water inflow and hence the rate of development of the eggs and fry. Heating to 14−15°C in particular enables fry to be produced several weeks in advance of batches incubated under ambient or ground water temperatures. This may be an economic proposition even if the heated water is allowed to discharge to waste. Much better, however, to recirculate the small volumes of water through a biological filter or use a heat exchanger or heat pump. The rate of development of the eggs can also be slowed down by cooling the water supply (see chapter 4; section 4.3.3). Again this can be more economically achieved by recirculating the cooled water.

A major disadvantage of the large 100 l vertical incubator is the enormous number of eggs which are carried in a single unit. All vertical incubators are more vulnerable to the spread of fungal infection because of the close, three-dimensional packing of the eggs; much better to spread the risk by using smaller, 15−20 l systems. It is also recommended that farmers use vertical incubators constructed from clear glass or perspex, rather than the opaque plastic which is commonly used, so that the eggs can be seen at all times and any sign of incipient infection of blockage remedied as it appears.

Whatever system of incubation is used, at eyeing the eggs should be 'shocked' by pouring or siphoning the eggs from one container to another placed approximately 50−60 cm below the first. Unfertilised or damaged eggs are killed by this shocking or addling and their yolk proteins turn white and opaque. In contrast healthy eggs are quite resilient at this stage and remain undamaged by the shocking. The dead eggs can be removed by manual picking with a suction pipette or by salt or sugar flotation. In the latter treatment salt or sugar is added progressively to a container in which the eyed eggs have been placed. At a certain concentration, which differs with each batch of eggs, the dead ones float and the viable ones remain at the bottom, thus allowing the former to be removed. However, most hatcheries now usually buy or rent egg-picking machines which can sort dead and healthy eggs at rates upwards of 1000 eggs per minute. During processing all eggs are passed across a beam of light and the white, opaque dead eggs, through which the light cannot pass, are rejected by a photo-electronically operated air jet mechanism. After shocking, picking, and volume and size measurements, as required, the eyed eggs are returned to the tray or jar incubator so that they may complete their development up to hatching.

3.3.3   Fry systems and first feeding

After hatching in conventional hatchery trays, the yolk-sac fry or alevins (Plate 3.1) fall through elongated perforations in the base of the tray into the trough below, leaving behind any dead eggs, discarded eggshells and deformed fry. After all of the eggs have hatched, the egg trays and their contents are removed leaving the fry to continue their development and growth in the trough. Some Atlantic salmon farms cover the floor of the troughs with sheets of plastic media or gravel to provide a refuge for the yolk-sac fry. Although the presence of a medium does give improved growth, there is, however, a danger it may also provide a reservoir of infection for the fry. More recent salmon hatcheries tend to have larger trays or baskets with two layers. Eggs are placed on the higher mesh layer and the yolk-sac fry drop through to the substrate layer beneath. The tanks of fry should also be covered because, although less affected than eggs, exposure to strong sunlight may produce abnormalities and additional mortality. Yolk-sac fry also grow faster if they are maintained in dim light or darkness.

When vertical stacks of hatchery trays, egg boxes or true vertical/ jar incubators are used, the yolk-sac fry have to be removed as soon as the first of the batch show any signs of 'swimming up', i.e. they leave the bottom of the tank and swim towards the surface of the water in order to take food. Transfer of the fry from incubator to tank or trough should always be accomplished by floating or immersing the tray and fry in the new tank and allowing the alevins to swim out under their own volition. On no account should the fry be tipped out or netted. At the swim-up stage it is important that the water depth in the tank is no more than 15 cm to allow the fry to reach the surface of the water easily. Water inflow should be of the order of 1 l/min for each 1−1.3 kg of fish.

After hatching probably the most significant hurdle in the development of young salmonids is the time of first feeding. It is essential that the young fish get off to a good start. This means that the diet, its timing and formulation and the frequency of feeding must be optimised. Fry prevented from taking food when they are willing to feed quickly lose weight and die. More importantly, they seem unable to recover their health or feeding response if subsequently food becomes available.

There is a considerable range of opinion as to the most appropriate time to begin feeding. Some farms begin when 50% of the fry have shown signs of swimming up whilst others start as soon as 70–90% of the yolk sac has disappeared. Realistically the best time to feed is when the fry are willing to consume food; this point can easily be established by spreading a little food on the water surface once or twice a day as the expected date of swim up approaches.

As soon as feeding commences, the fry should be fed for 20 hours each day using automatic feeders. As the numbers of fry which are feeding and the amount of food fed increases, so the inflow of water should be increased. At this stage of development the fry will be consuming 5–10% of their bodyweight per day in food and doubling their weight every week or so. With continual growth of the fry, upward adjustments in the size and amounts of the diet will ensure maximal feed intake. However, on no account should fry be moved on to a diet with particles which are a little large for some of the population. There is often a temptation to do this to save buying further bags of the smaller sized diet, a decision which will lead to stock losses and poor growth. By continually monitoring growth and performance and feeding optimally, at 10°C the best rainbow trout farms will get their fry up to 4–5 g (approximately 200 fry/kilo) 120–130 days after fertilisation. This is the stage when many hatcheries sell on their stock to production farms. Faster growth can be achieved by artificially warming the water and for many salmonids 14–16°C is considered the optimum temperature for growth.

After fry have reached 500 fish/kg they can be stocked in any form of tank or raceway. By this stage any potential risks of infection with infectious pancreatic necrosis virus (IPN) ought to have passed, whereas fry which are to be moved to earth ponds should remain in concrete or fibre-glass tanks or raceways until 16–18 weeks post-fertilisation to reduce the chance of infection with the protozoan parasite *Myxosoma cerebralis*, which causes whirling disease (chapter 6). There should also be nets or covers over all the fry tanks because fish of such small size will be eaten by many different predators.

Water inflows for each kg of fish can be reduced as the fry get larger. Thus 1 l/min of water inflow will support 1.0−1.3 kg of fry up to 1 g in size, 1.5 kg of 2.5 g fry, 2 kg of 5 g fry and 2.5 kg of 10 g fry. In each case double these flows should be available for tank cleaning and for emergencies.

3.3.4 Fingerling and smolt production

After hatching, most trout fry are stocked in concrete troughs (Plate 3.6) until they are moved out to larger tanks or earth ponds. As far as the farming of trout in freshwater is concerned, fingerling production relies heavily on the principles and techniques used in the culture of fry and with on-growing. Adjustments of feed rates and pellet sizes and the use of larger tanks and other enclosures are often the only alterations in methods employed by farms. By contrast, for the Atlantic salmon and several other farmed salmonids, it is the period of development between the fry and production-sized fish which is arguably the most important in the life cycle. During this period the young fish, which is known as a parr, becomes a smolt and is able to move from a freshwater to a seawater existence.

Atlantic salmon smolts are produced approximately one or two years after hatching, being known as S1s or S2s respectively (see section 3.2). Most farms strive for the maximum number of S1s

Plate 3.6 Concrete troughs for trout fry (note automatic feeders using compressed air).

because S2s occupy the smolt production tanks for two years rather than one year. Although at present it is not fully established why some fish become S1s whilst other from the same batch become S2s, it is clear that size is an important determinant. Attaining a body or fork length of 8–12 cm in September/October appears to be a prerequisite for smoltification the following spring, particularly if the water temperature remains at 10°C or above. Parr which fail to reach this size invariably become S2s. Increasingly, farms are looking for improved diets and feeding methods, together with the incubation of eggs and culture of fry and parr in heated waters, to improve growth and hence the overall percentages of S1s produced. Greater numbers of S1s can also be achieved by using earlier spawning strains of salmon or by photoperiodic advancements of maturation (chapter 4; section 4.3.3). By making eggs and fry available at earlier times, these methods provide longer periods of growth to enable the pre-smolt to attain the necessary minimum size to become an S1.

During the period up to smoltification, parr are usually maintained in circular tanks of 4–5 m or more in diameter, although there is some evidence that (at least for the first 6 months or so of development) they grow better in raceways. Just before release S1s would probably be in 10 m diameter tanks whereas the larger S2s may be over-wintered in even bigger tanks. All tanks should be partially covered because parr are shy by nature. Parr are usually graded in the Autumn or Fall and the different growth modes placed in separate tanks. This often helps the smaller fish because they take food and grow better once their larger siblings have been removed.

One of the major difficulties of smolt production is deciding on the optimum time for the transfer of the smolts into salt water. Smoltification involves marked changes in behaviour, body shape and colour and the development of a tolerance for seawater. All of these characters have been used to judge the timing of transfer. However, some of these changes are sudden in occurrence while others take several weeks to develop. Thus, the condition factor (weight (g)/fork length (cm^3)) × 100 decreases progressively for a month or so before smoltification from values of 1.3 or more down to 1.0. The characteristic silvery colour of smolts also develops over a similar time scale (Plate 3.2). In contrast, changes in behaviour and salinity tolerance occur more quickly. From a bottom-living existence as parr, smolts become pelagic or surface-swimming and gregarious, often forming shoals. The direction of their swimming is also reversed from being against the current to being with the current. These changes often occur within a few days. Smolts also tend to swim less actively than parr, a trait which in the wild would tend to carry them downstream, i.e. towards the sea.

Sometimes even with these changes salt water tolerance is not absolute. Consequently farms with supplies of both fresh- and saltwater adapt their presumptive smolts to saltwater by introducing progressively increasing levels of seawater in the freshwater inflows to the tanks. Increased levels of salt in the diet may also help adaptation. Similar techniques are also used to acclimate trout which are to be stocked in sea cages. Trout fingerlings weighing 30−50 g can be transferred successfully, although the best survivals are achieved with 60−70 g fish.

Smolts should also be adapted to lighter conditions before transfer to sea cages otherwise they may attempt to burrow away from the much stronger light in the cage into the mesh of the net; this can cause serious damage to the heads of the fish and scale loss. Extreme care should be exercised in the handling of smolts. In fact direct handling should be avoided and any transfer of smolts to sea cages or transport tanks should be carried out via water channels or by using a fish pump (section 3.5.4).

3.4   ON-GROWING

3.4.1   Introduction

On-growing fish to market weight traditionally takes place using excavated earthen ponds into which fingerlings are stocked at a unit weight of not less than around 5 g, which equates to a unit length of about 3 in, i.e. 6−8 cm. Such ponds still represent a very cost-effective means of rearing fish where land is fairly flat, inexpensive and easily excavated. The highly efficient Danish trout farming industry relies almost entirely on earthen ponds and typically each farm comprises groups of ponds arranged in a herringbone pattern around a central outlet channel (Fig. 3.4). A dam is installed across a suitable water course to give sufficient head of water, which is then diverted down an inlet channel, from which it passes, via wooden inlet monks or sluices, into the fish ponds (Plate 3.7). The ponds are in parallel and drain into the outlet channel which

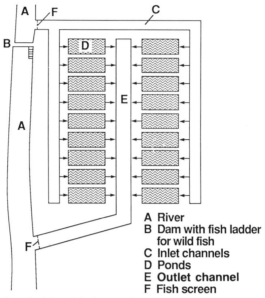

A River
B Dam with fish ladder
   for wild fish
C Inlet channels
D Ponds
E Outlet channel
F Fish screen

Fig. 3.4 Plan of typical Danish farm using excavated earthen ponds.

Plate 3.7 Excavated earth pond for trout production with wooden inlet screen in foreground.

also contains fish; thence through an outlet screen back into the river. Such earth ponds are usually 1–2 m in depth and, although water flows continuously through them, the rate of flow is sufficiently low that fish do not need to expend energy in keeping station, which in turn favours efficient conversion of feed to flesh.

By contrast with the traditional excavated earthen pond, it was found that narrow ponds had the advantage that fish could be more easily observed and certain husbandry procedures were facilitated (e.g. chemical flush treatments; grading and harvesting). Provided such ponds were lined or constructed in material that did not erode (e.g. concrete), much faster flow rates could also be used, enabling correspondingly higher stocking densities of fish. So-called 'raceways' (Plate 3.8) became a popular means of rearing trout and salmon, particularly in the USA, and these more intensive conditions allowed considerable economy in the surface area of land needed for fish farming. Another approach has been to use circular ponds (Plate 3.9) with a central outlet, so that fish are more evenly distributed in the water rather than congregating at the top end of a raceway unit. The circular flow also gives a self-cleaning effect with metabolic wastes being flushed in the resulting vortex towards the central drain. However, circular ponds can be criticised for causing greater contact between the fish, disease organisms and waste materials than in raceways and also for being less suited to undertaking chemical treatments.

More recently cage farming of salmonids has come to the fore, particularly for marine use. A simple arrangement of bag nets attached to a floating frame represents an economical rearing unit (Plate 3.10). Provided water circulation is sufficient for the quantity of fish stocked, such units can be attached to form a pen group for greater overall stability and ease of servicing.

Clearly a wide variety of different salmonid on-growing systems

Plate 3.8 Concrete raceways for trout production (note fish being fed on pellets blown from a mobile hopper).

Plate 3.9 Cleaning outlet screens on a trout farm which uses circular tanks (note 'demand' self-feeders).

Plate 3.10 Close-up view of floating net cages for salmon production (note walkways around perimeter).

has evolved from the original earthen pond. In many cases, the choice of system will reflect local factors, such as the amount of water available, surrounding terrain and access. This usually comes down to the most cost-effective mean of exploiting site potential in terms of capital investment, commensurate with good husbandry and low operating costs.

3.4.2 Efficiency of utilisation

Efficient use of water for fish farming presupposes that all the water available is being exploited, whereas in practice there are usually legal constraints on the quantity which may be used. For example where water is being abstracted from a river, the rights of other users have to be considered and there may also have to be sufficient residual flow in the river to permit wild fish stocks to thrive. In certain countries the right to abstract is governed by the terms of a licence which may also take into account the likely deterioration in water quality due to fish farm effluent. Although a river-fed trout farm normally returns its water to the river un-diminished in volume, it will impose an organic loading and BOD which can be detrimental to the river immediately downstream, particularly if the effluent water is insufficiently diluted when too high a proportion of water flows through the farm relative to its size and polluting capacity. This can become an even more severe problem in the case of floating cage or enclosure systems if metabolic waters are not carried away by currents but accumulate beneath the farm and threaten fish survival. It may be necessary to pump water to exploit certain sites because of insufficient gradient or site topography (e.g. shore-based marine farms) and some cage farmers install submersible pumps to increase water circulation. This entails capital investment in pumps and standby generators, together with increased energy utilisation in pumping costs.

In order to maximise the utilisation and carrying capacity of a site it is important firstly to ensure the incoming water is fully saturated with oxygen. If this is not the case, then splashboards and aerators should be installed to rectify the situation. By the same token, it is usually possible to re-oxygenate the water using aeration systems after it has passed through the fish ponds. By this means, multiple use can be made of the water as it passes through a series of ponds. If the dissolved oxygen level can be increased above 5.5 mg/l after initial use, then clearly further use may be made of the water, but care must be taken to ensure that faecal wastes and suspended solids are rapidly flushed away. The next limiting factor is usually ammonia, especially under conditions of alkaline pH, whereas under acidic and soft water conditions, other factors, such as carbon dioxide, may become limiting. Although recirculation systems are used in some hatcheries, to date their application to commercial scale on-growing of salmonids has been considered prohibitively costly because of the large quantities of water involved. Sedimentation tanks collect faeces and waste feed, but it is also necessary to install massive biological filters to strip out the ammonia. Nonetheless experimental recirculation units are available which operate on not more than 10% input of fresh water, at which rate it can become worthwhile heating the water depending on power costs, as described for intensive eel cultivation (section 3.6).

Increased stocking density enables a reduction in the surface area of land needed to rear fish, together with advantages in fish husbandry. As with battery chickens, production of fish per unit area becomes a meaningless index and the intensive fish pond, tank, raceway or cage should be considered as a three-dimensional rearing unit. Provided water quality is maintained, stocking density of salmon or trout growers can be increased to over 32 kg/cubic metre (about 2 lb/cubic foot) without loss in performance. Figure 3.5 indicates how the required rearing volume increases for fish below approximately 50 g in unit weight. Although smaller fish have a higher oxygen requirement than larger fish, their main problem at high stocking densities is excessive energy expenditure in swimming to keep station against the high water flow rates needed to ensure satisfactory water quality. Too high rates of flow will quickly cause fry or fingerlings to become exhausted and the resulting mass of fish may then suffocate and die pressed against the outlet screen. There has been insufficient research on the

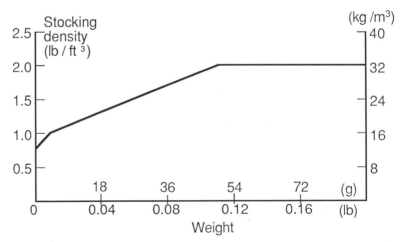

Fig. 3.5 Upper limits of stocking density recommended for Pacific salmon parr at different body weights (after Mayo, 1971).

effects of high stocking density upon performance, but it seems obvious that efficiency of feed conversion must decline as fish are forced to swim more actively. The corollary may be that enforced exercise is likely to improve muscle tone and hence flesh texture in the final product.

It follows from the above that in evaluating the efficiency of utilisation of water for fish farming, stocking density must be related to water flow. This may be illustrated by comparing the carrying capacity of different fish farming systems. Consider the hypothetical example of two rainbow trout rearing units with the following characteristics:

Pond A can carry 10 g rainbow trout at the rate of 1.7 kg of fish per l/s inflow.
Pond B can carry 10 g rainbow trout at the rate of 0.3 kg of fish per l/s inflow.

On this basis pond A appears more efficient than pond B, but the preference is reversed when more information is available as follows: Pond A is a large earthen pond with a rearing volume of 920 cubic metres. The inflow is 8 l/s and the capacity is 2 tonnes of 10 g rainbow trout at a stocking density of 2.4 kg/cubic metre of water. Pond B is a raceway unit with a rearing volume of 68 cubic metres. The inflow is 380 l/s and the capacity is 2 tonnes of 10 g rainbow trout at a stocking density of 33 kg/cubic metre of water. The differences in carrying capacity reflect the considerable difference in water exchange rate between the two units. The water in the raceway unit is exchanged every three minutes, whereas the water in the earthen pond is exchanged not more than every 32 hours; indeed the periphery of the earthen pond may remain largely stagnant with effective water exchange only occuring down the centre of the pond. As a result the stocking density of trout in the raceway unit at 33 kg/cubic metre is approximately fourteen times higher than in the earthen pond.

3.4.3 Rearing unit design

Capital expenditure for on-growing systems is required for land, excavation, rearing units, dams, pipelines and valves, buildings, plant and machinery (such as pumps and processing equipment). Capital cost is directly proportional to rearing volume and there is immense diversity in the design of rearing units for on-growing salmonids and other fish.

Excavated earth ponds (Plate 3.7) typically have dimensions 30 × 10 × 1.5 m deep, hence a rearing volume of about 500 cubic metres. Flow rates of 10–15 l/s enable average stocking densities of over 2 kg/m³ with at least three to six water exchanges per day. If land is cheap and easily excavated, such ponds are constructed in parallel. They are usually unlined, but if the ground is porous, the earth bottom may be covered with a layer of clay, and butyl sheeting used around the edges to prevent erosion. If ponds are fully stocked, most of the solid wastes will remain in suspension and drain out. However, there may be insufficient fish present to churn up the water and flush out solid wastes, which therefore accumulate on the pond bottom. Provided the pond is not overloaded, biological processes in the mud break down this waste material and a degree of self-cleaning is achieved. Fish are collected for harvesting by lowering the water level and spreading a seine net across the pond before pulling it down the entire length. Smaller fish may be flushed via the outlet monk into holding pens within the outlet channel before being graded and then transported to the appropriate ponds for further on-growing. Bar grading units on an earth pond farm were traditionally moved up and down the central outlet channel for convenience of operation, although such systems have increasingly been replaced by automatic fish pumps (section 3.5.3). Many earth pond trout farms have a concrete holding tank in which fish are kept for one to two days before marketing to reduce the risk of a muddy off-flavour in the final product. In order to improve access to individual ponds, some Danish earth pond farms have installed a simple railway along which feeding carts are propelled by hand.

Raceways (Plate 3.8) were pioneered in the USA and represent a more intensive approach to salmonid farming. They generally comprise parallel sets of narrow channels and are often constructed in sequential blocks with two or three raceway sets in series, so that a splash-down occurs from one set to the next in order to reaerate the water. Typically concrete raceways are about 30 m long, 3–10 m wide and less than 1 m deep. Fast flow rates down these more narrow rearing units mean that water is exchanged three times per hour or more, enabling stocking densities of at least 32 kg/m³ or more. Unlike earth ponds there is no risk of eddying with solid wastes settling out in low flow areas. Raceway design ensures homogeneous water quality across the width of the unit, although there is a progressive deterioration down its length. Compact design means that vehicular access is usually possible at any point, chemical flushing can be readily undertaken to control diseases, each unit can be quickly drained, and grading can be undertaken conveniently within the raceway.

The gradient on raceway farms is commonly 15 cm in 30 m and by allowing the water to drop 15 cm or more between individual raceway units, up to eight units can be operated in series on a water supply of 85 l/s. Some raceway systems are constructed with sloping sides and by reducing the depth of water, each unit can be converted from holding a large volume and surface area of water with a slow current to a virtual mill race of much smaller rearing volume.

Circular tanks are usually constructed in concrete, corrugated metal or fibre-glass with a peripheral water inlet (Plate 3.9). Water flows in a circular fashion around the tank, draining out through a central outlet comprising a standpipe surrounded by a screen. A typical circular tank is 4 m in diameter with a water depth of 0.75 m and an inflow of about 4 l/s will readily enable 200 kg of trout to be carried. Water quality is fairly homogeneous in a circular tank encouraging a more uniform distribution of fish compared with ponds or raceways. In theory pond hygiene is superior due to the self-cleaning effect of the central drain, but in practice this relies on the vortex created by flow rates which may be too strong for the fish. Also, circular ponds are not so well suited to automated fish handling methods.

Floating pen or cage systems (Plate 3.10) have become very popular for farming certain fish, such as Atlantic salmon, in seawater. However, floating pens also represent a highly cost-effective means of rearing fish in freshwater lakes, gravel pits, streams etc., given the appropriate water circulation and sufficient depth underneath the pens to discourage accumulation of wastes, (i.e. 3 m plus). Typically they comprise 1−2 cm square mesh nylon bag nets suspended from a square or circular flotation collar. The latter may have a timber or steel frame surrounded by expanded polystyrene floats, empty oildrums, etc. A square pen with dimensions of 10 × 10 × 4 m deep could be expected to carry up to 1 tonne of salmon if moored at a site with good tidal exchange. Recently there has been an upsurge of interest in the use of far larger marine cages in Japan and now also in Europe. For example, the Bridgestone cage is of flexible rubber construction with a capacity of 6500 m³, i.e. 10–20 times larger than traditional floating pens. The main advantages claimed are an all-weather capability allowing exploitation of more exposed sites and lower stocking densities resulting in enhanced growth rates, together with reduced capital and maintenance costs for a given level of production.

Walkways around each pen facilitate fish observation and handling and a group of pens may be conveniently arranged around a central service platform and feed store. Like other fish farming systems, it is advisable to discourage bird predation by means of top nets or overhead wires, but most floating pens are usually also surrounded by underwater predator nets to discourage seals, diving birds, etc. Water exchange is slowed down by the growth of algae and marine organisms on the netting and it may be necessary to use anti-fouling chemicals on the nets or to change nets if this becomes a problem. The main disadvantage of pen systems is the need for boat access unless connected to the shore by means of a walkway or pier. In the case of marine farms, floating pens are often vulnerable to storms unless moored in a sheltered cove or bay. Various other rearing options are also available to the marine farmer (Fig. 3.6) including shore-based tanks using pumped seawater, fixed enclosures, mid-water pens and seabed units. Increasing attention is also being focused on the scope for using large submersible pens in order to avoid the worst effects of storms, typhoons, etc.

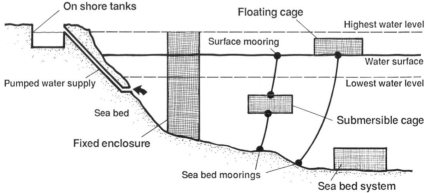

Fig. 3.6 Range of rearing options open to the marine farmer (after Milne, 1972).

3.5 HUSBANDRY PROCEDURES

3.5.1 Cleaning and routine maintenance

Providing holding tanks are suitably designed, fully stocked with fish and receive appropriate water flows, most concrete, plastic, butyl-lined and fibre-glass tanks are mainly self-cleaning. In contrast, many earth ponds and also some tanks of earlier design and construction build up deposits of uneaten food and fish faeces, which must be removed as part of the routine maintenance of the farm if the general health of the stock and/or the water quality are not to deteriorate. Every six months or so earth ponds should be emptied, allowed to dry and silt accumulations removed; alternatively excessive build-up of sediment can be removed with a sludge pump. The shape of the earth pond, the slope of its banks and the velocity of water flowing through it also affect the amounts of sludge deposited. A temporary lowering of the height of external standpipes or removal of the top board or monks, which are the usual methods of adjustment of water levels in earth ponds, will by increasing water flow tend to flush out sediment into the effluent channels. Feeding fish close to the effluent screen of earth ponds also helps sludge removal because fish, like many other animals, produce faeces immediately after feeding; their proximity to the effluent flow ensures that the faeces produced are swept out of the pond rather than falling to the bottom as they would if the fish were fed at other parts of the pond.

Probably the most important of all the routine maintenance procedures on the farm involves the checking and cleaning of the main intake screens for the water supply to the farm. This becomes especially important in the Autumn or Fall during times of heavy leaf fall. Some farms with recurring problems install automatic devices for removing the leaves and other debris present in their water supply. Farms overhung by trees may suffer leaf fall into all tanks and during certain periods of the year all tanks will need to be checked at least once or twice a night, particularly during concomitant high winds. Motor-driven revolving circular screens to the water inflow are more trouble-free than static screens, although they should not be relied upon either to remove leaves or to avoid problems due to obstruction by larger objects. Motorised equipment is also subject to breakdown and must be regularly checked and serviced. Floating booms can also be used to prevent larger debris from reaching the main intake screens. All farms should have float alarms which warn of reduced and/or increased water flows to the farm; these can either ring alarms if staff are on site or be programmed to page personnel by telephone. Some farms have similar low water devices behind the screens of all their larger and/or important tanks and ponds.

All outflow screens, whether centrally placed in circular tanks, or at the end of ponds or raceways, should be cleaned as often as is necessary and at least one to three times a day with a stiff brush, otherwise water flow will be restricted and still water areas develop (Plate 3.9). Dropping the standpipe or top boards of monks during this cleaning will help to ensure that all the debris dislodged from the screen is swept from the tank and not back into the stock. If present, inflow screens should be subject to the same cleaning. Mesh screens with oblong rather than round holes help with cleaning. For earth ponds wooden screens are easier to manoeuvre and less subject to jamming after frost than metal ones. At regular intervals (weekly), all the internal surfaces of fabricated tanks and raceways (i.e. concrete, metal or fibre-glass) should be brushed to remove deposits of food and faeces and any algal growth.

All dead, dying or unhealthy looking stock should be removed. Although dead fish are found amongst healthy stock even on the most well-run farms, it is essential that they are removed as soon as possible if they are not to become a potential source of infection for other fish. Their removal is particularly important in earth pond systems because they may sink into the mud at the bottom of the pond and remain a possible hazard until the pond is drained.

Effluent channels or pipes from tanks and ponds should also be kept clean and free from fish. Accumulations of debris within these channels may constitute a source of infection or, by decomposition, result in a worsening of water quality. Settlement lagoons should also be kept free of fish (which might otherwise stir up solids) and at least once a year the settled material should be removed mechanically or with a sludge pump. Because effluent channels and settlement ponds are close to the main water outlet from the farm, deterioration in water quality at this point may lead to difficulties. This may be a special problem for farms which use effluent or back channels and settlement lagoons to stock fish.

Walkways between tanks should be maintained, any stagnant pools should be drained or filled and the grass between earth ponds kept short, otherwise there may be access problems during grading, harvesting, bad weather and emergencies. Vermin, in particular rats, should be removed as their nests may cause bank erosion and leakage, particularly if they are sited near the outlet monks or pipes.

All cleaning and maintenance operations on the farm are made more acceptable to the husbandry staff if they have appropriate waterproof and all-weather clothes and boots. To avoid any possibilities of introducing disease organisms to the farm, these clothes should be used exclusively for work on the farm and not taken elsewhere. Entrances to the farm and especially hatcheries should have footbaths containing a suitable germicidal solution which everyone should use. Vehicles should be parked outside the area of the farm delimited by these baths.

Check lists should be available on tanks, ponds or raceways so that the husbandry staff can record that a particular cleaning or other maintenance procedure has been completed. The proper recording and collation of stock records and other information is an important duty of farm staff and should be encouraged. Often this activity is considered superfluous by some staff because at a particular moment of time they know exactly what stock is in what tank or pond and what treatments have recently been undertaken. With the passage of time and during problems out of normal working hours, or after staff changes or absences, however, such information is lost or not easily retrieved. Microcomputers and inventory software systems are now so cheap and easy to use that there is no reason why even the smallest farm should not be able to keep accurate stock and husbandry records. This provides a bank of information on which disease diagnosis and treatment and also longer-term decisions relating to feed levels and stocking strategy can be based. This is in addition to the provision of an immediate picture regarding the current stock position. Farmers are constantly faced with problems and decisions which might have been more easily solved had adequate records been taken of similar occurrences on earlier occasions.

If there are unexpected stock losses on the farm it is advisable for the staff on duty immediately to take water, food and fish samples and store them in suitable stoppered containers in the refrigerator or freezer. Farmers can benefit from knowing other sources of water discharge in their area, likely pollution hazards, and also from making their existence known to these other parties.

For cage sites it is essential that the structure of the cage, its angle joints, netting and moorings are checked at regular intervals. Generally, checking the moorings will require employment of a skin diver. Netting should be lifted regularly to check for predator and poacher damage. Nets should also be removed at intervals from each cage and air/sun dried for removal of weed, molluscs and other materials which block water exchange through the mesh. Walkways should be checked and any organic growth removed so that they do not become excessively slippery. All boats which are in constant use should be regularly serviced and equipped with back-up motors and emergency flares.

Whatever the type of farm, all electrical equipment, stand-by generators, aerators, feeders, float alarms and motorised screens should be checked and serviced at regular intervals. Maintenance should not be left until breakdown demands a repair. All this checking should become part of the routine procedures on the farm. Again check lists should be available to ensure that these duties have been completed; this also gives an accurate picture of the intervals between successive services of equipment.

Although most routine work can be conducted during normal working hours, it follows from the above procedures that all farms should arrange manning rosters so that at least one member of the farm staff is on call 24 hours a day, 365 days a year.

3.5.2 Feeding

When it is realised that feed costs constitute 40% or more of total production costs on most salmonid farms, then the importance of optimising all aspects of the feeding and nutrition of the stock can be appreciated. Feeding rates (amounts of food fed each day) will vary depending on the size of fish and the temperature of the water supply.

First-feeding or swim-up fry are generally fed to satiation which at 10°C is 7–10% of their bodyweight each day. As fish grow they take proportionately less ration each day. Thus, in the growth from fingerlings and yearlings, fish are fed 1–6% of bodyweight per day at 10°C, whereas at 5°C and 15°C they would take 0.5–2.5% and 1.5–7.5% bodyweight/day respectively. All these levels can be increased, possibly by 20–30%, in order to improve rates of growth although as satiation is approached, there is generally a reduction in the food conversion ratio or FCR (calculated by dividing the weight of dry food fed by the gain in wet fish weight). There is considerable variation in the FCRs achieved by trout farms although most would expect to average an FCR of at least 1.5. Under ideal conditions FCRs of 1.0 or less are possible. Diets which are poorly digested by the fish or low in protein will produce FCRs of 2.0 or more.

In addition to feeding a ration of appropriate size, different sizes of fish will require pellets of different physical dimension. Feed manufacturers are often the best source of information about the most appropriate daily ration, diet formulation and size and methods of fabrication of pellet for particular stocks and sizes of fish and for feeding under differing environmental conditions. They should always be consulted because of their experience of similar farms and parallel conditions.

Methods of feeding and frequency of feed administration are also important in achieving nutritional effectiveness. Generally, first-feeding fry, whether they are trout or salmon, should be fed as soon as they will take food and they should be fed a little and often. In fact, many farms feed early fry stages continuously using automatic feeders (Plate 3.6). Larger fish need only be fed two or three times a day. In each case the recommended daily ration (% bodyweight per day) is divided by the number of feeds to be employed. Although the different feeds can be of equal amounts, some farms prefer to feed a disproportionately larger meal at the first feed of the day.

There are three major methods of feed administration: hand, automatic and demand feeding. Smaller farms often administer feed by hand thus saving some capital expenditure. Hand feeding also has certain other advantages over automatic feeding. Firstly, hand feeding enables the feeder to gauge the health of the stock from the strength of the feeding response, i.e. the avidity during the first moments after feeding and the general appearance and behavour of the fish (Plate 3.11). In trout farming the absence or reduction of the characteristic frenzied movements or 'boiling of the water' immediately after the pellets hit the water, together with a darkening of some fish and/or abnormal movements (e.g. flashing) are good indications of a water quality or disease problem developing.

Although salmon show a much less pronounced feeding response than trout, with experience their behaviour can still provide important information. Hand feeding also prevents unnecessary wastage of food because feeding can be stopped immediately any pellets

Plate 3.11 Checking feeding response on a rainbow trout farm.

remain uneaten by the stock (Plate 3.12). Expanded pellets can be helpful in assessing the feeding response during hand feeding because they remain on the water surface for longer than conventional pellets. Some farmers who have turbid or cloudy water supplies use feeding platforms, mounted a foot or so under the water surface, to show when food is not being eaten.

On cage farms for Atlantic salmon in particular, the temporary suspension of a sheet under the cage may also enable the amounts of uneaten food to be ascertained. Although such a system cannot be used every day, similar enclosures are used routinely for disease treatment without adverse consequences to the stock.

Although there is considerable advantage to be gained from hand feeding, for manpower reasons this method is not feasible for larger farms and most of these resort to automatic feeders driven

Plate 3.12 Hand-feeding salmon (note anti-predator net above cages).

by clockwork, water, compressed air, or electric or battery mechanisms. In general feeders powered by mains electricity are to be preferred because of their reliability irrespective of how they are motorised. Automatic feeders generally incorporate a timeclock device. This enables the exact number and timing of the feeds to be specified. Of course they enable the fish to be fed many more times each day than is possible with hand feeding. Daylight or solar-operated devices have been used effectively to time feeders on cage sites which are too remote or possibly too small to justify manning throughout the day.

Automatic feeders are particularly useful for feeding fry. First-feeding or swim-up fry benefit greatly from continuous feeding. Used in combination with extended daylength provided by artificial lights in the hatchery, fry may be fed 24 hours each day, although in practice it is better to feed for only 18−20 hours because all-day feeding may provoke gill problems in the stock. As the stock increase in size, so the frequency of feeding can be reduced. Thus division of the daily ration into eight and five portions are necessary for fish of 0.5 g and 1 g size respectively, whereas for larger fish the ration can be fed in only two or three feeds. Broodstock fish are able to take their complete daily ration in a single feed. The more feeds there are each day, the more constant are the levels of production of ammonia, suspended solids, BOD, etc., by the stock, a single feed generally producing a peak in effluent production some six to eight hours later. A major disadvantage of early automatic feeders was the small area over which the food was dispensed. Feeders now, however, are able to distribute pellets over much of the surface of tanks, ponds and cages. Models are also available which disperse moist diets without clogging.

Because of the advantage of hand feeding in providing information on the general health of the stock, in practice many farms (even the largest) feed only 60—70% of the daily ration of their stock automatically, with the remainder being fed by hand. Neither hand nor simple automatic feeding is sufficient to cope with the largest production units. On these farms food is distributed over the ponds or raceways by compressed air blowers usually mounted on a lorry or tractor (Plate 3.8). Compressed air systems are also used to deliver feed through pipes to feeders on cages which are remote from the shore.

Some farms use a third method of feeding, known as 'demand' feeding. Demand feeders are devices in which the appetite of the fish is used to define the amounts of food eaten. Usually, demand feeders consist of a food hopper with an aperture whose opening is controlled by a movable gate. Attached to this gate is a pendulum whose tip extends down into the water where it can be nudged by the fish (Plate 3.9). Lateral movements of the pendulum cause the gate mechanism to open, allowing food to flow out of the food hopper into the water. The fish are thus able to demand food. Disadvantages of demand feeders include the gate mechanism sticking open and/or inadvertent or unnecessary operation by fish, both resulting in food waste. Often farms using demand feeders, like those primarily using automatic feeders, combine these methods with a proportion of the feed being administered by hand.

Whichever method of feed administration, whatever feeding frequency or daily ration size is employed, the effectiveness of these regimes should be checked every one or two weeks for every batch of fish on a farm. These checks are carried out by weighing a representative sample of fish in a particular tank or pond. Knowing the total number of fish in the tank and hence their total weight, and the total amounts of food fed, enables the growth and FCR to be calculated and any deficiences in performance from the expected norm remedied accordingly. Careful recording of weights, growth, FCR etc., enables most farms to predict what performance is to be expected from their stocks under a variety of environmental conditions.

3.5.3 Grading and harvesting

Most stocks of fish and especially salmonids and other carnivores exhibit hierarchical or peck-order patterns of feeding behaviour. This means that some fish regularly receive more food than others in a population. In time these feed differences in intake of food are reflected by disparities in the growth of different fish. If stocks are poorly or irregularly fed, then these weight differences may become even more divergent. These ranges in size have important implications for commercial practice. Firstly, it is not possible to feed either the most appropriate ration size or size of pellet to a population or tank of fish where the individuals widely differ in size. Secondly, the presence of some fish, possibly twice the size of others, in the tank leads to bullying, tail and fish nipping by the larger fish and sometimes to cannibalism. To avoid this aggressive behaviour and enable optimum rations and pellet sizes to be fed, ideally all the fish in a tank or pond should be the same size. This is achieved by a sizing or grading of the stock. Larger and smaller fish from a population are combined with similar sized grades from other stocks in separate tanks leaving the medium grade or grades to continue. Three to five grades may be achieved at a single grading, although most farms only select three.

Unnecessary grading, however, is to be discouraged because inevitably it stresses the fish, they go off their food and lose growth for a day or so; some fish may be damaged and/or die. There is no hard and fast rule regarding the frequency of grading. Grade when size disparity becomes an obvious problem or when the fish have to be handled or moved for any other purpose, e.g. disease treatment, thinning of the stock, harvesting or selling on to other farms. In general, farms often grade once or twice in the hatchery, the first when the fry reach 500 fish to the kilogram and two or three times as fingerling, yearling and production stock. All fish to be graded should be starved for at least 12 hours before handling, except of course fry.

Grading can be carried out by eye and/or measuring boards, but this is inaccurate and impractical for larger farms. Hand-held grids consisting of a series of parallel bars or slits appropriately spaced to grade a specific size range of fish are commonly used (Plate 3.13). Some devices are available where the spacing can be adjusted to suit the fish being graded. Alternatively a series of graders is required. Larger, mechanised graders work on the same basic principle as the hand-held ones, although the bars are often tapered and able to revolve. The tapering allows fish of several different grades to be separated at the same time, the fish sliding along the length of the bars until their girth dimension enables them to slip through and be channelled along pipes into separate holding tanks (Plate 3.14). Water is constantly pumped over the rollers to reduce

Plate 3.13 Bar grader for salmonid fry.

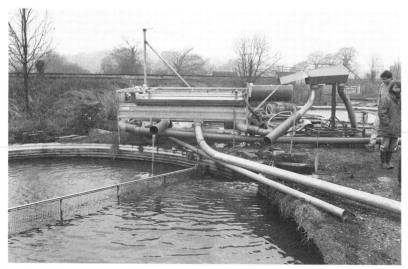

Plate 3.14 Fish pump and grader in use on trout farm.

frictional damage to the fish. Some graders use motorised rotating belts instead of rollers; these are better if water temperatures are very low (and the fish lethargic) because they do not depend on the movement of the fish to rotate the rollers. Whatever their construction, these mechanised graders are usually mounted on wheels so that they can be moved around the farm as required.

An important part of the grading process, particularly at harvest, is the parallel weighing of representative fish from each of the different grades. This enables the physical dimensions of the grader to be calibrated to the individual weights of the fish. At present a machine which can directly weigh live fish is not commercially available. However, it is likely, with the current rapid progress in microprocessor development, that such a device will soon reach the market. We have performed some development work with an imaging system which gives a very accurate read-out of fish weight. It remains to be seen whether such a grading device can achieve the speed of throughput required by most commercial farms for their grading operations.

Transport of the fish to the grader can be by hand but increasingly farms are using fish pumps (Plate 3.14) to move fish around the farm in 4, 6 or 8 in (about 20 cm) diameter pipes. The impellors on these pumps are rounded and cause no damage to the fish. Although the wide-bore pipes are not full of water, sufficient water is pumped through them to enable the fish to pass along, again with little or no damage, apart from the stresses of confinement and crowding. Some commercial graders incorporate a fish pump as part of their design.

If sufficient differences in head of water are available on the farm site, then it may be possible, during construction, to build a second series of pipes from the individual stock tanks to a central and lower (head) placed grading unit. Removal of standpipes in the tanks allows the fish to be directly moved or drained to the grading area. This system of pipes, of course, has to be totally separate from that which supplies and removes the water flow to the tanks. Much of the improvements in production per man-year which have recently been made in trout culture are solely due to the reduction in the handling and netting of fish. The most cost-effective farms now use central grading areas to which the fish are transported by fish pumps or a gravity-fed piped system. Similar systems can also be used during fish harvesting.

In general, before any grading or movement of fish it is necessary to concentrate or crowd the fish. This can be achieved by reducing the amounts of water in an enclosure, i.e. by dropping the external standpipe or the height of the monk. In tanks and raceways various types of movable screens are used to partition off sections of the tank and hence 'crowd' the fish into a smaller volume of water. This makes it easier to net the fish. When crowding fish in circular tanks, a second screen is required to retain the fish. It is from the crowded sections of the tank or raceway that the influent pipe to the fish pump is placed. Some farms, particularly in the USA, incorporate vertically-placed grading bars into these screens so that, as the crowder is moved up, the fish grade themselves by swimming between the bars. The larger fish are, of course, retained by the screens. In practice this method of grading, although less stressful to the fish, is much less effective as a proportion of under-grade fish still remain with the larger stock.

Probably the most important grade to be conducted by a farm is that which is carried out at harvesting. Increasingly, chain/multiple retail outlets, because of packaging specifications, are demanding a narrow size range in the fish provided by the production farms. Undersize fish are rejected by the retail companies who also pay no additional monies for over-weight stock; both aspects represent significant losses in revenue for the farm, losses which can only be minimised by improvements in the accuracy of the final grade.

In preparation for harvesting all fish should be starved for 24 or preferably 48 hours. Stock which has been grown in earth ponds should be maintained in fibre-glass or concrete tanks during this period and receive a good flow of water so that any muddy or earthy taints, mainly due to the presence of algae in the water, are removed from the flesh before they are sacrificed. Samples of fish should also be regularly cooked and tasted to check for any taints or 'off flavours'. Some farms also feed with a lower quality or low fat diet one or two weeks before harvest to reduce carcass fat and the incidence of 'soft belly'. Trout which are to be sold with pigmented flesh should, of course, have been on an appropriate diet for at least 6 weeks before harvest, preferably commencing at half the final harvest weight. Generally, Atlantic salmon are fed on a pigmented diet throughout the period of salt water growth. Harvesting salmon and other large fish is labour-intensive and made much easier by the use of automatic hoists, particularly in the case of cage farms (Plate 3.15 (a) and (b)).

Whether they are asphyxiated or electrocuted in a proprietary killing tank, fish should be placed on ice immediately after sacrifice. Processing staff even on the smallest farms should be instructed in the methods of gill and gut removal and the absolute necessity for maintaining proper hygienic measures and controls. The shelf-life of fish is considerably extended by gutting, gilling and handling fish in a clean environment. A well prepared and fresh fish can only enhance the image of the product. In contrast a poor product will reflect on every other production unit. Atlantic salmon destined for smoking may need to be bled before sacrifice so that the blood vessels are not as discernible in the finished product. Again it is essential that this is carried out under the most hygienic conditions. Generally, the retail outlets will insist on certain standards of hygiene in the processing unit and will regularly check that these and the quality of the product are satisfactory. Larger farms will wish to perform their own quality control by conducting organoleptic trials and sending out samples for bacteriological analysis.

3.5.4 Transport of fish

Live fish are transported in a variety of systems and vehicles ranging from the small tank or other container in the back of a car, through larger tanks in pick-up trucks and vans, to even larger

(a)

(b)

Plate 3.15(a, b) Winching salmon from farm cage into service boat at harvest.

tanks on flat-bed lorries and lastly to custom-built tankers (Plate 3.16). Whatever system is used the tank should be fully insulated to ensure as far as possible that the temperature of the water remains the same as that in which the fish were being maintained on the farm.

The major difficulties of transporting fish relate mainly to the provision of adequate supplies of oxygen, the removal of ammonia and carbon dioxide, and the maintenance of an acceptable temperature. At no time during transport should the oxygen levels in the water be allowed to fall below 6 mg/l. Oxygen can easily be provided by pumps or gas bottles which aerate or oxygenate respectively. Either system can be used, whether the holding tank is small and transports only a small number of fish, or large on a transporter lorry carrying several tonnes of stock. For the latter, a pump which continually circulates water is probably to be preferred. The pump should be of sufficient capacity to circulate all the water in the system every three to five minutes. By spraying the water during its circulation or passing it through a venturi device, it will be continually saturated with oxygen. Auxiliary aeration or oxygenation systems should also be installed both to ensure oxygen saturation and to act as a back-up or stand-by should the water circulation fail for any reason. The shape of the holding tank can also help with the mixing of the water with oxygen, thus ensuring all the water in the tank is fully saturated. For this reason a round tank rather like those used in the transport of milk, beer or concrete is considered the most suitable.

Suitable aeration and circulation will also serve to blow off excess carbon dioxide providing there is adequate ventilation. However, carbon dioxide usually does not constitute a problem to salmonids until levels reach 15 mg/l.

Provided the pH of the water is not too high (less than 7.5), ammonia, which is only toxic in its unionised (NH_3) form (see chapter 2; section 2.3.1), should not be a problem, particularly on short journeys of a few hours. For longer hauls or where higher stocking densities of fish are being carried it may be necessary to remove the ammonia with an ion exchange or a klinoptilite column. The latter material is cheap and can be easily reconstituted, after combination with ammonia, by flushing with a solution of common salt.

During hot weather it may be necessary to add ice to the water to maintain a reasonable temperature. However, under no circumstances should the fish be allowed to come into contact with

Plate 3.16 Live fish transportation: a custom-built road haulage system for live eels.

the ice. Proper insulation of the tank will minimise any increases in temperature during transport. For larger systems, particularly on long journeys with higher air temperatures, a refrigeration system may be required to cool the water continually. For salmonids the lower the temperature the less active are the fish, and the lower their consumption of oxygen and production of metabolites. Lower temperatures also increase the absolute levels of oxygen in the water at saturation.

Although the weight of fish which can be transported is dependent on many factors, it is suggested that at a temperature of 10°C, 0.5–1.5 kg of 8–11 inch (220 g) rainbow trout can be carried in each gallon of water provided the water quality conditions itemised above are satisfactory. Halving the length of the fish halves the weight of fish which can be transported. A reduction in temperature of 5°C enables 30% more fish to be carried. In general, other salmonids travel less well than rainbow trout and farmers should start with weights approximately 50% of those quoted above. All tanks, pipes and connections should be properly cleaned and disinfected after each shipment. On no account should the washings be allowed to pass into lakes, streams or other watercourses.

Small quantities of fish can also be transported in plastic or polythene bags under an atmosphere of pure oxygen. The secret is to have a small volume of water and approximately three times the volume of oxygen. Bags of 45 l (10 gallon) volume with 11 l (2.5 gallons) of water can support up to 2.5 kg of 5 inch (about 12 cm) rainbow trout at 10°C for journeys up to 12 hours in length. Placing the bag in an insulated iced package can prolong the duration of the journey.

Before transport all fish should be starved for 24–48 hours depending on their size and should, of course, be in a healthy condition. Fifteen minutes after the fish have been loaded, their behaviour should be checked. Similar checks should also be made at periodic intervals during the journey, especially one hour after loading, at which time it is suggested a major portion of the ammonia is released. On arrival the temperature of the delivery and receiving waters should be measured to make sure that the fish are not exposed to a temperature shock on transfer. Differences in temperature can be equalised over a few hours by floating the bags containing the fish in a tank of the receiving waters or by slowly running in small amounts of the new water.

Salmonid eggs are also routinely transported both within countries and internationally. Generally, it is best to transport them at the eyed stage, although some farms move uneyed eggs within 24 hours after fertilisation and also move those which are still in ripe broodstock. We find high losses in transporting newly-fertilised eggs and, as disease considerations often prevent movement of broodfish, we prefer to transport eggs in the unfertilised state if it is not possible for them to be moved at the eyed stage. Unfertilised eggs should be transported in ovarian fluid and under no circumstances should they be allowed to come into contact with water. Eyed eggs are transported in a moist atmosphere but should not be immersed in water. Generally, a hatchery tray maintained over a sealed compartment of ice and enclosed in a styrofoam or polystyrene box constitutes the most suitable container for egg transport. All of the packaging should be burnt on delivery and the eggs treated with an iodophor preparation (see chapter 6; section 6.6.1).

3.5.5   Effluent treatment and management

During growth and metabolism fish take in and utilize oxygen and nutrients and produce a number of waste products which are then released into the water via the urine and faeces and from the gills. The waste products released include ammonia, suspended solids and organic phosphates and nitrates and, because these and other organic materials excreted may be subject to oxidation, they have a biochemical oxygen demand, or BOD, which removes oxygen from the water into which they are discharged. A high BOD can produce anaerobic conditions and kill off most animal and plant life, except some bacteria and fungi.

Ammonia is the chief excretory product of protein metabolism and is actively passed out of the fish into the surrounding water through the gill epithelium. The remaining waste products represent those materials left in the faeces and to a lesser extent in urine after the bulk of the useful components have been assimilated from the ingested food. Useful materials such as phosphates are also excreted when their concentrations exceed those required by the fish. Ammonia, suspended solids and BOD are toxic to fish and other organisms although the level at which adverse effects are produced varies with different species. In contrast organic phosphates and nitrates are not toxic but in water they lead to eutrophication; this enrichment with nutrients results in an excessive growth of many plants and other organisms which again have a high demand for oxygen. Removal of oxygen will encourage the development of anaerobic (low oxygen) life forms rather like waters with a high BOD. Under anaerobic conditions nitrates may also be converted into nitrites which are extremely toxic.

All of these waste products, together with any food that is uneaten, pass out in the effluent from fish farms and are discharged into streams, rivers and other watercourses. Many countries strictly control the levels and concentrations of agricultural and industrial discharges. Usually this control is based on the differences in levels of specified materials found between the influent and effluent waters to the site of discharge, i.e. the addition or change made by the farm or industry. Sites on spring water or near the sources of rivers may be subject to an absolute consent. However, in Britain discharge consents are usually based on the increases in concentration of ammonia, suspended solids and BOD and the reductions in volume of water flow and oxygen tension which occur during the passage of the water through the farm. Less commonly discharge conditions include pH, phosphates, presence of sewage fungi, temperature, colour, turbidity, formaldehyde, malachite green, free chlorine, total phenols, copper, antibiotics and oils, as well as unspecified items likely to be toxic. In some other European countries there is more concern about eutrophication and in particular the phosphate levels in the effluent. Certain countries (e.g. Sweden) place limits on the tonnage of fish which farms are allowed to produce, mainly because of the close correlation between food fed and the amounts of waste produced. Similar controls of total production are often imposed on cage farms because of the difficulties of measuring effluent.

Although varying environmental and farm conditions can affect the levels of waste materials produced, in general 25–50 g of ammonia, 200–300 g of suspended solids, 100–200 g of BOD, 5–15 g of phosphate and 30–60 g of nitrate are produced for each kg of dry pelleted food fed to trout; moist and trash fish-based diets produce much higher levels of waste products. If the amounts of water flowing through a farm are known then the levels of waste product in the effluent may be calculated. If these levels fall outside the consent conditions for the farm, then some form of treatment of the water must be carried out before discharge.

On average oxygen levels fall by 2–3 mg/l during the passage of water through a farm. If necessary this reduction in oxygen tension can be remedied by aeration or splash boards in the water flows.

The levels of suspended solids and to a lesser extent the BOD of the effluent from the production tanks can be treated by settlement, preferably in purpose-built ponds or lagoons. Settlement is the most usual and often the only form of effluent treatment practised by farms. Settlement ponds must be of sufficient size, in relation to total water flows, to allow time for the particles to settle; the shape of the lagoon is also of importance. As a rule of thumb, maintaining the surface hydraulic loading of the effluent into the settlement pond below 40 m^3/m^2 surface area of the pond per day, together with a ratio of pond length to retention time below 4 m/minute (a measure of water velocity) will maximise settlement. Inflow and outflow channels from the settlement pond should also be of sufficient width to reduce the velocity of entry and exit of the water. Baffles can be provided to ensure that all parts of the settlement pond are utilised.

Reductions in suspended solids and BOD levels in the effluent can also be achieved by modifications in the formulation of the diet of the fish. Replacement of complex carbohydrate, which in general is poorly digested by carnivorous fish, by increased amounts of more digestible protein or fat reduces the amounts of faeces and hence suspended solids produced. Unfortunately, increasing the levels of protein or fat results in corresponding increases in the ammonia excreted or in the amount of lipid deposited in the fish carcase respectively. Recently, some improvements have been made in the digestibility of the carbohydrates used in diet formulations, although salmonids generally are unable to tolerate digestible carbohydrate levels which exceed 15% of total food input.

Ammonia and phosphate constitute the most difficult components of the effluent as far as treatment is concerned because of their solubility. Less than 10% of the total ammonia excreted is settleable. The remainder can only be removed by biological filtration which, because of the large volumes of effluent to be treated, is uneconomic and impracticable for salmonid production farms. Biological filtration can be used to advantage with high value stocks maintained on relatively small or recirculated water flows (section 3.6). In such systems, colonies of nitrifying bacteria, maintained in trickle or submerged filter beds or activated sludge units, convert ammonia to nitrate. Nitrate is not toxic to fish or other life until very high concentrations are reached; in fact high levels are already naturally present in most waters as a result of the agricultural run-off of fertilisers. Klinoptilite filters may be used as a short-term remedy for high ammonia, but because the ammonia is only bound on the filter column rather than converted to a non-toxic form, it still has to be disposed of by the farm.

The last remaining waste product of intensive fish production over which there is concern is phosphate. At present there is no practical method for removing this material from the effluent of production farms with their high volumes of water flow. Phosphate stripping by ion-exchange is possible but expensive. Absorption and assimilation by aquatic plants has been tried but many of these die in the winter and consequently return the phosphates and other materials back to the effluent. Currently, the reduction of phosphate by dietary means is the only control available to the farmer. At present most commercial diets contain 10–20 g of phosphate in every kg of feed. Clearly the formulation of these diets includes a large safety margin, because experimentally the ionic demands of the fish are fully met with diets which contain only 6–8 g phosphate/ kg feed. As intakes in excess of 5 g/kg would appear to be directly excreted, farms with difficulties in achieving phosphate discharge consents should request a special low phosphate diet from their feed company.

3.6 RECIRCULATION SYSTEMS (WITH SPECIAL REFERENCE TO EEL CULTURE)

3.6.1 Introduction

Eel culture under controlled conditions has been practised since the 19th century in Japan and Europe where, traditionally, eels are a delicacy. Interest in the intensive culture of eels has been stimulated in more recent times due to steadily increasing demand from the market, coupled with increasingly limited supplies of wild eels, both of which have served to maintain high market prices. This has created the right climate for the intensive culture of eels which has been made possible by recent advances in technology and husbandry skills.

In Europe, the main supplies are of the native European eel (*Anguilla anguilla*), in Japan and Taiwan both European and Japanese (*Anguilla japonica*) eels are favoured, while various other eel species are farmed or caught in the Far East, Australasia and

North America. Cultured eels are mainly produced by extensive and semi-intensive techniques, but intensive systems have gradually developed over the last 15 years and in Europe intensive culture probably accounts for around 15% of total production. Under conditions of culture, growth rates can be considerably speeded through the use of warm water. For example, the optimum temperature for the growth of European elvers has been shown to be 26.5°C. Such conditions can be attained with constancy through the use of warm waters from industrial or geothermal sources in open-flow, or alternatively in recirculation systems which conserve the heat of the water. In these circumstances, eels can be grown to a weight of 250 g in 18 months from glass-eel intake, which is approximately three to five times faster than would occur in the wild population.

3.6.2   Eel husbandry

The life cycle of the European eel may take 15−20 years to complete naturally and comprises four distinct stages: leptocephalus, elver, brown or yellow eel, and the sexually-mature silver eel. The freshwater eel is a catadromous fish, migrating from fresh to sea-waters in order to breed. So far, artificial propagation of eels has not been achieved commercially and, as a result, one constraint imposed upon the eel farmer is a total dependence on wild-caught elvers. Until this is resolved the industry will always be subject to the uncertainty which attaches itself to any catch of wild fish. In addition, there will never be any prospect of stock improvement by genetic manipulation.

  The first feeding of elvers is still considered to be one of the most difficult phases in the cycle and mortalities can be very high, mainly due to a proportion of the stock failing to learn to take artificial feeds with the result that mortalities over the first 6 months can be in excess of 30%. In order to minimise the extent of non-feeding, eel farmers have to present a highly stimulatory feed to the elvers for several days after intake in order to ensure that the stock come on to feed as they develop physiologically. This 'weaning' feed is often based upon moist raw materials bound into a sticky paste which is presented on a mesh-bottom floating tray, allowing the elvers to feed through the mesh (Plate 3.17). After 7−10 days feeding on the weaning paste, the elvers are

Plate 3.17 Farmed eels feeding on paste diet.

gradually converted to compound feeds of either moist or dry varieties. Moist feeds must be prepared on-farm by mixing the commercial meal with water and, sometimes, fish oil. Dry feeds in granule or pellet form offer the following advantages:

1. No on-farm preparation.
2. Automatic or demand feeders can be used.
3. Less pollution from uneaten food.

The last point is often critical when considering recirculation systems as unnecessary organic loading on the filter must be avoided.

Elvers are predominantly olfactory feeders until they reach a size of 10–20 g and will readily accept granulated feed provided it is presented on the tank bottom or solid-based feeding tray. Larger eels feed visually and will accept either floating or slow-sinking pellets on the water surface or in mid-water. Fingerling and larger eels can be fed by hand or from automatic or demand (pendulum) feeders. The feeding rate for elvers at optimum temperatures is 5.0–7.5% bodyweight per day (dry feed basis) and for large eels is

0.5–1.5% bodyweight per day whereas feed conversion rates are typically 1.6–2.0 at optimum temperatures.

Initial stocking densities are typically in the region of 10 kg/m^2 for elvers but can be steadily increased as the eels grow to around 100–150 kg/m^2. The growth rate within one year class varies widely between individuals, and is mediated by size, sex and 'stress'. Hence, it is usual to grade eels at two month intervals using net meshes or bar graders. Specific growth rates of first-feeding elvers can be as high as 3.1% per day under optimum conditions. These rates reduce progressively with size such that eels of 400 g would exhibit growth rates of 0.1–0.2% per day. Male and female eels exhibit parallel growth rates until a size of 60–90 g is reached. After this point, male eels grow at slower rates than females and the former rarely reach a size in culture greater than 200 g.

Female eels show little signs of a check to growth over the size range cultivated, which can be 150–500 g. Eels are harvested at sizes between 100 and 500 g (all females). Usually, market-sized eels are graded out and removed from the farm alive in road tankers.

At the elevated temperatures associated with eel culture, there are increases in the risks of disease problems as the reproductive rate of pathogenic organisms will be increased. Such risks should be minimised when considering recirculation systems as a disease outbreak in one tank may spread rapidly throughout the system. This is achieved by introducing to the system only elvers shown to be free of disease and by employing water-sterilisation methods such as ultra-violet equipment or ozone treatment. The treatment of disease outbreaks also has significance when the design of a recirculation system is considered. Provision must be made for isolating the eel tanks from the biofilter during which time water-dispersible chemicals can be employed.

3.6.3   Recirculation and environmental control

If the water supply is insufficient to meet the requirements of fish, fish-holding systems may be employed which involve water re-use. This may comprise relatively straightforward procedures, such as aeration, or progressively more complicated procedures involving greater degrees of water re-use until, as a final measure, recirculation systems are employed. These may achieve water re-use rates of more than 90% by employing biological or ion-exchange filtration. The operation of recirculation systems for aquaculture offers a number of distinct advantages over open-flow systems which can be listed as follows:

1. Water and heat conservation.
2. Environmental control.
3. Disease control.
4. Management of production.
5. Freedom from site limitations.

In many cases, however, these advantages are outweighed by one serious disadvantage: additional capital and operational costs are incurred. Even the simplest form of water re-use involving only aeration and/or solids removal will increase the total volume of water in the system by 20–40%, which will involve increases in capital costs. Greater degrees of recirculation will require significant increases in capital outlay (Fig. 3.7). Additional capital expenditure is usually required for the provision of water pumps, aerators, air compressors, comprehensive alarm systems, water sterilisers, heat exchangers, etc. Operational costs are usually increased due to the necessity to run water pumps and air compressors. Often, extra operatives are required in order to maintain complicated systems. Also, expensive water heating may be desirable, but it must be remembered that this should result in improved growth rates which will improve profitability. Therefore, the development of

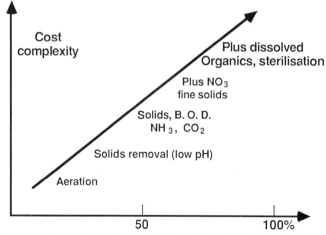

Fig. 3.7 Degree of recycle and system design (after Muir, 1981).

recirculation systems has been restricted to specialised cases, mainly where water economy or heat conservation is paramount and the value of the end-product is sufficiently high to justify the extra costs. Such cases might include Atlantic salmon (*Salmo salar*) smolts, marine flatfish such as Dover sole (*Solea solea*) and turbot (*Scophthalmus maximus*). The practical application of such systems to eel culture will now be described in more detail.

The reconditioning facility of a recirculation system has to reduce or remove waste feed and metabolic by-products as well as return the water to an acceptable quality before re-use (Fig. 3.8). The reconditioning facility may comprise several units, each with a specific task, for example:

1. Sedimentation and/or solids filtration.
2. Biological filtration and/or ion-exchange filtration.
3. Water sterilisation.
4. Aeration/oxygenation.

Effluent from the eel holding tanks should first be settled and/or filtered to remove organic material deriving from feed and faeces which might be detrimental to the operation of a downstream biofilter. Under special conditions, where a trickling biofilter employing high void space plastic media is used, prior settlement may be omitted and the settlement unit may be placed downstream of the biofilter. This arrangement has the advantage of removing settleable organic material deriving from the biofilter as well as from the crude effluent. Settlement is fairly readily achieved by employing lagooning techniques or 'swirl' separators. On smaller-scale units, mechanical filtration has been employed with success.

The effluent from the sedimentation/filtration unit may now be directed to a biofilter or ion-exchange filter with the aim of reducing the circulating level of ammonia, the main end-product of nitrogen metabolism of fish. Designs for biofilters are numerous, involving

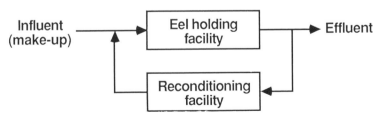

Fig. 3.8 Simplified diagram of water re-use system.

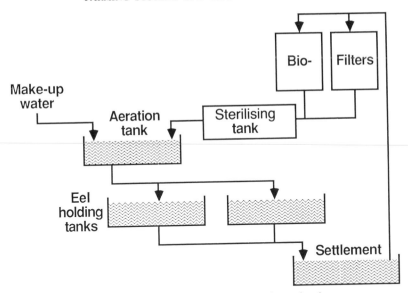

Fig. 3.9 Layout of generalised recirculation system for eel culture.

trickling or submerged filter, upflow and downflow filters, activated sludge and many others. Usually the biofilter contains a medium which has a high surface area to volume ratio, so providing a large area for bacterial attachment. The principle involved in biofiltration is the nitrification of ammonia to nitrites and nitrates by the bacteria attached to the substrate of the filter.

These end-products are relatively less toxic than the unionised ammonia in the effluent. In most practical applications, the levels of nitrites and nitrates are controlled by regulating the amount of new make-up water such that circulatory levels of nitrites are maintained below a critical level. The rate of make-up water addition is often the method employed to prevent the water pH from descending too low. Were no make-up water to be used, it would be possible under certain conditions for pH to fall below 6.0 due to the release of NO_3 and CO_2 into the water. Occasionally, a buffer material, such as crushed limestone or sea shell, is employed as part of the biofilter.

After passing through the biofilter, the effluent may be sterilised, usually by employing ozone or ultra-violet methods. Then, the water is vigorously aerated or oxygenated to replace the dissolved oxygen removed by the eels and the biofilter. This step is not quite so critical if a trickling biofilter is employed as a degree of aeration can then occur. The cleaned effluent is then mixed with a proportion of make-up water and is circulated back to the eel tanks. The layout of this generalised recirculation system is shown in Fig. 3.9 for eels. Where technically and commercially feasible, variations on such systems may be adopted for rearing other species, particularly at the juvenile stage when total fish biomass and hence demand for water is relatively low.

4 Propagation and stock improvement

4.1 INTRODUCTION

In the intensive farming of fish, an ability to control the reproductive cycle of the species under cultivation is most important. The artificial induction of spawning enables supplies of eggs and fry (i.e. seed) to be made available even from those fish which do not normally spawn in captivity, thus avoiding the problems of either collecting broodstock or harvesting supplies of fry from wild populations (Fig. 4.1). For species which are able to mature and spawn under conditions of intensive culture, controlled reproduction can provide seed at precisely those times required by on-growing farms and not just during the few months of the year when natural spawning occurs. Effective seed production also demands a thorough understanding of the special husbandry and nutritional requirements of

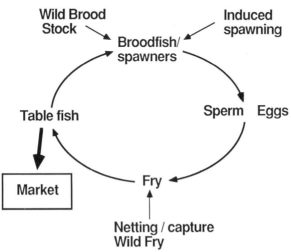

Fig. 4.1 Induced spawning methods provide continuity of supply of fish seed.

broodstock fish and the newly-hatched and developing fry. Combined with breeding and broodstock selection programmes designed to develop and maintain those traits which are best suited to intensification, this array of knowledge ensures that hatcheries are able firstly to maximise their production of eggs and fry and secondly to tailor this production to the needs of the farms which grow fish up to table size.

In addition to being able to produce good quality seed on demand, it is also essential that fish culturists are able to prevent or inhibit sexual maturation. During maturation many fish undergo profound changes in growth and flesh composition (Fig. 4.2), which at worst render the flesh unacceptable for market sale and at best offer poor performance from the stock and wide variations of flesh quality. Methods of sex control which depress or inhibit maturation ensure that the metabolic activities of the fish and more importantly the inputs of high-cost artificial feeds are channelled into somatic growth (i.e. the production of saleable flesh) and away from gonadal development and maturation.

The ability both to stimulate and prevent maturation assumes a greater importance with the intensification of the farming process. Generally, intensification demands greater tonnages of production in order to reduce unit costs through economies of scale. As a

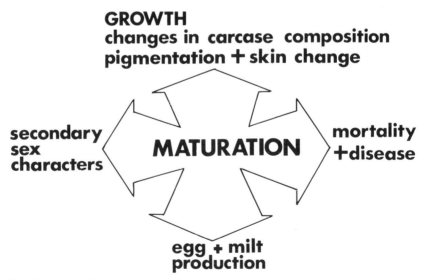

Fig. 4.2 Some of the physiological changes which accompany sexual maturation in fish.

consequence, the increased volume of product has to be retailed through the more extensive markets which are available to whole-salers and chain multiple and freezer stores with less reliance on local sales. The sophisticated nature of these outlets and their distribution networks, require that the product is properly processed and packaged and that it is of consistent size and quality and available throughout the year. Such a continuity of production is not possible if farmers are totally dependent on the natural spawning of broodstocks for their supplies of eggs and fry. Seasonal changes in flesh composition and fish size which arise as a consequence of maturation may also prevent farmers from meeting the quality and product specifications imposed by the various multiple retail outlets.

Clearly, problems associated with maturation and the provision of supplies of eggs and fry constitute important constraints as far as further development of intensive fish farming is concerned. In this chapter, after considering the process of reproduction and the maintenance and spawning of broodstocks of the major cultivated species, some of the methods used to control size or gender, to induce spawning artificially, and to alter its timing are described. Finally stock improvement is discussed in relation to cross-breeding and strain selection.

4.2 REPRODUCTIVE CYCLES AND MATURATION

For most wild stocks of fish, spawning is characteristically an annual event with the mature eggs and sperm being produced at a time of the year when external conditions and the supplies of available food are most favourable to the survival of the fertilised embryos and developing fry. Most fish use seasonal patterns of changing daylength and temperature, the presence of food or the onset of rainfall to time gonadal maturation and spawning accurately. Gonadal maturation, however, irrespective of whether it is ovary or testis, involves many months of preparation and growth with the release of gametes generally only comprising a brief period in the developmental process. All of the stages of maturation, from the inital differentiation of the sexes to the final ripening of the ovary and testis, must be closely controlled if the production of eggs and sperm is to be successful. Although factors from the external environment are responsible for the initial induction and subsequent course of reproduction, it is the hormones produced by the endocrine system which directly control this process. Many hormones are involved in the control of maturation and spawning but those produced by the hypothalamus, pituitary and the gonad itself are of primary importance. The major components in the control of reproduction and some of the steps which are amenable to artificial manipulation are itemised in Figure 4.3. Clearly, an understanding of the relationships between the various components involved in the control of gonadal function is essential if the reproductive cycle is to be modified for aquacultural advantage.

Briefly, the elements of this control are as follows: In response to changing conditions in both the internal and external environments, a part of the brain known as the hypothalamus secretes small peptide hormones called releasing hormones. These peptides pass a short distance to the ventral surface of the brain, just above the roof of the mouth of the fish, where they control the activity of specific gonadotropic cells in the pituitary gland. In turn these cells secrete gonadotropic hormones which pass in the blood to the gonad of the fish where they control all structural and functional changes in the testis and ovary. An important gonadotropic influence is the induction of hormone secretions from the gonads. It is these gonadal hormones acting either independently and/or in concert with the gonadotropic hormones which directly control all aspects of gonadal growth and maturation. In male fish the major testicular hormones produced, often generically referred to as androgens, are testosterone and 11-ketotestosterone. In female fish two principal classes of hormones are produced: the oestrogens which include oestradiol-17β and oestrone and the progestagens 17α-20β-dihydroprogesterone and 17α-hydroxyprogesterone. All of these gonadal hormones have a steroidal structure which is the same or similar to that of the corresponding hormones of higher vertebrates.

Amongst the fish which are of commercial importance there is a remarkable range of reproductive strategies and tactics. Some fish spawn once a year with a single batch of eggs (e.g. salmonids), others produce a series of batches of eggs over a protracted breeding season (e.g. flatfish). Some spawn in the Winter or Spring whilst others produce eggs in either the Summer or Autumn months. Some ovulate a few hundreds or thousands of relatively large eggs whereas others produce millions of very small ones. Usually there are separate male and female sexes although some species start off life as males or females and later reverse their sex (e.g. gilthead bream and grouper).

Artificial Control

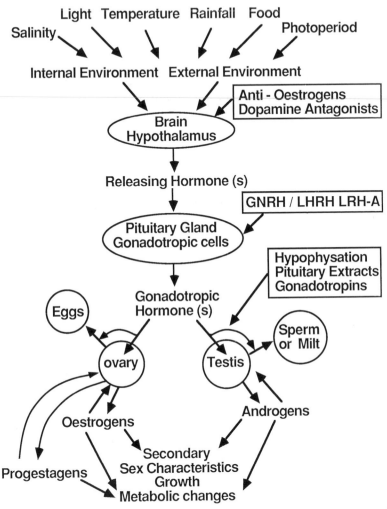

Fig. 4.3 The stepwise control of reproduction and some of the hormones used to induce spawning artificially.

Despite this diversity, as far as is known from the species which have been studied, all have their reproductive cycles controlled by the same cascade of endocrine controls briefly described above. We shall see later that exploitation of this control, by hormone administration and environmental manipulation, allows fish culturists to alter the times of spawning of a wide range of fish species for commercial advantage.

Aside from the changes in the ovary and the testis which culminate in the formation and release of sperm and eggs, sexual maturation also has profound effects on other facets of the metabolism of the fish (Fig. 4.2). One of the most important of these effects, as far as fish farming is concerned, is the diversion of nutrient materials from somatic into gonadal growth. Many fish experience reductions in growth and food conversion efficiency during the later stages of maturation. In salmonids both sexes suffer from depressions in growth just before spawing, although this effect is shown especially by male fish (Fig. 4.4). Similar reductions also affect male eels and turbot, whereas in tilapia and catfish it is female fish which show the poorest growth during sexual maturation.

In addition to the modifications of growth, reproduction often requires the profound reorganisation of tissues or the development of quite new structures specifically for the purposes of reproduction. These secondary changes, often described as secondary sexual

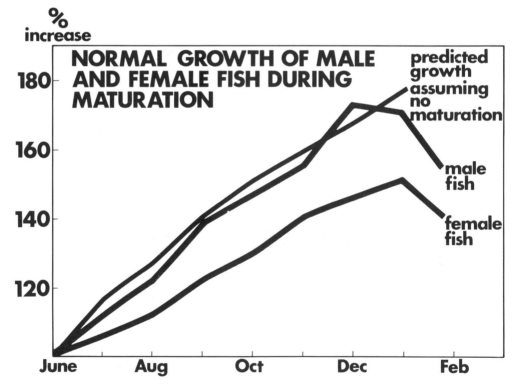

Fig. 4.4 Percentage weight increase of male and female rainbow trout during the months up to spawning in January/February.

characteristics, supposedly prepare fish for spawning. However, they often diminish the acceptability of fish for commercial sale. Again, many of the changes are especially well demonstrated in salmonid fish. Thus, during maturation in trout and salmon there are changes in carcase composition with reductions in muscle (i.e. flesh) protein and fat and increases in water. The flesh becomes progressively less firm as maturation approaches, and from the point of view of the consumer, much less inviting. In addition there are alterations in both flesh and external colouration which result in the characteristic paling of the muscle and the darkening of the skin (Plate 4.1). There are also changes in the thickness of the skin, modifications in mucus secretion and often the development in some male salmonids of a protruding upper or lower jaw which is known as a kype. These changes in skin structure, together with the additional energy demands of reproduction, increase the susceptibility of fish to bacterial and fungal attack. This increased vulnerability to infection is often exacerbated by the physical damage which many fish inflict upon each other during courtship and spawning. Catfish of both sexes and male salmonids in particular take part in aggressive displays and fighting.

In parallel with the diversion of resources into the support of maturation-related processes, food is also used to grow gonads which for many commercially important species are removed from the carcase during processing and before sale. Gonad growth is of particular significance in female fish where the ovary may constitute 20–30% or more of the body weight. Even in male fish the testes often comprise 10% of the total weight. Clearly, the diversion of

Plate 4.1 Mature female Atlantic salmon showing dark coloration.

expensive feeds into the growth of ovaries and testes, which are then thrown away, constitutes a significant drain on the profitability of fish farming.

For many fish the trauma of reproduction is such that they die after spawning. In rainbow trout there are mortalities amongst both sexes at maturation but generally most female fish regain condition to spawn again successfully. By contrast many post-spawned male rainbows show poor growth and die. Extreme examples of the mortality which is associated with maturation are shown by the different species of Pacific salmon and eel, all of which die after their first and only spawning. A similar fate also awaits maturing Atlantic salmon, although some fish regain condition to spawn again as kelts.

Clearly, the changes which are associated with maturation have important implications as far as the marketing and sale of fish are concerned. For this reason many farmers aim to sell their fish before they show signs of maturation. This strategy is, however, not possible when maturation occurs in advance of the fish reaching table- or market-size. One of the most important problems, for example in tilapia farming, is the early maturation and high fertility of the fish; this precocity results in large numbers of small, stunted and sexually-mature fish which are generally too small for sale. In Atlantic salmon some male fish reach maturity as parr or smolts. Both Atlantic and Pacific salmon also experience early maturation at the sea water stage. Thus, Atlantic salmon grilse mature after only one rather than two years' growth in sea water; similarly Pacific salmon jacks begin maturation in advance of the normal spawning time. Similar problems also affect rainbow trout production; here spawning of the female fish occurs at 2 or 3 years of age whereas all of the males mature by 2 years and a proportion of these will have become precociously mature as underyearlings. Whilst most rainbow trout in Europe are sold at the 180−280 g size range (Fig. 4.5), clearly only male maturation constitutes a management problem, and as will be seen later, this difficulty can be largely avoided by producing all-female or sterile fish. If rather than producing portion-size fish, farmers grow rainbow trout up to larger sizes, then all fish, irrespective of their sex, will mature before reaching market-size. This is a problem which can only be avoided by the cultivation of sterile fish. Some of the methods used to produce sterile or single-gender stocks will be considered in section 4.4.

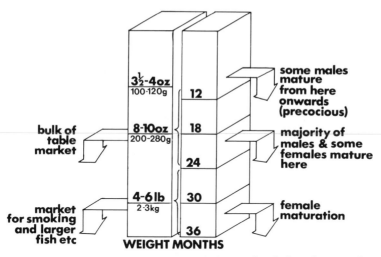

Fig. 4.5 Age and size of rainbow trout in relation to the timing of maturation and marketing.

4.3 GAMETE PRODUCTION AND THE CONTROL OF SPAWNING

In this section the technique which uses pituitary extracts to induce spawning, known as hypophysation, will be described, together with some of the procedures involved in the handling of broodstock fish and the artificial production of eggs. Also described are methods in which environmental manipulations are used to extend the availability of gametes (eggs and fry) beyond the natural spawning times of the fish being cultivated.

4.3.1 Induced spawning, broodstock handling and fertilisation

Techniques of induced spawning have been used to modify the reproduction of fish for almost 50 years. Traditionally, such methods have been employed to induce spawning in fish which do not breed in captivity (e.g. carp). However, recently it has become more widely recognised that methods for inducing spawning might be used to manage more effectively the production of eggs and sperm from fish which to breed under farm conditions (e.g. salmonids) as regards the timing of spawning.

Most of the commercially used methods of induced spawning

involve the injection of gonadotropic hormones which have been derived from the pituitary gland, generally in the form of dried gland extracts. However, with improvements in our understanding of the importance of the hypothalamus in the control of pituitary function, there has been a growing interest in the use of gonado-tropin releasing hormones to induce spawning (Fig. 4.3), particularly as these peptide molecules can be easily synthesised artificially. Furthermore, the relatively small size of the releasing hormones means that treatment with these materials does not provoke the immunological reactions which sometimes occur in recipient fish in response to the injection of pituitary extracts or gonadotropic hormones. This reaction, which involves the development of anti-bodies against the gonadotropic principles injected, considerably reduces the effectiveness of subsequent hypophysations; in some cases the injected extract is totally inactivated by the antibody reaction of the recipient fish. It would also appear that the differ-ence between species and groups of fish, as far as the molecular structure of the releasing hormones is concerned, are less important than the corresponding differences in structure, between groups, of the pituitary hormones. Thus, the same releasing hormone is likely to be used more successfully in a wider range of fish groups than the pituitary extracts. One of the biggest problems as far as the use of pituitary extracts is concerned, is that they are taken from fish of variable maturity and hence the potency of the extracts shows considerable batch variation. Despite these limitations, at present it is the pituitary extract which is the most widely available and consequently this is the agent most commonly employed for the induction of spawning of broodstock on commercial farms.

The first trials on induced spawning in fish were carried out in South America in the 1930s. Subsequently, considerable efforts were devoted to the development of methods for the induction of spawning in the Chinese and Indian carps. Carp are of considerable importance to the economies and nutrition of many Asian peoples but unfortunately the majority of these fish fail to reach full maturity under farm conditions. Initially, pituitary hormones from cattle and sheep were used to induce ovulation in the Indian major carp, the mrigala (*Cirrhina mrigala*), in the late 1930s. However, it was not until the 1950s that the techniques of induced spawning became more widely established. As a result a wide range of species, in addition to the carps, has now been successfully spawned using these hypophysation techniques (Table 4.1).

Table 4.1 Common and specific names of fish on which techniques of induced spawning have been successfully used

---

Catfish
    Channel catfish − *Ictalurus punctatus*
    African or Sharptooth catfish − *Clarias gariepinus* (*lazera*)
    Indian catfish − *Heteropneustes fossilis*
    Also bullhead, blue, black, yellow, brown, wels, and flathead catfishes −
       Ictalrus, *Pylodictis* and *Silurus* spp.
Chinese carps
    Grass carp − *Ctenopharyngodon idella*
    Silver carp − *Hypophthalmichthys molitrix*
    Bighead or spotted silver carp − *Aristichthys nobilis*
    Black carp − *Mylopharyngodon piceus*
Other carps
    Common carp − *Cyprinus carpio*, includes the scale, line (Zeil), mirror,
       leather and koi carps.
    Goldfish − *Carassius auratus*
    Medaka − *Oryzias latipes*
Eels
    Japanese eel − *Anguilla japonica*
Flatfish
    Japanese flounder − *Limanda yokohamae*
Indian major or king carps
    Catla − *Catla catla*
    Rohu − *Labeo rohita*
    Mrigala − *Cirrhina mrigala*
Milkfish − *Chanos chanos*
Mullets
    Grey mullet (flathead) − *Mugil cephalus*
    Grey mullet (thinlip) − *Liza ramada* (*Mugil capito*)
Pikes and perches
    Northern pike − *Esox lucius*
    Walleye − *Stizostedion vitreum vitreum*
    Zander − *Stizostedion* (*Lucioperca*) *lucioperca*
Salmonids
    Rainbow trout − *Salmo gairdneri*
    Coho salmon − *Oncorhynchus kisutch*
    Atlantic salmon − *Salmo salar*
    Steelhead trout − *Salmo gairdneri*
    Chinook salmon − *Oncorhynchus tschawytscha*
    Chum salmon − *Oncorhynchus keta*
    Cutthroat trout − *Salmo clarki*
    Ayu − *Plecoglossus altivelis*
Sparids
    Gilthead bream − *Sparus aurata*
    Red seabream − *Pagrus major*
Sunfish
    Bluegill − *Lepomis macrochirus*
    Largemouth bass − *Micropterus salmoides*

To be successful, hypophysation requires firstly the use of pituitary materials from suitable donor fish. The donors should be mature pre-spawned (i.e. unspent) fish, preferably of the same species as the ones which are to be spawned (i.e. the recipient). Pituitaries can be preserved in alcohol or acetone; the acetone-dried material is easier to handle and is now available from a number of reliable suppliers in the USA, Israel and Germany. In the past these pituitaries have been obtained almost exclusively from the common carp, but material from other cyprinids is becoming increasingly available, including very recently a source of salmonid pituitary extract. The methods involved in the preparation of acetone-dried pituitary glands are, however, easy to carry out and consequently many farms may prefer to produce their own extracts. Either a ventral or dorsal approach is used to remove pituitary glands from a suitable donor fish. The pituitary gland lies on the underside of the brain above the roof of the mouth and both approaches involve dissection. A useful landmark in doing this is the presence of the pair of optic nerves (optic chiasma), which cross from the right and left sides of the brain to the left and right eyes respectively, just anterior to or in front of the pituitary gland.

Although pituitary glands can be used in the fresh state to induce spawning, generally it is more convenient to produce an acetone-dried preparation. If many glands are processed at the same time, then a year's supply of material may be produced on one occasion. The pooling and mixing of many glands of variable gonadotropic activity has the advantage of producing a single extract which has a consistent potency. The large quantities of material available in such a supply will also allow samples to be assayed in order to ascertain the exact potency of the preparation. In order to preserve the glands, they should be placed in absolute acetone which has been refrigerated or placed on dry ice. The acetone should be changed several times over a 24 hour period to ensure that the glands have been properly dried and defatted. Desiccated glands are then dried on a filter paper or under vacuum and stored, either intact or after being ground to a powder, in sealed light-proof containers in a refrigerator, or preferably a −20°C freezer. Glands stored under these conditions retain their potency for many years. For example, 15 year-old acetone-dried carp pituitary extracts have been used to spawn various cyprinid fish in this laboratory.

Unfortunately, hypophysation is only effective in inducing spawning in recipient fish which have reached the final stages of maturation. In scientific parlance for female fish this means that the nucleus must have begun or preferably have completed its migration towards the edge of the egg cell. Generally, pituitary injections have failed to advance satisfactorily the earlier and much more extended periods of maturation (i.e. the stages of early egg growth and the incorporation of yolk). Catastrophic reductions in the quality or survival of the eggs occur if injections are made too early in development. Because hypophysation techniques can only be used successfully with mature eggs, they are unable to alter the timing of spawning significantly and, at most, ovulation may be advanced by only a few weeks by injections of pituitary extracts. Thus, the major use of hypophysation is to enable farmed broodstocks to mature and spawn under conditions of intensive cultivation at approximately the same time and in the same manner as fish in the wild. For fish which do spawn in captivity, induced spawning, although not offering full control of spawning time, does allow the production of eggs and fry to be adjusted as required to procedures on the farms. Similar constraints also apply to the induction of milt release from males, although generally the spawning of male fish is far more tolerant of the farm environment than that of female fish. For this reason much lower doses of pituitary materials are required in the induction of spawning by hypophysation in male fish.

A major difficulty with hypophysation is the identification of female fish which will be able to respond to pituitary injection. Secondary sexual characteristics, such as tubercles or roughening of the surface of the body, changes in colouration, softening and rounding of the abdomen and a reddening and protrusion of the anal papilla and vent have all been used, but each is notoriously unreliable. Increasingly, many farmers monitor development by taking small samples from the ovary using either a glass or plastic catheter or cannula or a wide-bore biopsy needle (Gauge 11) and examining these under a microscope at approximately 20 × magnification. To aid examination, the eggs should be cleared in a suitable fluid, such as Stockard's solution, which contains glacial acetic acid, formalin (40% formaldehyde) and glycerine. This solution enables the position of the nucleus to be seen easily; eggs can also be stored permanently in this solution. Pituitary extract injections should only be given to fish in which the nucleus is on the periphery of the egg cell. Less mature fish should be returned to the holding ponds or tanks for re-examination at a later date.

Of all the fish on which induced spawning has been practised (Table 4.1), it is the carp which has been most widely used. The hypophysation of the common carp will be considered in detail here and reference made to modifications in the techniques which are necessary for other species of fish. Common carp have to be maintained at 20°C for 1000–2000°C · days (degree-days) in order for the fish to reach the stage of maturity when hypophysation will be effective, 1000°C · days is sufficient if the fish are maintained permanently at temperatures above 20°C (e.g. 40 days at 25°C). After male and female fish have been identified at appropriate stages of maturity, they are injected with carp pituitary, generally in the form of an aqueous extract of the acetone-dried material. Appropriate amounts of the entire or dry powdered glands are weighed and then homogenised thoroughly with a small quantity of sterile water or 0.6% saline. Further amounts of the same solvent are then added to achieve a concentration of 20–50 mg of dried pituitary material per ml of water or saline. The exact concentration is unimportant, although it is essential that this is accurately determined in order to administer the appropriate dose for spawning induction. Although on occasions complete homogenates have been used, usually the insoluble fragments from the whole pituitaries are removed by sedimentation or centrifugation at 3000 revolutions per minute (rpm) before administration of the aqueous phase which contains the water-soluble gonadotropic principles. Fresh material may also be used, as mentioned above, although the dosage subsequently injected must be increased to take account of the six-fold reduction in weight of the pituitary which occurs during acetone-drying.

As regards the procedure for hypophysation, female fish are injected with two doses of the pituitary material (Fig. 4.6). The first injection should contain 0.3 mg of the acetone-dried powder per kg fish body weight followed 6 hours later by a second injection of 3 mg/kg. Ovulation occurs approximately 6–12 hours after the second injection. Male fish should be injected with 0.3 mg/kg at the same time as the female receives the second injection. Two males should be available for each female just in case sperm production has failed or the sperms are infertile. Injections are usually made intramuscularly below the dorsal fin, although sometimes the intraperitoneal route (into the abdominal cavity) is employed. Other fish species may be spawned with different dosages and injection regimens of the pituitary extracts (Table 4.2).

Other hormone preparations have also been used to induce

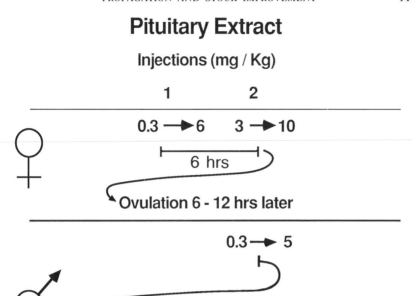

Fig. 4.6 Regimen for induction of spawning (hypophysation) in female (♀) and male (♂) carp.

spawing in fish, including mammalian hormones like human chorionic gonadotropin (HCG) or the hypothalamic gonadotropin releasing hormones (e.g. LHRH, GnRH) and their synthetic analogues (e.g. LRH-A/LHRH-A). HCG especially has been widely used, mainly because it is easy to acquire, relatively cheap and of known and consistent potency. On the debit side fish often become unresponsive to HCG injections because of immunological reactions; there are also reports of regressive ovarian changes following HCG treatment, possibly as a result of overstimulation. Sometimes HCG injections are more effective when combined with doses of pituitary extracts. However, in general similar responses are achieved with both materials and 100 International Units (IU) of HCG are said to be equivalent to 0.5 mg of acetone-dried pituitary. Some of the dosages of HCG used to induce spawning are shown in Table 4.2.

For reasons described earlier, in recent years many fishery scientists have been looking towards the releasing hormones and their synthetic analogues as a solution to the problems of inconsistencies in the purity and supply of pituitary extracts. Injections of small quantities of these materials elicit release of gonadotropic hormones

Table 4.2 Dosages of carp pituitary extract (CPE), human chorionic gonadotropin (HCG), luteinising hormone releasing hormone (LHRH) and an LHRH analogue (LHRH-A) required for induced spawning of different fish species in the first (1) and second (2) injections

Fish species	CPE in mg/kg		HCG in IU/kg		LHRH in mg/kg		LHRH-A in µg/kg	
	1	2	1	2	1	2	1	2
African catfish			4000					
Atlantic salmon							125 µg (implant) ♂750	
Ayu					6			
Bighead carp	6	6 ♂5	500–2000 or 50	200 + 4(CPE)	5	5	2 ♂1	9
Black carp	2 (Salmon P.E) 0.1 SGG100						10 + 1–2(CPE)	5
Coho salmon		6			0.1 SGG100 + 2 (×3)		200	200
Chum salmon			1000–1800				♂50	
Channel catfish	4	4	800–1000					50
Common carp	0.3	3 ♂0.3						
Gilthead bream	50 + 3300 HCG		150	150				
Grey mullet	6	6 ♂5	17 000	34 000			20 (CPE)	200
Grass carp			1500 or 100	850 + 1(CPE) 100	0.5	6.5	5–10	63 ♂5
Goldfish			50	100	0.1–1.0 (×5 daily)		10–100	100
Indian carp	2	8	400	1600				
Medaka					1 (6 × biweekly)			
Sea bass			250					100
Striped bass			250–330 ♂100–150	250			100	
Silver carp	6	6 ♂5	800–900 or 50	200 + 4(CPE)	2		2	10 ♂5
Walleye	5 (×4)		140	140			2	

NB All units expressed per kg bodyweight. SGG100 is a purified SPE. All dosages for female fish except as specified (♂).

from the pituitary gland and it is the increasing concentrations of these hormones which bring about final maturation and spawning. Despite their obvious advantages there have been some difficulties in the use of these hormones in commercial culture, possibly because insufficient investigation has been carried out. From the data available it would appear that the naturally-occurring releasing hormones are less effective than the synthetic analogues, probably because the former are more rapidly metabolised or removed from the fish. In general the synthetic analogues like LHRH-A are up to 1000-fold more potent (Table 4.2).

For the carp, and possibly all fish, there does seem to be some advantage in pretreating those which are to receive LHRH-A with a dose of the synthetic dopamine antagonist pimozide (10 mg/kg) as this appears to inhibit the production of other hypothalamic hormones which actively reduce gonadotropin production by the pituitary gonadotropic cells. Further work, however, is clearly required on appropriate schedules of treatment for these compounds. Similar observations might also be applied to the two anti-oestrogens tamoxifen and clomiphene citrate (Fig. 4.3), which have both been shown to induce spawning.

During hypophysation all handling of broodstock should be carried out under anaesthesia, for this will avoid damage to the scales or surface of the fish and the internal organs, especially the ovary or heart. A range of anaesthetics is available (see Table 4.3), all with various advantages and disadvantages as far as their use and efficacy in different species is concerned. Many workers use MS-222, although this is expensive. A cheaper and equally effective alternative, 2-phenoxyethanol, is a liquid at temperatures above 10°C and hence much more easily adminstered. All anaesthetics should be used with caution as they may be toxic to workers after long-term use. All fish should be rinsed in water after anaesthesia because it has been reported that some anaesthetics may reduce sperm motility and egg quality. When trying a particular anaesthetic for the first time, it should be used at a low dilution and with the least valuable specimens of a particular stock, as pH, water hardness and fish size are known to have significant modifying influences on the potency of anaesthetics.

In some cases the fish which have received the pituitary injections are allowed to spawn naturally in a tank or aquarium. However, more usually farmers wish to strip their fish so that the eggs and sperm may be properly fertilised and processed through suitable

Table 4.3 Commonly used anaesthetics

Benzocaine (ethyl-4-amino-benzoate): 50 mg/l (dissolve in small quantities of
  alcohol or acetone and add to water)
Chlorbutol: 1 g/l
Hydroxyquinaldine: 10 mg/l
Metomodate: 0.2–0.8 mg/l
Methyl pentynol: 0.5–3 ml/l
MS-222 (tricaine methanesulphonate): 50–100 mg/l (buffered to pH 7)
2-Phenoxyethanol: 0.3 ml/l
Quinaldine: 50 mg/l
Tertiary amyl alcohol: 1–10 ml/l
Urethane: 100 mg/l

Sodium amobarbitol (85 mg/kg of fish body weight), sodium amytal (50–80 mg/kg)
sodium barbitol (50 mg/kg) and novocaine (50 mg/kg) may be used for
anaesthesia or sedation, although all must be administered by injection

---

NB   These concentrations of anaesthetic have been used successfully on farms.
However, when used for the first time, all should be tested because each species
responds differently to anaesthesia. The speed and depth of anaesthesia for each
species also shows considerable variation with different water supplies and fish
size and at different temperatures.

incubation and hatching containers. For some fish species it may
be necessary to put a stitch in the vent at the time of the first or
second pituitary injection to prevent premature release of eggs.
This is then removed when the farmer is ready to strip the eggs.
Generally, stripping is carried out in a similar fashion in all fish.
Between 6 and 12 hours after the second hormone injection the
fish are anaesthetised again and re-examined. The females should
be grasped around the caudal fin with the hand and, with the
weight of the fish supported by the arm which is holding the tail,
slight pressure applied to the abdomen (Plate 4.2). If ovulation has
occurred then this will produce a steady stream of eggs. If only a
few or no eggs appear then the fish should be returned to a holding
tank and examined again. Fish from which eggs run freely from the
vent are ready for stripping and the eggs are collected into a
suitably-sized container. By gently massaging the fish from the
front of the abdominal cavity backwards, starting in the mid-line
between the two opercula (i.e. stripping) all the eggs which have
been released from the ovary can be removed from the body of the
fish.

Different fish show considerable variation in the numbers of
eggs which are produced, i.e. their fecundity varies. Although the

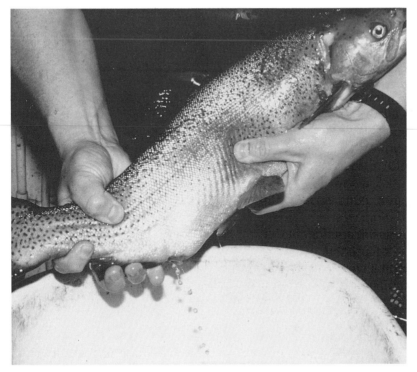

Plate 4.2 Manually stripping hen rainbow trout of eggs.

total fecundity of each fish species varies with its weight, expressing fecundity in terms of the average number of eggs per kg of fish is helpful in calculating the total weight of broodstock which must be held in order to produce a specific number of eggs. Amongst fish, which as a group are extremely fecund animals, the salmonids produce relatively few eggs per kg body weight, whereas the carps and flatfish produce very large numbers (Table 4.4). This means that the production of salmonid eggs and fry constitutes a considerable commercial undertaking whereas cultivation of many of the other groups of fish can be achieved with relatively small broodstocks. The dietary intake of broodstocks can also have profound effects on their fecundity. Trout maintained to half the recommended ration produce only 75% of the numbers of eggs of those maintained on full rations. Reductions in ration during the second half of the reproductive cycle have also been shown to decrease the size of eggs produced. There is a tendency for broodstock fish to be fed only maintenance rations. However, in view of

Table 4.4 Fecundities of farmed fish (number of eggs per kg fish body weight)

African catfish	80 000
Atlantic salmon	1 800
Ayu	210 000
Bighead carp	120 000
Bluegill	110 000
Brook trout	2 600
Brown trout	2 100
Channel catfish	7 000
Chinook salmon	770
Cod	680 000
Common carp	150 000
Coho salmon	990
Cutthroat trout	1 600
Goldfish	110 000
Grass carp	80 000
Grey mullet	800 − 1 200 000
Lake trout	1 750
Pike	25 000
Pink salmon	900
Rainbow trout	2 200
Red seabream	300 000
Silver carp	160 000
Smallmouth bass	17 000
Sockeye salmon	1 100
Striped bass	220 000
Turbot	1 200 000
Walleye	55 000

the known effects on fecundity, egg size and also egg quality, it is recommended that all broodstocks are maintained on diets with high levels of fat and protein (see also Chapter 5).

Stripping should be carried out as soon as possible after ovulation, otherwise an over-ripening of the eggs will occur with a loss of viability. Different species vary in the time at which the over-ripening process begins. Trout eggs lose their quality if left in the abdominal cavity more than 10 days after fertilisation at 10°C. In salmon over-ripening does not generally become a problem until 14−18 days after ovulation. Turbot and halibut eggs become over-ripe if not stripped from the fish within 10 and 1−2 hours of ovulation respectively. In contrast, striped bass eggs and those of many of the carp species and other cyprinid fish start to become over-ripe less than 60 minutes after ovulation. For these species it may be necessary to make repeated checks on the broodstock after the second pituitary injection for the first signs of ovulation.

4.3.2 Fertilisation and larval development

Following the collection of eggs from the female, two male fish should be stripped in the same way. From the resulting pool of milt a small amount should be added to each batch of eggs. Two different methods of artificial fertilisation may be used. The first, known as wet fertilisation, involves the stripping of sperms and eggs into a container filled with water. Water activates the sperm and promotes fertilisation but, depending on the species, this activity may only last for between 15 and 30 seconds. Hence, there is the likelihood that some of the eggs will not be fertilised. Generally, the preferred procedure is the dry method which was first developed in the late nineteenth century. In this method both the male and female fish should be carefully dried before they are stripped. The eggs are then stripped into a dry pan, the sperm added and the mixture gently stirred with a finger or feather. Approximately 1 ml of milt is used for each 10 000 trout eggs. Reasonable fertilisation rates may be achieved with far less milt but excessive amounts of milt should be avoided as this may lead to a blocking of the micropyle (the opening in the egg through which the sperm enters to fertilise the egg). After 5 to 10 minutes water should be added and from here onwards procedures for the wet and dry methods of fertilisation are much the same. The excess sperm and yolk debris should be rinsed off the eggs using further changes of fresh water. Over the next 30 minutes to 2 hours, depending on temperature and species, the eggs imbibe water and swell. This process is known as water-hardening. In some species the eggs become sticky with the addition of water and if the eggs are to be conveniently handled in the hatchery, this stickiness has to be removed. This is a particular problem for the common and Indian carps and also for many other cyprinids.

Inactivation of the sticky layer is carried out as follows: A solution comprising 0.4% sodium chloride and 0.3% urea is gradually added to the eggs over a period of approximately one hour during water-hardening. This has the effect of temporarily inhibiting the development of adhesion while prolonging the fertilising capacity of the sperm. For this treatment to be effective it is essential that the eggs are continually stirred. The supernatant liquid is then decanted and the eggs washed twice with fresh hatchery water before treatment with a 0.16% solution of tannic acid. This solution is removed after approximately 10 seconds and the eggs washed three times with fresh hatchery water. A second tannin rinse of similar duration is followed by further thorough washings of the eggs with fresh water. Subsequently, the eggs are transferred to appropriate incubation containers which may be conventional hatchery trays or zouger jars (chapter 3; section 3.3.2). Provided stirring is continued throughout this treatment, discrete eggs result and incubation can proceed without any problems. Suspensions of textile starch, kaolin and clay and mixtures of milk and full cream powdered milk at concentrations of 10−25 g/l have also been used to render the eggs non-adhesive. These can be useful for the eggs of fish species in which the tannin proves to be especially toxic. All of these methods rely on the various particles forming a thin coating around the eggs during the water-hardening process. After this has been completed the particles should be rinsed off the eggs.

After fertilisation, eggs pass through a series of well-defined stages of development, some of which are of particular importance in hatchery management. The first landmark is eyeing, when the embryo begins development of the pigmented retina in the eye so that eye spots become visible in the egg. At this stage eggs can be safely moved and for salmonids this is the point when eggs are often sold and transported to different parts of the world. After eyeing further development occurs and often one can see the young fish moving inside the egg shell (Plate 4.3). At hatching some of the yolk still remains in the hatchling which is now called a yolk sac fry or alevin (Plate 4.4). The fry derives all its nourishment from this yolk sac which is an expansion of the abdominal region. Gradually, the yolk sac is used up by the fry (Plate 4.5) and near the time when it is fully resorbed the fry begins actively to swim and take its first feed. Hatching and first feeding are particularly vulnerable periods in the development of fry and difficulties at these times can lead to high mortalities.

There is considerable variation between species in the time at which eyeing, hatch and swim-up or first feeding occur (Table 4.5). Usually, the timing of the stages after fertilisation is determined primarily by the size of the egg and the amounts of yolk available; development also proceeds faster the higher the temperature. In general the larger the egg, the longer is the period of embryonic development and the bigger the size of the hatched fry. The size of the fry is of considerable importance to the choice of hatchery procedure and diet. If the fry is large then it can be fed on an artifical diet immediately. However, each of the diet particles

Plate 4.3 Trout egg compared with smaller carp eggs.

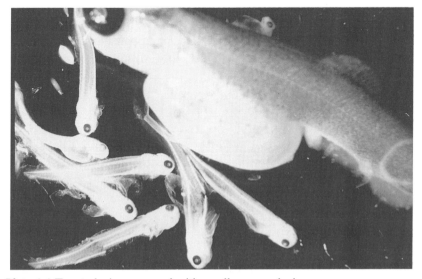

Plate 4.4 Trout alevin compared with smaller carp alevins.

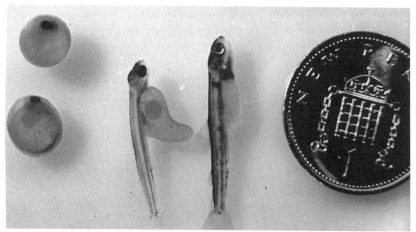

Plate 4.5 Larval development in salmonids (egg, alevin and older alevin showing yolk sac resorption; note diameter of coin is 2 cm).

taken in by the first-feeding fry must be large enough to contain all the necessary components of a balanced diet and retain these without excessive leaching. Salmonid fry are particularly amenable to cultivation because they are large enough to consume artificial feeds immediately after the supplies of nutrients from their yolk sacs have been exhausted. Unfortunately, the fry of most other intensively-farmed fish (e.g. turbot; Plate 4.6) are very small and consequently have to be fed on live organisms because only these are of a small enough size to be eaten by the hatched fry and yet provide all the nutrients required in a balanced diet.

Live organisms of course are also not subject to the leaching

Plate 4.6 Larval development in turbot (day 1 larvae, day 10 larva and day 25 larva; note diameter of coin is 2 cm).

Table 4.5 Approximate development times to eyeing, hatch and first feed (in temperature units/degree days at 10°C)

	Eyeing	Hatch	First feed	Preferred spawning temperature in °C
African catfish		45		23−27
Atlantic salmon	250	430	800	3−12
Ayu		220	270	13−20
Bluegill		42	153	18−27
Brook trout		440		4−12
Brown trout		410		5−12
Channel catfish	130	200	280	21−29
Chinese carp		40	120	21−29
Chinook salmon	250	420	890	5−12
Chum salmon	420	560	960	5−12
Coho salmon	250	420	970	5−12
Common carp		80	130	17−25
Cutthroat trout	150	300		5−12
Goldfish		80	130	20−26
Grey mullet		55	130	18−24
Grouper			110	22−28
Lake trout		490		5−12
Pike		100	183	4− 8
Pink salmon	420	500	805	5−12
Rainbow trout	160	310	500	5−12
Red seabream		40	100	15−22
Sea bass		50	135	12−18
Smallmouth bass		70	172	15−22
Sockeye salmon.	500	670	1000	5−12
Striped bass		55	180	15−22
Tilapia spp.	30	100	260	24−28
Turbot		90	140	10−18
Walleye		170	190	9−14
Yellowtail		60		18−25

NB   These developmental rates have been calculated at a temperature of 10°C (50°F) to aid comparisons between species. Appropriate temperatures for the culture and spawning of each species are shown in the right hand column. Gaps appear in the table where data are not available.

which affects artificial feeds. A whole series of live foods has been used for different fish larvae including phytoplankton (e.g. *Dunaliella* and *Chlorella*), rotifers (e.g. *Brachionus* spp.), copepods, flagellates (e.g. *Isochrysis*), the brine shrimp, *Artemia salina*, nauplii and the cladocerans *Moina* and *Daphnia*. The organism(s) chosen for each species of fish is determined by the size of the larval food particles relative to the size of the mouth of the hatched fry being cultivated. Thus, turbot fry can be started off on rotifers (Plate 4.7) and then progress to brine shrimp nauplii (Plate 4.8) when their mouths have become larger. Carp can be fed initially on brine shrimp and then on cladocerans (0.8−1.2 mm).

All of these live foods must be cultivated in the hatchery and in recent years considerable efforts have been devoted to the development of methods for continuous culture. Probably the most popular live food is the brine shrimp nauplius. However, the cysts have shown considerable variations in quality depending on their source and the time of the year. Recently methods have been developed to decapsulate the cysts as a further processing step before general sale and distribution. Because the cyst is removed together with various surface contaminants, the decapsulated nauplii are a more consistent and reliable product. This development can only improve the production of many cultured fry in the hatchery. Some dried products and micro-encapsulated foods are now becoming available; at present these diets are being widely-used in crustacean culture. However, their efficacy compared to live foods remains to be demonstrated for a wide range of fish species.

After fry have reached a size where they might reasonably take

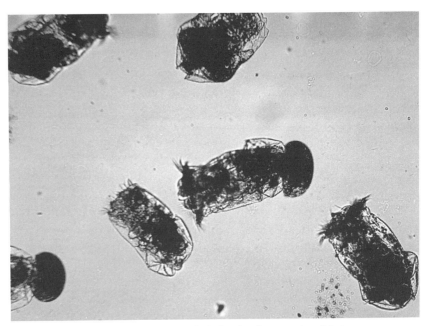

Plate 4.7 Rotifers (70−80 microns size) for feeding marine fish larvae.

Plate 4.8 Nauplius of brine shrimp (*Artemia salina*) (150 microns size).

particles greater than 1 mm in size, they are slowly weaned on to artificial diets.

4.3.3   Environmental control of spawning

Under natural conditions reproduction in fish is timed by changes in the external environment (Fig. 4.3). This enables the production of young fish to be synchronised with seasonal variations in climate or food supplies. Generally, fish which inhabit higher and lower latitudes spawn in response to changing daylength (e.g. salmonids and flatfish), whereas sub-tropical and tropical species like the carps are much more dependent on cycles of temperature or the timing of onset of rainfall (i.e. monsoons). Although seasonal reproduction is of considerable adaptive significance to wild stocks of fish, it is a positive disadvantage in intensive farming where supplies of eggs are required throughout the year. However, the precision and reliability of the reproductive responses of fish to specific environmental cues offer the means by which the timing of spawning can be adjusted to provide eggs and fry on demand.

Carp were probably the first fish on which the environmental control of spawning was practised. Farmers concerned with the production of carp eggs and fry are well aware of the importance of temperature in the control of maturation. In the common carp maturation is stimulated by temperatures in excess of 17°C and preferably of 21−25°C. Under such conditions maturation of the eggs takes 1000−2000°C·days (i.e. 50−70 days). Grass carp need 1400°C·days and the temperature needs to be over 22°C. The sensitivity of this response is well demonstrated by the timing of natural spawning of the common carp in the UK. Generally, spawning occurs in late May but in years like 1985, when the temperature of still water rarely exceeded 20°C, most common carp failed to mature fully and spawn. By contrast, if carp are maintained permanently at temperatures over 20°C, then they continually come into and out of spawning condition at approximately 3-month intervals. Similar cycles of reproduction are also seen in other species of carp and the Indian catfish maintained at constant high temperatures.

Temperature can also be used to modify spawning time in salmonids. Reduction in water temperature to 1−2°C during the Autumn for rainbow trout delays spawning by up to 3 months when compared with fish maintained throughout the year in constant 10°C spring water. This is probably the reason why supplies of rainbow trout eggs from many hatcheries in Denmark, Sweden, the Faroe Islands and parts of Scotland only become available during the late Spring. Similar delays in maturation are also seen in bass and catfish in response to colder waters. Clearly, only farms with ready supplies of both spring water and low temperature river water can use this method commercially, for the artificial cooling of water is an uneconomic proposition. There may also be other difficulties as a result of the temperature reduction. Certainly, many fish species resorb their eggs if exposed to sudden reductions in water temperature during the later stages of maturation.

Low temperature waters can also be used to slow down the early embryonic development of the fertilised egg. For the rainbow trout temperatures between 1 and 2°C from 10 days after fertilisation onwards delay the time of hatching of the eggs by 50 days (Fig. 4.7). Provided eggs are incubated at higher temperatures for the first 10 days after fertilisation there are no reductions in egg quality. If supplies of low temperature water are not available from natural sources, then the low flows required for incubation can be aerated and recirculated past the eggs after chilling in a simple refrigeration unit. Low levels of malachite green can be maintained in the closed system to inhibit fungal infection and, as necessary, a klinoptilite filter, with by-pass, to remove the ammonia.

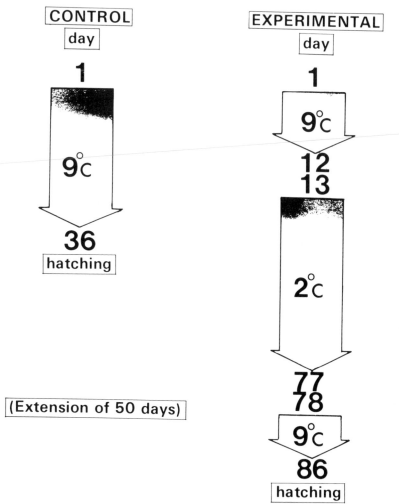

Fig. 4.7 Effects of low temperature on the rates of development of rainbow trout eggs (experimental), compared with 9°C spring water (control) (from Bromage and Cumaranatunga, 1987).

Of greater practical application for altering the rate of maturation and the time of spawning is the use of modified light or photoperiod regimes. All fish whose reproduction is cued by light spawn at specific phases of the annually changing cycle of daylength. Most salmonids in the Northern hemisphere spawn in the Autumn and Winter months under decreasing or short daylength (Fig. 4.8). However, the early growth of the egg and the incorporation of yolk will have taken place 6–9 months earlier under a long or increasing daylength, whereas other photoperiodic species spawn

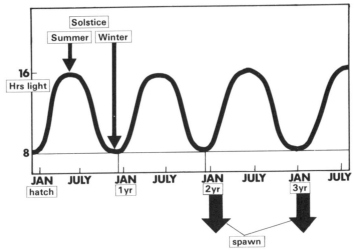

Fig. 4.8 Timing of spawning of rainbow trout in the UK in relation to daylength (from Bromage *et al.*, 1984).

at different times. For instance, most flatfish spawn in the Spring under long days whereas the gilthead bream and sea bass reach maturity early in the year on short but increasing daylength.

With trout, adjustments of the seasonal rate of change of photoperiod into shorter and longer periods than a year (Fig. 4.9) have produced advances and delays of 3 to 4 months respectively in the timing of spawning. Generally, the eggs produced earlier are somewhat smaller and those spawned later a little larger than the eggs from natural spawnings, but there are no effects on the quality of the eggs. Constant long days (18 hours light and 6 hours dark each day) from January until May followed by constant short days (6 hours light and 18 hours dark each day) have produced similar 3 to 4 month advances in spawning time. Delays of equivalent duration have been produced by exposing fish to constant short days throughout the year or short daylengths until September followed by constant long days. These constant photoperiod regimes are obviously less complex than modified seasonal light cycles and thus easier to manage and use on commercial farms. Once the spawning time of a stock has been modified in this way, the fish must be maintained permanently under controlled light otherwise their spawning will revert to its timing under natural ambient conditions. Usually, after the spawning of a stock has been advanced or delayed by 3 to 4 months, then the fish are maintained on a light

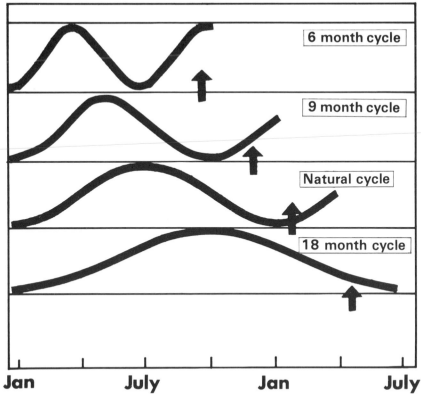

6 month cycle

9 month cycle

Natural cycle

18 month cycle

Jan   July   Jan   July

Fig. 4.9 Alteration of spawning time in rainbow trout using daylength manipulation (from Bromage and Duston, 1986).

cycle which will give out-of-phase yearly spawnings. In this way the eggs produced will be the same size as those produced by natural spawning.

All of these photoperiod manipulations require that the tanks are totally blacked out, with light provided by white light fluorescent tubes and adjusted weekly as necessary with 24 hour cycle electronic timers. It is important that no light is allowed to interrupt the artificial dark periods as even the shortest periods of illumination may influence spawning time. Similar techniques have also been used successfully with the coho and pink salmon, the brown and brook trout, the sea bass, gilthead bream and turbot, although only with turbot and rainbow trout have these methods achieved commercial realisation. The use of environmental manipulation to change spawning time is a particularly attractive proposition for flatfish, bass, bream and carp because the very high fecundities of these fish mean that only small numbers of broodstock need to be maintained under artificial conditions.

Using photoperiod control it is possible to spawn fish more than once a year. Thus rainbow trout maintained under constant illumination or long days for a month before and at least a month after spawning mature again after an interval of approximately 6 months. There is, however, considerable variability in this response, with some fish spawning after 5 months, others after 6, 7 or 8 months, and still others failing to mature until the normal spawning time 1 year later. Similar results have been reported for flatfish, although it would appear than this technique has not been used commercially. There are also strains of rainbow trout in both Japan and the USA which spawn every 6 months apparently without photoperiodic interference, although at this stage it is not clear whether this characteristic will be maintained over successive generations.

4.3.4   Storage of eggs and sperm and cryopreservation

Most farmers would find it useful to store eggs and sperms for short periods of time. This enables separate batches of eggs to be fertilised together or more effective use to be made of limited supplies of sex reversed or out-of-season sperm. Eggs may be stored in the refrigerator for 4−5 days at 2−3°C without any consequent changes in quality. Longer periods are possible with sperm although it must be maintained in a high-oxygen environment. On no account should either be mixed with water as this will lead to activation of the sperm and water-hardening of the eggs.

Much less successful is the long-term storage of gametes and embryos. In fact for the commercial farmer only the storage of sperm currently offers any real prospect of commercial application. Storage of sperm is achieved by cryopreservation, which is storage at −196°C, the temperature of liquid nitrogen. Problems occur during the freezing and thawing which must be carried out at optimum rates in order to avoid damage due to osmotic change and formation of ice crystals; these rates vary considerably according to the species, the material to be frozen, its size, etc. Rates of 5−1500°C/min have been used although those approximating to 30−50°C/min would appear the most effective. Often, cryoprotectants like glycerol, DMSO in combination with methanol, and egg yolk or milk powder, are used to minimise some of the problems consequent on the freezing and thawing. Moderate success has been achieved with the sperm from a number of salmonids, the gilthead bream, sea bass, cod, plaice, some tilapia and Indian carp. However, there is still much painstaking research to be carried out before cryopreserved sperm are routinely used on fish farms.

4.4 SEX CONTROL

Because of the deterioration in consumer appeal and the unpredictable nature of the various changes in growth, food conversion efficiency, behaviour, body and flesh colour and disease susceptibility which occur during reproductive development (section 4.2), many farmers would prefer to inhibit or delay maturation and therefore employ predominately immature or sterile stocks in their production of table fish. Often these problems are more severe in one of the sexes, generally the male, and under these circumstances a reversal of sex and the production of single-sex stocks may be the preferred option.

A number of methods of controlling sex are now available (Fig. 4.10), some in routine farm use, some still at the research bench, but all requiring an understanding of the way in which sex is determined. For most fish, in line with other vertebrates, sexual differentiation occurs early in development, often immediately after

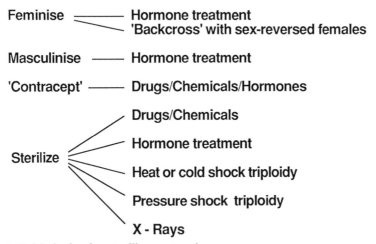

Fig. 4.10 Methods of controlling maturation.

hatching, and is controlled by the chromosomal (genetic) configuration of the nucleus of the fertilised egg. In man the X and Y pair of sex chromosomes determines gender or sex with the female or homogametic sex having a pair of like X sex chromosomes and the male or heterogametic sex having dissimiliar X and Y sex chromosomes in the pair (Fig. 4.11). Most fish are the same and all the gametes produced by the male and female fish at spawning (i.e. sperms and eggs before fertilisation) contain half the total number of chromosomes found in the egg after fertilisation by the sperm and also present in the cells of the developing fish. The gametes are said to have a 'haploid' set of chromosomes and the fertilised adult complement is described as 'diploid'. In the egg the haploid set of chromosomes includes a single X sex chromosome, whereas the haploid set in each sperm will include either an X or a Y. At fertilisation the diploid genetic constitution is re-established by the fusion of the haploid sperm and egg nuclei. Thus all the adult cells formed by division during embryonic development contain two sets of chromosomes, one derived from the egg and the other from the sperm. Fertilised eggs which contain the XX sex chromosome configuration will become females and those with the XY will develop into males. Most fish so far studied would appear to have similar mechanisms of sex control with the female being the homogametic sex. Only *Oreochromis aureus* of the commercially important species is thought to have the male as the homogametic sex.

DETERMINATION OF SEX

chromosomes

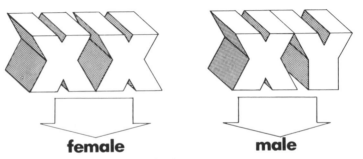

Fig. 4.11 The sex-determining mechanism.

Although genetic sex is established at fertilisation, sexual differentiation generally does not occur until after hatching. At this point modifications in development, in response to the genetic constitution of the embryo, programme the primordial germ cells to become either ovary or testis. Subsequently, with further growth of the gonads and the production of sex-specific steroid hormones, the fish undergo full maturation and realise their genetic sex. Fortunately for aquaculture, the sex-determining mechanisms in fish are extremely 'labile' or malleable processes offering considerable potential for artificial manipulation. In this section a range of methods for sex reversal and sterilisation will be considered, together with the respective advantages and disadvantages of each.

4.4.1 Sex reversal

Methods involving direct hormonal feminisation and masculinisation and single-sex broodstock production are as follows:

(a) Direct hormone sex reversal

Sex reversal has been accomplished in a number of fish species by the administration of steroid hormones before and/or during the period of sexual differentiation. Generally, first-feeding fry treated with male hormones or androgens develop testes and male sexual characteristics at maturity, whereas treatment with oestrogens produces fish with ovaries and female characters. The striking feature is that both changes occur irrespective of the genetic constitution of the fish which have received the hormonal treatment. The decision taken by the farmer regarding the sex of fish to be produced will, of course, be determined by the particular sexual maturation characteristics of the two sexes during their growth cycle on the farm. Generally, farmers would wish to exclude male salmonids, flatfish and eels from their farms, whereas in tilapia and catfish culture the aim would be to produce all-male stocks.

An important advantage of the techniques of sex reversal which use direct hormone treatment is that the hormones are administered to the fish in their feed. Consequently, the treatment is easy to manage and carry out on farms. Generally, the hormones are administered for only a 40–60 day period, rarely 100 days, from the time the fry take their first feed (Fig. 4.12 shows a protocol) and hence hormonal residues will have disappeared from the flesh of the fish long before they are marketed. In fact the levels of the two hormones, oestradiol-17β and 17α-methyl testosterone, which are the two most commonly used for sex reversal, are reduced to unmeasurable quantities only 5 days after their withdrawal from the feed. These hormones are also very similar or identical to the naturally-occurring hormones. Consequently, if used at the correct levels they will cause no adverse changes in the fish under treatment. All hormones, however, should be handled with the same care and respect as any other chemical or medicine as they can cause problems if mishandled or improperly used. In many countries there is legislation governing the addition of materials to animals which are to be sold for human consumption and it is essential that before usage the farmer seeks the advice of his veterinarian or fishery scientist. In the UK, hormone additions to feeds fall under the auspices of the Medicines Act and so they can only be incorporated under veterinary prescription. In the USA the Food and Drugs Administration controls the additions of hormones and other materials.

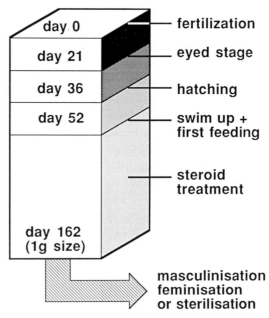

Fig. 4.12 Protocol for the steroidal masculinisation, feminisation or sterilisation of rainbow trout (at 9°C).

Hormonal supplements to the diet are simply prepared. First the appropriate quantities of steroid are accurately weighed out and then dissolved in a known quantity of ethyl alcohol. Because of the small quantities of hormone involved it is essential that all is dissolved in the alcohol. Appropriate quantities of fry food are also weighed and the hormone−alcohol solution thoroughly mixed, ensuring an even distribution throughout the measured amount of food. More consistent mixes can be produced by spreading the food thinly over a polythene sheet and then spraying the hormone solution over the food with a hand-held electric or mechanical paint sprayer. Plastic gloves should be used for all these procedures and if the spray method is used it should be conducted in a fume cupboard. Farmers may prefer to have the diets made up for them by the local veterinary surgeon or fishery scientist. After the addition of the hormone solution, the food is thoroughly dried to evaporate the alcohol, leaving the hormone as a coating around all the food particles. The dried diet should be bagged, suitably labelled and stored in the freezer until use. The same procedure is used to produce both oestrogen and androgen-supplemented feeds for feminisation and masculinisation respectively.

Feminisation

Feminisation is achieved by the administration to first-feeding fry of oestradiol-17β. Other oestrogens have been used for feminisation (e.g. ethynyl-oestradiol which is a longer-acting preparation), but oestradiol is the steroid of choice because oestradiol and oestrone are the naturally-occurring oestrogens in fish; oestradiol is generally more effective then oestrone in inducing sex reversal. A range of oestradiol dose levels and treatment periods have been used and the recommended levels for a number of different fish are shown in Table 4.6. Levels of 20 mg of hormone per kg of food fed feminise most salmonid fry.

In all the treatments described in this section it is essential that the first-feeding fry receive the appropriate levels of hormone treatment throughout the period of sexual differentiation. In salmonids this is probably for the first 50−70 days after first feeding at 10°C. With the rainbow trout our laboratory has produced 100% all-female stocks in some commercial trials but this has not been a consistent result. To achieve 100% feminisation rates the hormonal diet must be fed for at least 16 hours each day at the ration level

Table 4.6 Dosages of 17β-oestradiol (mg/kg food fed) and treatment period required to feminise different fish

Fish	Dosage in mg/kg of food fed	Treatment Period in days from first feed
Atlantic salmon	20 (+2 immersions of alevins in 250 μg/l)	120
Brook trout	20	60
Channel catfish	60	21
Coho salmon	10 (+4 immersions of eyed eggs and alevins in 50 μg/l 2 hours each)	70
Goldfish	100	
Masou salmon	Alevins immersed in 5 μg/l for 18 days	28 (starting 5 days after hatching)
Medaka	125	
Rainbow trout	20	60
Tilapiids		
Oreochromis mossambicus	50 (ethynyl-oestradiol)	20
Turbot	5	60

recommended by the feed manufacturers' tables. Incomplete feminisation occurs if the level or duration of feeding falls.

With the Atlantic salmon and two species of Pacific salmon, levels of feminisation have been shown to be improved by immersion of the alevins and/or eyed eggs in a solution of 5–50 μg/l of oestradiol, possibly indicating an earlier onset of sexual differentiation in these fish.

With oestrogens one has to take care not to exceed the recommended dose. At high levels oestrogens cause liver damage and mortalities in salmonids and flatfish. In contrast cyprinids are far more tolerant of oestrogens. Possibly for this reason they require higher levels for feminisation. With the exception of *Oreochromis mossambicus*, where a dose of 50 mg ethynyl-oestradiol/kg induced feminisation, tilapia would also appear not to respond consistently to oestrogens.

At the levels required for total feminisation, all oestrogens cause some growth depression. However, as soon as the period of hormone administration is completed, the treated fry regain lost ground when compared to the normally-fed stocks.

Masculinisation

Masculinisation of fish is usually carried out with 17α-methyl testosterone which is a derivative of the naturally-occurring hormone testosterone. Although testosterone and methyl testosterone have similar effects on the development of male secondary sexual characteristics, methyl testosterone is more effective in inducing sex reversal, possibly because it is a longer-acting preparation. Ethynyl-testosterone has also proved effective but it is not very soluble in alcohol. Recently, 11-ketotestosterone, the other naturally-occurring fish androgen, has been used to masculinise tilapia. However, the levels required for masculinisation were extremely high making the use of this androgen in sex reversal a prohibitively expensive procedure.

Generally, much higher levels of androgen can be used on fish without fear of toxicity. In salmonids and flatfish, levels of only 3 mg of hormone per kg of food fed are effective in inducing masculinisation. Most tilapia require treatment levels of 30−60 mg/kg (Table 4.7). Higher levels of androgen cause a form of sterilisation, particularly in salmonids (section 4.4.2). Again in the salmonids there does seem to be some advantage in immersing alevins and/or eyed eggs in a solution of the same androgen. A positive feature of methyl testosterone and androgens in general is that they are anabolic and therefore stimulate growth. However, on the negative side, and despite the fact that these direct hormonal sex reversals of fish are inherently safe and effective procedures, it should be pointed out that there is considerable consumer resistance to the use of hormones in animals which will be consumed. For this reason, in particular, it is probable that the main uses of hormonal sex reversal in the future will be to prepare stocks for subsequent genetic manipulations.

(b)   Genetic sex reversal and single sex stocks

Because of the various objections to the hormonal treatment of fish which are subsequently to be used as a source of human food, there have been many attempts to produce single-sex stocks by genetic means. The first step in this procedure is the sex reversal of fish which are of the XX homogametic sex and (as explained in

Table 4.7 Dosages of 17α-methyl testosterone (mg/kg food fed) and treatment period required to masculinise different fish

Fish	Dosage in mg/kg of food fed	Treatment period in days from first feeding
Atlantic salmon	3	90
Chinook salmon	3	60
	(+2 × 2 hour immersions in 400 μg/l of alevins)	
Coho salmon	1	60
	(+ 2 × 2 hour immersions in 400 μg/l of alevins)	
Common carp	100	from 40 until 88 days after hatching
Goldfish	25	60
Medaka	25	28
Rainbow trout	3	70
Steelhead trout	5	220
Tilapiids		
Oreochromis aureus	60	25
Oreochromis mossambicus	60	28
	or 30	60
Oreochromis niloticus	40	60
Tilapia macrochir	40	60
Tilapia zillii	50	45
Turbot	3	60

section 4.4 above) this is usually the female. Thus for salmonids, first-feeding female fry are masculinised with methyl testosterone and develop testes and male secondary sexual characteristics but still have the female genetic constitution. The fertilisation of normal eggs with sperm from these sex reversed or masculinised females produces fish which all have the female XX sex chromosome configuration (Fig. 4.13).

Initially one has to masculinise a mixed male and female stock for there is no way of distinguishing the sex of first-feeding fry. Subsequently, when these fry have reached sexual maturity, back-cross techniques, involving determinations of sex ratios in broods derived from different male fish, may be necessary in order to distinguish masculinised females (XX genotype) from normal males (XY genotype). In rainbow trout this differentiation is helped because the masculinised females have blocked or incomplete sperm ducts and thus cannot be stripped of milt at full maturity. Any

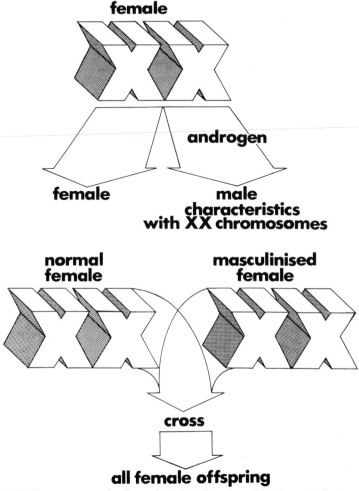

Fig. 4.13 Two-stage production of all-female salmonids. Stage 1 (upper figure) involves the masculinisation of genetically female fish. Stage 2 (lower figure) involves using the sperm from these to fertilise eggs from a normal female.

male fish which is able to release milt is not a sex-reversed female and must not be used in the production of all-female broods. At one stage it was hoped that by treating fish with the lower doses of androgen, or the same dose for a shorter period of time, sex reversal would be accomplished with normal development of the sperm ducts. The results of such trials have, however, been largely equivocal.

In order to obtain sperm from the masculinised females (i.e. those with incomplete sperm ducts), the fish have to be killed, the testes removed and placed in a dry container. Only fish with nodular testes and incomplete or no sperm ducts should be used. One drawback with this technique is that it is difficult to judge whether the fish is fully mature or ripe. In normal fish the ability to strip milt is the criterion used. Sperm maturity can only be ascertained by checking under the microscope the motility of a small sample from each and every testis.

First the testis is sliced using a razor blade and from the milky fluid which is usually released, a small drop is placed on a microscope slide and covered with a coverslip. To the edge of the coverslip is added a drop of water, or preferably a similar quantity of ovarian fluid. The water or ovarian fluid produces an intense swimming activity in mature sperm. Only testes which have these highly motile sperm should be considered suitable for fertilisation. Motility only lasts for approximately 30 seconds and thus it is important to have the microscope properly focused ($\times 50$ magnification) and the preparation located before addition of the activating liquid. Motility should not be confused with the eddy currents which sometimes occur with the addition of water.

With practice this technique produces 100% all-female stocks, at least as far as rainbow trout are concerned. Similar rates of fertilisation and survival of the eggs and fry are achieved with the sex reversed as with the normal spawnings. The process becomes easier after the first all-female stock has been produced because all of the first-feeding fry will be female and consequently when they reach maturity all the male fish are certain to be sex-reversed females.

4.4.2 Sterilisation

Although the production of single-sex stocks may avoid a number of the more important problems associated with sexual maturity, maturation-related losses in production will still occur if the all-female or all-male stocks are not marketed before they become mature. Clearly, these difficulties will only be avoided by sterilisation. Some of the methods of inducing sterility are considered below.

(a) Hormonal sterilisation

If salmonid first-feeding fry are treated with higher doses of methyl testosterone than those required to achieve masculinisation during sex differentiation (Fig. 4.12), then a form of sterilisation is induced. Even with the doses of methyl testosterone which are recommended for masculinisation, a certain proportion of sterile fish is usually produced, probably as a result of differences in the amounts of the hormone-supplemented food consumed by the first-feeding fry. Sterilisation has been achieved in rainbow trout, turbot and Atlantic, coho and chinook salmon (Table 4.8). Treated trout have little gonad development up to 2 years of age but subsequently very low rates of maturation can occur. The technique seems much more effective in the Pacific salmon where immersion of the eggs and alevins in 250 μg/l of methyl testosterone, followed by dietary inclusion at 20 mg/kg, produces 100% steriles. These fish migrate to sea as smolts but do not return as mature adults. As all Pacific salmon die after spawning, this sterilisation represents a considerable enhancement of the fishery. If maintained in sea cages, the sterile fish grow a little more slowly than normal females up to the time of normal maturation, at which time the latter stop growing, spawn and die whereas the steriles continue their growth.

(b) Triploidy

Another method of producing sterile fish is by the induction of chromosomal change. The change most commonly employed is triploidy, which is achieved by subjecting eggs to an environmental shock (e.g. heat, cold, pressure or chemical) during a critical period shortly after fertilisation. In most fish the final division of

Table 4.8 Dosages of 17α-methyl testosterone (mg/kg food fed) and treatment period for sterilisation of different fish

Fish	Dosage in mg/kg of food fed	Treatment period in days from first feed
Atlantic salmon	30 (+2 immersions of eggs and alevins in 250 μg/l)	120
Chinook salmon	9	63
Coho salmon	20 (+4 immersions of eggs and alevins in 250 μg/l)	63
Rainbow trout	30	110
Tilapia macrochii	40	90
Turbot	25	80

the chromosomes in the egg is induced by the penetration of the sperm at fertilisation. This produces an egg cell which contains half the genetic material possessed by adult somatic cells. The egg cell has one rather than two sets of chromosomes and it is said to be haploid (see section 4.4). The effect of the environmental shock is to block the final division of the egg nucleus; it in fact blocks the extrusion of the second polar body (Fig. 4.14). Thus, the egg nucleus is diploid and contains two full sets of chromosomes. Fertilisation by the haploid sperm produces an egg which has three sets of chromosomes, hence the name 'triploid'. The effects of triploidy on reproduction vary greatly with the sex of the fish. Female triploids appear the most strongly affected, with the presumptive ovaries often never developing beyond the earliest stages and never producing female sex hormones. In contrast male triploids appear unaffected by the treatment. Although they are unable to produce fertile sperm, the testes develop, produce male hormones and in consequence the fish show secondary sexual characteristics

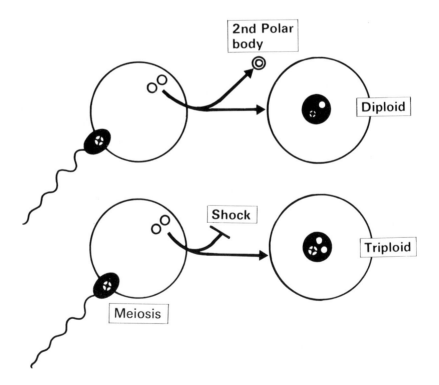

Fig. 4.14 Production of triploid embryos by shocking fertilised eggs.

and other maturation-associated changes. Clearly then, triploid-induced sterilisation is only of use to commercial production if it is performed on all-female stocks.

Various methods of triploid induction are in use. In rainbow trout triploidy is induced by raising the temperature in which the newly-fertilised eggs are incubated, at 28°C for 10 minutes, beginning 20−40 minutes after fertilisation. This produces a triploidy rate of 90%+ of the treated eggs and usually only a 10−20% additional mortality in comparision with normally fertilised and incubated eggs. Temperatures higher than 28°C produce higher triploid rates but lower survivals, whereas temperatures below this (e.g. 26°C) produce fewer triploids and fewer mortalities. Only eggs of the highest quality should be used and only simple facilities are required. After initial incubation of the fertilised eggs for 20−40 minutes in 10°C water in a meshed hatchery tray, the eggs are moved in the tray to an insulated egg box or water bath containing water at 28°C. A thermostatically-controlled water bath is recommended if large numbers of eggs are to be processed.

In Atlantic salmon, heat shocks of 5 minutes at 32°C or 10 minutes at 30°C, beginning 20−30 minutes after fertilisation, have been successfully used. Ovarian development does progress a little further in this species although triploid females are sterile. No sperm are produced in triploid males but the testicular production of hormones is unchanged. High pressure shocks of 3−6 minutes at $7.0 \times 10\,000$ kiloPascals, administered 20 minutes after fertilisation, have also been used effectively in Atlantic salmon and in this species this method produces better survival than with the high temperature technique of shocking. Triploidy has also been successfully induced in flatfish (including plaice and turbot), the common carp, channel catfish and *Oreochromis aureus*.

Much of the interest in triploidy is based on the assumption that the triploid homogametic sex is sterile and that it grows better overall with no development of secondary sexual characteristics. Certainly, this is true for flatfish, tilapia and catfish. In rainbow trout there was some suggestion that triploids did not perform as well as their diploid siblings (i.e. brothers and sisters), even though they continued to grow throughout the period when maturation would depress growth in normal fish. Subsequently, it was found that when diploid and triploids are present in the same cage or tank, the triploids do not compete as aggressively for the available food. Consequently, in a mixed population overall they only grow marginally better. If maintained in a separate tank, however, they show much improved performance and of course no spawning and no maturation-associated mortality.

(c) Irradiation

There have been recent reports of encouraging results from radiation-induced sterilisation. Treatment of young fish with carefully regulated gamma radiation from a cobalt-60 source sterilises fish with few mortalities and other side effects. Further validation of this technique is required, but if it proves successful it will have the advantage of inducing sterilisation in both sexes and hence there will be no necessity to use only all-female eggs. Whether there will be a market resistance to the consumption of irradiated fish similar to that of hormonally-treated animals is unknown.

4.4.3 Hybridisation and gynogenesis

Hybridisation and gynogenesis are two further methods of chromosome set manipulations which result in modifications of sex and maturation.

(a) Hybridisation

The crossing of different species or genera of fish, a procedure known as hybridisation, sometimes produces progeny whose performance in terms of growth, food conversion efficiency and resistance to disease and adverse environmental conditions is better than that of either of the parental fish. This improved performance, which is called 'hybrid vigour' (or heterosis), parallels the production benefits which invariably accrue from interbreeding different strains of the same species.

Hybrids are, however, often sterile offering further advantages for intensive culture. Thus, hybrids of brown trout females and brook trout males (known as tiger trout) are sterile, as are crosses between brown trout and chars. Hybridisation has also been used to produce all-male stocks in the tilapias. Thus, crosses between male *Tilapia hornorum* and female *Oreochromis niloticus* or *O.*

mossambicus and also *Tilapia macrochir* males × *O. niloticus* females are all reported to produce all-male progeny. Unfortunately, it has not always been possible to repeat these results, possibly because the sex-determining mechanisms in tilapias show many race, strain and species differences. The presence of only 5% females in otherwise male stocks is unfortunately sufficient to nullify the possible advantages of this method of all-male production.

An important difficulty with most hybrids, irrespective of their advantages in intensive culture, is that they will have to be accepted by the consumer. It remains to be seen whether there is sufficient interest in hybrids to overcome this possible market resistance.

(b)   Gynogenesis

This is a technique which allows all the genetic material in the progeny to be derived from the mother. Aside from producing all-female stocks, its major use will be in breeding programmes because it allows a much more rapid selection of traits to be made than is possible with conventional male–female spawning.

In gynogenesis the eggs are fertilised with irradiated sperm and then heat- or cold-shocked; irradiation of the sperm prevents any genetic contribution but does not affect its motility. When a sperm enters an egg, it induces final division of the egg nucleus. Normally, the haploid egg nucleus thus formed receives the haploid sperm nucleus and a diploid fertilised egg is produced. Although the irradiated sperm activates the division processes in the egg, no fusion of the genetic material in the nuclei occurs. If there was no further interference with the fertilised egg, then a haploid embryo would result and most if not all of these die. However, shocking of the eggs at this stage blocks the final division of the egg nucleus (section 4.4.2(b)) which remains in the diploid condition. Some of these embryos are viable and all have derived their genetic constitution exclusively from their mother. For similar reasons all of the progeny have XX set chromosomes and hence are all-female. Gynogenesis has been carried out in the rainbow trout, brown trout, coho salmon, rohu, catla, flounder and grass carp. However, survivals from most of these experiments have been very poor. As mentioned earlier, it is likely that this technique will find its greatest use in decreasing the time required to achieve in-breeding in selection programmes.

4.5 SELECTION AND BREEDING

Generally, when considering long-term methods of stock improvement most farmers turn firstly to selection and secondly to out-breeding or line crosses, which encourage heterosis or more accurately inter-strain vigour. Artificial selection in particular has provided the basis for many of the improvements in agricultural productivity encountered over the past five thousand years. Simply choosing the best animals and plants as parents for the next generation has largely resulted, over a period of a thousand or so generations of selection, in the highly-developed varieties of cattle, sheep, chickens, arable crops, etc., on which we all depend today. Stock improvement is of no less importance to aquaculture, although for the majority of fish under cultivation appropriate breeding and selection programmes have not been carried out and the fish being farmed are the same or very similar to those which were initially taken from the wild. Probably only amongst the different species of carp, and to a lesser extent a few of the salmonids, is there any evidence of successful selection for a particular trait or performance characteristic. There is, however, considerable interest in the commercial benefits which might accrue from (amongst others) improvements in growth, food conversion efficiency, disease resistance, tolerance of poor quality or acid waters, and egg production and survival; also from changes in body shape, flesh quality and time of first spawning, particularly if these are shown to be inherited.

When considering the implementation of a stock improvement programme it is important to realise that the physical appearance of a single or a series of characters (known as the phenotype) is determined in part by the genetic make-up (genotype) of the animal or plant and in part by the environment in which it lives. To add to the complexity, the relative importance of the genotype and environment in the determination of each individual characteristic of the phenotype of an animal shows considerable variation. This means that some characters can be passed easily on from one generation to the next, whilst others are modified much more profoundly by the environment in which an animal is placed. Superior fish may only be superior because of a better water supply or diet, or because they are well looked after. The extent to which any particular character is passed on from parent to offspring is termed its heritability; this figure represents the ratio of selection pressure imposed to phenotypic response achieved in the progeny.

Unfortunately, at present there is little reliable information regarding the heritability of characters in fish. Understandably, there has been much interest in growth. However, the heritability of growth is said to be low, with most of the reported gains in growth of farmed fish being probably due to improvements in culture conditions. Notwithstanding this it has been suggested that there are good opportunities for genetic improvements in growth because, although heritability may be low, the enormous range of individual fish size found in most populations would allow very high selection pressures to be imposed. Similar optimism also prevails with regard to the timing of first spawning (especially in salmonids), although there is little experimental evidence to suggest that selection for this characteristic would be effective. But one note of caution regarding fish growth: the fastest growers may be the most aggressive and possibly not the most suitable fish for further selective improvement.

Fecundity and disease resistance are both said to have significantly higher heritabilities, but again this is largely unproven, particularly with regard to improvements under commercial conditions. A good example of successful selection concerns body shape in the carp. The cultured varieties have much deeper bodies and smaller heads than the wild form; some variants also have different scale patterns or markedly reduced scaling. They are also said to have a lower bone-to-flesh ratio, although this change is by no means clear. For the future it is essential that fish breeders have more information about the heritabilities of commercially-important characters; otherwise there is no chance that selection programmes will prove successful.

Before starting any programme of selection, farmers should ensure that their broodstocks represent the best strains or stocks which are available. Differentiation between strains is more difficult with fish because the strain types are not as well-defined as with other domesticated animals, e.g. dogs, cattle, etc. However, a population of fish is now generally recognised as a strain when (1) the population has been separated from the original source for at least two generations, (2) the population has been shown to differ significantly in one or more performance characteristics from the original source, and (3) the population is sufficiently large to be able to provide supplies of eggs or fry to other farms. In order to choose the most appropriate stocks, the farmer must first undertake a survey of the major strain types available and, as far as is known, their important characteristics. This might involve a regional or country-wide assessment, or possibly even consideration of stocks in other countries. Surveys of this kind are particularly important when cultivation of a species of fish which has not been previously farmed in a particular area or country is being considered. If eggs or fish are to be imported from other countries, farmers should make doubly sure of the health status of the stocks. If a health accreditation is not available from an experienced government agency, then it is recommended that the farmer seeks an alternative supply. For many countries imports are prohibited unless the brookstock has been tested regularly over the previous 2 to 3 years, although this safeguard does not universally apply (see chapter 6; section 6.7.1).

Once suitable stocks have been established, farmers beginning a programme of selection and line breeding should carefully document the characteristics of their fish and how these change with each generation. Any variability between strains must be maintained, wherever possible, otherwise future opportunities for selective improvement may be lost forever. There should be a minimum of 50 pairs of male and female fish, preferably 100 in each breeding population. This will ensure a reasonable degree of phenotypic variation in the population and, providing seed is derived from at least 50 of the pairs, it will also avoid any risk of in-breeding. All of the broodstock for each of the available stocks should be marked or tagged so that they can always be distinguished from other strains. On no account should farmers rely merely on separate tanks to isolate the different stocks as only a few escapees can negate even the most successful breeding programmes. Freeze-branding or marking with dyes or numbered tags may be used, although there can be problems because brand marks grow out and tags can be lost. Farms currently considering large-scale breeding programmes might utilise microtags; these are small, binary-coded metal studs, a few millimetres in length, which are implanted into the nose of the fish and can be read with a suitable external scanner without killing the fish. Such devices are used routinely to mark smolts before their release in sea ranching programmes.

Each year a group of one to two hundred fish should be chosen as the parents for the next generation. These should be chosen because they exhibit the best examples of the character on which selection is to be based. Selection for one character in each line offers the best and most rapid chances of improvement. However, it is likely, even if only one character is selected, that five or six generations may have to pass before there is any real genetic improvement. Gynogenesis (see section 4.4.3) may speed up this process, although there are likely to be complications as a result of the viability of some of the offspring. An unselected line should always be maintained as, in addition to allowing an easy assessment to be made of the efficiency of selection, it can also be used at a later stage in the programme to improve the vigour of the stock by heterosis. It can also provide a fresh supply of genetic 'raw material' should a particular line of selection prove unfruitful. Many trout hatcheries only sell eggs derived from crosses between strains because often these show the best performance; they also prevent other farms from acquiring pure strains on which extensive programmes of selection have already been carried out.

5 Nutrition and growth

5.1 INTRODUCTION

As a result of recent developments and improvements in the techniques of fish culture, particularly those concerned with the production of eggs, fry and juvenile fish aquaculture production has been growing at the rate of 10% per annum (chapter 1; section 1.1). Up to the present time more than 300 different species of finfish have been cultivated, all with different feed and nutritional requirements. The continued expansion and improvements in efficiency of aquacultural production require continued improvements in nutritional formulation and feed technology. However, to be effective feed developments must consider on the one hand the nutritional requirements of the cultured species, in terms of their energy, protein, lipid, vitamin and mineral requirements and on the other hand the range of available feed ingredients, their cost and digestibility, and the nature and regularity of their supplies.

Formulation of well-balanced diets and their adequate feeding are two of the most important requirements for successful aquaculture. Without intake of suitable feeds, fish are unable to remain healthy and productive, regardless of the quality of their environment. The production of nutritionally well-balanced diets for fish requires research, quality control, and biological evaluation. Figure 5.1 outlines the various factors which affect successful feed formulation and the development of feeding programmes for culturing fish. Nutritionally deficient diets impair fish productivity and result in a progressive deterioration of health until overtly recognisable diseases ensue. The borderline between reduced growth and diminished health, on the one hand, and overt disease, on the other, is always very difficult to define. Therefore, the ability to recognise a deterioration in growth and health performance during its initial stages and take corrective action will remain

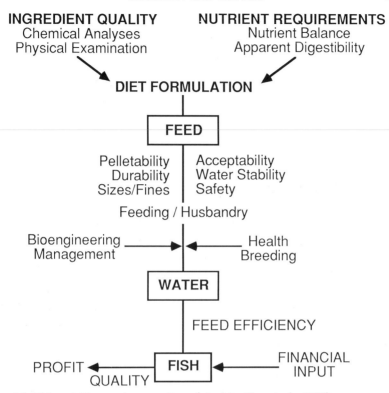

INGREDIENT QUALITY
Chemical Analyses
Physical Examination

NUTRIENT REQUIREMENTS
Nutrient Balance
Apparent Digestibility

DIET FORMULATION

FEED

Pelletability | Acceptability
Durability | Water Stability
Sizes/Fines | Safety

Feeding / Husbandry

Bioengineering | Health
Management | Breeding

WATER

FEED EFFICIENCY

PROFIT | FINANCIAL
| INPUT
QUALITY | **FISH**

Fig. 5.1 Fish nutrition and aquaculture (cited in Cho *et al.*, 1985).

an essential part of the skill of the fish culturist. Diets can also negatively influence the healthy growth of fish by inducing nutrient deficiencies, imbalances, or toxicoses, or by introducing infective agents to fish. A well-balanced diet containing all the essential nutrients in suitable proportions, however, not only results in higher production, but also provides the means to promote recovery from diseases or help the fish in overcoming the effects of environmental stress. Hence, nutritionally balanced and quality-controlled diets are of critical importance for aquaculture.

Fish growth (productivity), together with feed conversion, is greatly influenced by various factors, such as quality of feed, feed intake and water temperature. These factors ultimately affect the nutritional requirements and dietary levels of nutrients. The main factors influencing the intake of feed that is organoleptically acceptable to fish are water temperature and energy density of the diet. Water temperature influences metabolic rate and energy expenditure. Thus growth of fish, in terms of availability of nutrients and energy requirement, is greatly affected by water temperature due to fish being poikilothermic. Protein and energy requirements change as a result of varying water temperatures, and diet formulation should take this into account. The growth rate of rainbow trout, for example, is controlled by water temperature and the total feed requirement is proportional to the rate of liveweight gain (Fig. 5.2). Wherever possible, optimum water temperature should be maintained to ensure maximum feed intake and normal metabolic activity.

Many types of fish feed are currently available, including wet, moist and also steam- and plain-extruded dry pellets, as well as those made of raw trash fish and animal offal (Fig. 5.3). Most farms, however, use either dry or moist diets; in intensive fish culture both have similar formulations, although the nutrients in moist diets are provided by a formulated mash diet together with raw fish and/or slaughter-house offal and waste products.

Moist pelleted feeds have several disadvantages when compared to steam-pelleted dry feeds. The former are high in water content and, to avoid spoilage, must be transported and stored in a freezer until fed. In addition, there are also difficulties relating to the regular supply of fresh raw materials and raw fish and also the risk of introducing pathogens. Improper transportation and storage also

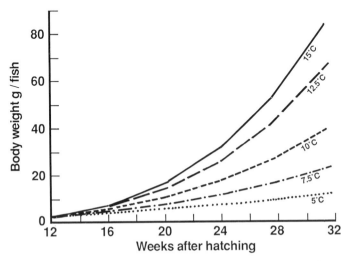

Fig. 5.2 Growth of rainbow trout at various water temperatures (cited in Cho *et al.*, 1985).

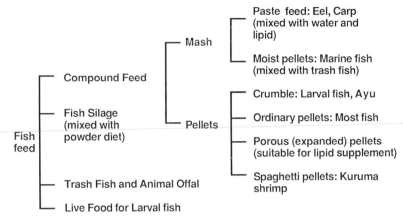

Fig. 5.3 Types of fish feed.

reduce the stability of certain vitamins and fats as well as increasing the growth of fungi and bacteria in such feed. Moist pellets do, however, have some advantages in coastal regions where fresh raw fish and their by-products are regularly available at low prices. The physical characteristics of moist pellets also make them more palatable to some fish species, such as yellowtail (*Seriola quinqueradiata*) which will not accept dry pellets.

By contrast, dry pellets are easier to manufacture, transport and store and the bulk purchase of quality feed ingredients ensures a continuous supply of feed. It is also possible to formulate and fabricate fish diets more precisely and easily using dry ingredients. Most of the nutrients in dry feeds, including vitamins and lipids (which are generally stabilised with antioxidants), can also be stored safely at room temperature. Feeding of dry pellets, either by hand or with mechanical feeders, is also much simpler and easier to achieve than with moist pellets.

Compared with our understanding of the diets of other domestic animals there is still a lot of work to be carried out on fish nutrition. At present fish diets (dry or moist) are not as precisely balanced for nutritional requirements. For example, they are not formulated on the basis of the true bioavailability or digestibility of the ingredients in the feed. Hence, there is a considerable wastage of nutrients due to the use of generous safety margins by feed manufacturers; this may increase the oxygen and other requirements of the fish and also the production and release of waste products into water courses. Feeding techniques and strategy are therefore a critical part of the daily activity at most fish farms.

5.2 DIETARY REQUIREMENTS

Growth of animals is usually defined as a 'correlated increase in the mass of the body in definite intervals of time, in a way characteristic of the species'. This statement implies that growth rate is characteristic for each species and its developmental stages. The maximum growth in terms of increase of weight and size is determined by nutrition and an optimum nutritional regime. True growth involves an increase in the structural tissues such as muscle and bone and also in the various organs. Thus growth is characterised primarily by an increase in protein, minerals and water. Energy-yielding nutrients such as lipid and carbohydrate are important to support the growth processes, and an adequate supply of vitamins is also required. Normal growth is somewhat difficult to define, but it usually indicates the state of productive performance of healthy growth, which varies with species and age. Consequently, feeding tables and standards for normal or optimal growth are invariably different for each fish species and consist of a series of values corresponding to the different ages and body weights of fish during the growth period.

Growth of fish, feed conversion and carcase composition are generally affected by species, genetic strain, sex, stage of reproductive cycle, etc., leading to different nutritional requirements. Growth is also greatly affected by quality of diets in terms of nutrient balance, energy content, bioavailability of each nutrient, etc., as well as environmental conditions such as water temperature, oxygen content, water flow rate, etc. The total requirement for a given nutrient during growth includes the amount needed for maintenance as well as the amount required for the formation of new tissues. The values given in feeding tables represent these combined requirements. Of the various nutrient needs for growth, the requirement for energy is by far the largest and primarily governs the total food allowance.

Thus, all types of formulated fish feeds must satisfy the nutritional requirements of the cultured species in terms of protein (essential amino acids: EAA), lipid (essential fatty acids: EFA), energy, vitamins and minerals. The quality of the feed ultimately will depend upon the levels of nutrients available to the fish. However, the nutritional quality of feeds is difficult to define because of the interactions which occur between the various nutrients during and after digestion and absorption This is also greatly affected by the energy value of the diets because fish adjust their food intake to satisfy their need for energy. Thus, the actual nutrient intake is regulated by the available energy level of the diets and the energy requirements of the fish.

The effectiveness of diets formulated on the basis of digestible energy and nutrients can be evaluated by measuring weight gain, feed efficiency (g weight gain/g feed) or feed conversion (g feed/g weight gain), and by the body composition of fish receiving the diets under particular culture regimes. The efficiency of each feed-stuff in the diets can be evaluated using the following scheme:

1. Analysing the feed ingredients.
2. Measuring the digestibilities of feed ingredients.
3. Formulating and reformulating balanced diets in combination with several ingredients or substituting with other ingredients.
4. Observing the resulting levels of productivity.
5. Measuring feed intake and weight gain and calculating feed efficiency.
6. Calculating energy and nutrient retention efficiencies (NRE) by analysing the fish carcases. NRE can be calculated using the formula:

$$NRE = \frac{\text{Nutrient gain in carcass}}{\text{Nutrient intake} \times \text{Digestibility of nutrient}}$$

In this chapter the dietary requirements for each nutrient are discussed, but for a more complete discussion of the chemistry and physiological functions of each nutrient, the reader is referred to more specialised biochemistry texts.

5.2.1 Protein

Protein is the principal constituent of the tissues and organs of the animal body; other nitrogen-containing compounds include the nucleic acids, enzymes, hormones, vitamins, etc. Thus a liberal and continuous supply is required in the food for growth and repair of decomposed tissue proteins. The proteins consist of carbon, hydrogen and oxygen, in common with the lipids and carbohydrates. In addition they contain a large and fairly constant percentage

(15−18%; 16% on average) of nitrogen. Most of them also contain sulphur, and a few contain phosphorus and iron.

When proteins are hydrolysed with acid or alkaline solutions or enzymes, about twenty different amino acids are obtained. These amino acids are also the end products of protein digestion. Thus proteins are polypeptides, the polymerised units of amino acids. Consequently the study of protein nutrition deals primarily with amino acids and the quality of proteins is defined by the quality, quantity and bioavailability of the amino acids.

Recent studies on essential amino acids (EAA) have demonstrated that fish require the following ten amino acids as essential for growth and maintenance of life:

1.
$$CH_3 \diagdown CH \cdot CH_2 \cdot CH \cdot COOH$$
$$CH_3 \diagup \qquad\qquad | $$
$$\qquad\qquad\qquad NH_2$$

Leucine

2.
$$C_2H_5 \diagdown CH \cdot CH \cdot COOH$$
$$CH_3 \diagup \qquad | $$
$$\qquad\qquad NH_2$$

Isoleucine

3.
$$CH_3 \diagdown CH \cdot CH \cdot COOH$$
$$CH_3 \diagup \qquad | $$
$$\qquad\qquad NH_2$$

Valine

4. $CH_3 \cdot CH \cdot CH \cdot COOH$
$\quad\quad\; |\quad |$
$\quad\quad OH\; NH_2$

Threonine

5. $C_6H_5 \cdot CH_2 \cdot CH \cdot COOH$
$\qquad\qquad\quad |$
$\qquad\qquad\quad NH_2$

Phenylalanine

6. $CH_3 \cdot S \cdot CH_2 \cdot CH_2 \cdot CH \cdot COOH$
$\qquad\qquad\qquad\qquad\; |$
$\qquad\qquad\qquad\qquad NH_2$

Methionine

7.
$$C \cdot CH_2 \cdot CH \cdot COOH$$
$$\| \qquad\quad |$$
$$CH \qquad NH_2$$
$$N$$
$$H$$

Tryptophan

8.
$$NH_2 \diagdown C \cdot NH \cdot CH_2 \cdot CH_2 \cdot CH_2 \cdot CH \cdot COOH$$
$$NH \diagup \qquad\qquad\qquad\qquad\qquad\qquad |$$
$$\qquad\qquad\qquad\qquad\qquad\qquad\qquad NH_2$$

Arginine

9. $HC = C \cdot CH_2 \cdot CH \cdot COOH$
$\quad\; |\qquad\quad |\qquad |$
$\; HN \qquad N\quad NH_2$
$\qquad \diagdown\; \diagup$
$\qquad\quad C$
$\qquad\quad H$

Histidine

10. $CH_2 \cdot CH_2 \cdot CH_2 \cdot CH_2 \cdot CH \cdot COOH$
$\quad\; |\qquad\qquad\qquad\qquad\; |$
$\; NH_2 \qquad\qquad\qquad\quad NH_2$

Lysine

Table 5.1 Requirements of some fishes for essential amino acids (percentage protein)

	Eel	Carp	Rainbow trout	Chinook salmon
Arginine	4.5	4.4	4.0	6.0
Histidine	2.1	1.5	1.8	1.8
Isoleucine	4.0	2.6	2.8	2.2
Leucine	5.3	4.8	5.0	3.9
Lysine	5.3	6.0	6.0	5.0
Methionine + Cys/2	5.0	2.7	3.3	4.0
Phenylalanine + Tyrosine	5.8	5.7	6.0	5.1
Threonine	4.0	3.8	4.1	2.2
Tryptophan	1.1	0.8	0.6	0.5
Valine	4.0	3.4	3.6	3.2

Source: Ogino (1985).

Different species of fish have different requirements for these EAAs (Table 5.1). The quality of proteins is usually evaluated by both biological and chemical methods, although the former methods which use body weight gain and nitrogen retention as criteria for protein quality are more accurate. Of the biological methods, protein efficiency ratio, biological value, and net protein utilisation are the most frequently used, and can be calculated using the formulae given below. Biological value is included, although insufficient data are available for fish due to difficulties in determining metabolic, faecal and endogenous nitrogen separately.

1. Protein efficiency ratio (PER)

$$PER = \frac{\text{Grams liveweight gain}}{\text{Grams protein fed}}$$

2. Biological value (BV)

$$BV = \frac{\text{True nitrogen retained}}{\text{Nitrogen absorbed}} \times 100$$

Also, $A = I - (F - Fo)$
        $R = A - (U - Uo)$
where I = Nitrogen intake
        F = Nitrogen excreted to faeces

Fo = Metabolic faecal nitrogen
U = Nitrogen excreted to urine
Uo = Endogenous nitrogen
R = True nitrogen retained
A = Nitrogen absorbed

$$\text{Thus,} \quad BV = \frac{R}{A} \times 100 = \frac{I - (F - Fo) - (U - Uo)}{I - (F - Fo)} \times 100$$

3. Net protein utilisation (NPU)

$$NPU = \frac{\text{True nitrogen retained}}{\text{Nitrogen intake}} \times 100$$

$$\left(NPU = BV \times \frac{\text{Digestibility}}{100} \right)$$

NPU is usually determined by the following formula:

$$NPU = \frac{\begin{array}{c}\text{Nitrogen increase in fish fed the test protein diet +} \\ \text{Nitrogen decrease in fish fed the protein-free diet}\end{array}}{\text{Nitrogen intake from the test protein diet}} \times 100$$

Of all the nutrients the level and type of protein inclusion is probably the most important, mainly because its high price on the commodity market greatly affects the cost of formulated diets. Protein is essential for normal tissue function, for the maintenance and renewal of fish body protein and for growth, as described above. The protein requirements of fish are influenced by various factors, such as fish size, water temperature, feeding rate, availability and quality of natural foods, overall digestible energy content of the diet and the quality of the protein. The protein requirement of fish is closely related to the optimum dietary protein levels, although these levels are not identical to each other. The former value is characteristic for each species under controlled rearing conditions and should be expressed as a value per unit of fish body weight per day. The requirement of carp and rainbow trout for protein is reported to be very similar at around 1 g/kg body weight per day for tissue protein maintenance and 12 g/kg body weight per day for maximum body protein retention. The efficiency of nitrogen utilisation for growth is, however, highest with a protein intake of 7−8 g/kg body weight per day. Consequently, the optimal protein level for practical feeds is determined by the protein requirement, bearing in mind the quality of the protein, the digestible energy levels, the rearing conditions, the manufacturing process for the feed, water temperature and market costs.

Investigations of the optimal protein requirement for fish have, somewhat surprisingly, shown few differences. Levels of 35−45% crude protein in the diet would appear to satisfy most of the species currently under cultivation (Table 5.2), although eels, red seabream, yellowtail and puffer require values of 45−55%. For all these species the dry feed ingredients are made into a moist pellet by mixing with water in a ratio of 50:50 or 60:40, their protein contents being reduced to 50−60% of the original protein inclusion level. Generally, these protein levels have been determined by using semi-purified or purified diets containing single high quality protein sources, such as casein, whole egg protein or fish meal, all of which have an NPU of more than 60 and a PER of 3.8 when determined in rainbow trout at a dietary protein level of 30% (Table 5.3).

When one attempts to formulate practical feeds, it is important to know whether it is possible to substitute the purified or high quality proteins used in experimental diets with other low cost protein sources; also whether dietary protein levels can be reduced

Table 5.2 Optimal dietary protein levels for certain fish species

Species	Protein source	Optimal protein level in %
Rainbow trout	Fish meal, casein	35−45
Common carp	Fish meal, casein	30−38
Channel catfish	Fish meal, whole egg	22−36
Japanese eel	Casein + arginine + cystine	44.5
Tilapia zillii	Casein	35−40
Tilapia nilotica	Fish meal, casein	30−35
Milkfish	Casein	40
Grouper	Fish meal, tuna muscle	40−50
Red seabream	Fish meal, casein	45−55
Yellowtail	Fish meal	55
Tiger puffer	Casein	50
White shrimp	Fish meal	28−32
Kuruma shrimp	Casein, squid meal	40−60

Sources: cited in Cho *et al.* (1985); NRC (1981); NRC (1983).

Table 5.3 Protein efficiency ratio (PER) and net protein utilisation (NPU) of some protein ingredients determined with rainbow trout at a dietary protein level of 30%. Biological value (BV) was determined with carp at a dietary protein level of 10%

Protein	PER	NPU	BV
Egg yolk	3.8	60	89
Whole egg	3.8	61	
Egg albumin	3.9	62	
White fish meal	3.4	62	76
Muscle protein (rainbow trout)	3.8	62	
Muscle protein (cuttlefish)	3.9	62	
Casein	3.5–3.6	56–58	80
Casein:Whole egg (1:1)	3.8	63	

without reduction of growth rate and/or feed conversion. The protein ingredients commonly employed in fish feeds at the present time are shown in Table 5.4. Their availability will differ with each geographical location. Protein quality is dependent upon the quantity and quality of essential amino acids and their bioavailability, as described above.

Table 5.4 The contents of protein, lipid and energy in some common ingredients in fish feeds

Ingredient	Crude protein in %	Crude lipid in %	Gross energy in kcal/g
Blood meal	80.0	1.0	5.21
Brewers dried yeast	45.0	0.4	4.31
Feather meal	85.0	2.5	5.36
Chrysalis meal	69.7	3.8	5.88
Meat and bone meal	45.0	8.5	4.50
Meat meal	66.8	12.0	4.89
Fish protein concentrate	80.0	0.3	4.37
Herring meal	70.0	10.0	4.94
White fish meal	62.1	11.1	4.57
Corn gluten feed	21.0	2.0	4.45
Corn gluten meal	60.0	2.0	5.07
Rapeseed meal	36.0	2.6	4.30
Soybean, full fat, cooked	38.0	18.0	5.34
Soybean meal, dehulled	48.0	0.5	4.30
Soybean protein concentrate	68.0	0.3	3.48
Soybean germ meal, defatted	25.3	1.8	4.07
Wheat flour	17.6	5.9	4.08
Rice bran, defatted	17.2	2.1	3.65

Proteins ingested by fish are utilised as a source of energy, for the renewal of tissue protein, and growth (increase of nitrogen). Because of the costs of the protein, the feed will be more cost-effective if all the protein is used for tissue repair and growth and little catabolised for energy. 'Energy-yielding or sparing' nutrients like lipids and carbohydrates can theoretically reduce the oxidation of protein to energy and hence improve the utilisation of dietary protein (protein-sparing effect). The beneficial effects of the incorporation of such protein-sparing nutrients have been extensively studied and optimal ratios between protein and energy have been proposed for many fish (Table 5.5). Unfortunately, most species of cultured fish are carnivorous and consequently the amounts of carbohydrate which can be handled are limited, with lipids playing a more important role as a source of dietary energy. Protein level can be decreased by approximately 15% in rainbow trout diets, if high-quality lipids, capable of satisfying the EFA requirement of the fish, are added at levels of approximately 18%. Furthermore low protein diets (35%) with high energy values (18% lipid-containing beef tallow as a main energy source and EFA) are effective for normal growth of rainbow trout at all stages of their development. Similar results have also been obtained with carp, although these and many other omnivorous fish can utilise carbohydrates as well as lipids as a dietary energy source. In these fish the optimal protein level in the diet can be held at 30–35% if the digestible energy content is more than 340 kcal per 100 g diet.

In Japan, fish farming diets rely heavily upon white fish meal as the major protein source. This is derived from demersal fish trawled in northern waters. With the establishment of 200 nautical mile fishing zones the supplies of these fish have decreased, making it necessary to optimise the levels of fish meal in the diets as far as possible. Recently, there have been increases in the proportion of so-called 'brown meal', prepared from coastal fish such as sardine and mackerel, as a substitute for white fish meal in fish feeds. There are high levels of lipid and red muscle in this meal and it has a protein quality which is almost comparable to that of white fish meal. However, most regionally-prepared domestic meals, especially whole meal containing fish solubles, contain high concentrations of histidine which is quickly converted to histamine by bacterial enzymes. The occurrence of gizzard erosion (GE) in chicken has recently been reported in Japan, and was found to be due to so-called 'GE substance' which is derived from the histidine or

Table 5.5 Protein sparing effect by dietary lipids in fish

Fish	Diet				$\dfrac{E(L) - E(P)}{E(P)}$ in %
	P:C:L* in %	Energy content in kcal/100 g	Calorie/protein	Protein required for 100 g wt gain E in g	
Rainbow trout	P 49:30:5	367	75	38	
	L 36:30:15	402	112	31	−18
Channel catfish	P 40:18:5	275	69	48	
	L 36:32:15	407	113	40	−17
Carp	P 42:25:6	316	76	39	
	C 32:45:5	356	110	33	−15
	L 32:30:15	384	125	31	−30
Eel	P 52:22:7	361	70	67	
	L 41:23:16	398	97	51	−24
Yellowtail	P 68:4:5	353	52	124	
	L 53:4:15	369	70	75	−40
Plaice	P 40:0:6	233	58	96	
	L 40:0:18	347	87	75	−22
	C 40:20:9	392	85	62	−35

*P, diets with high protein and low lipid content; L, diets with low protein and high lipid content; C, diets with high carbohydrate and low lipid content.
Source: cited in Watanabe (1982).

histamine in brown or whole meal. GE substance is also formed when histidine or histamine are heated with protein. Both GE substance and free histamine have been shown to produce pathological changes in the stomach of rainbow trout. Thus when utilising brown meal in fish diets, especially complete meals, it is important that as low a temperature as possible is used in the drying of the fish meal so that any oxidation of lipids is minimised.

5.2.2 Lipids

Both animal and plant ingredients contain a group of substances, insoluble in water but soluble in organic solvents, such as ether, benzene, and chloroform, which are generally referred to as lipids. Lipids are roughly classified into three groups: simple, compound and derived lipids. Simple lipids consist of neutral fat (triglyceride), glyceryl ester, wax (true wax, cholesterol esters, vitamin A or D esters), etc. Phospholipids and cerebrosides are the main constituents of the compound lipids. The derived lipids are the hydrolytic end-products of the simple and compound lipids, and fatty acids are their main constituents. As regards the composition of fish carcases and fish feeds, neutral fats are by far the most important members of the lipid groups, but the other lipids such as phospholipids play very significant roles in nutrition and cellular physiology. In common with the carbohydrates and proteins, the lipids contain carbon, hydrogen, and oxygen, and certain phospholipids contain nitrogen and phosphorus also.

Fishes also utilise lipids for energy, for cellular structure, and for maintenance of the integrity of biomembranes. The fluidity of biomembranes is regulated partly by the fatty acids of the phospholipids that control such processes as cellular transport and by the activities of membrane-associated enzymes. Most fish are different from terrestrial animals in that their tissues contain fairly high amounts of $\omega3$ highly unsaturated fatty acids (HUFA) such as $20:5\omega3$ and $22:6\omega3$. The degree of unsaturation in tissues increases when environmental temperature is lowered, thereby maintaining membrane fluidity to allow normal cellular function.

From the standpoint of practical fish culture, dietary lipids are an important source of energy and the only source of essential fatty acids (EFA) in fish. Besides these functions they are important carriers of certain non-fat nutrients, notably the fat-soluble vitamins

A, D, E and K (see section 5.2.5). Recent studies of lipid metabolism in fish have shown that the requirements for EFA differ
considerably from species to species (Table 5.6). One of the most
marked and characteristic differences in EFA requirements between
species is that between freshwater and marine fish. Rainbow trout
require fatty acids of the linolenic family as EFA, whereas carp,
eel and chum salmon require not only linolenic but also linoleic
acids for good growth. However, these particular fatty acids were
found to be non-essential for marine species such as red seabream,
plaice, and yellowtail, which instead required alternative HUFA,
such as $20:5\omega3$ and $22:6\omega3$.

$$CH_3 \cdot CH_2 \cdot CH_2 \cdot CH_2 \cdot CH_2 \cdot CH=CH \cdot CH_2 \cdot CH=CH \cdot CH_2 \cdot CH_2 \cdot CH_2 \cdot CH_2 \cdot CH_2 \cdot CH_2 \cdot COOH$$
linoleic acid ($18:2\omega6$)

$$CH_3 \cdot CH_2 \cdot CH=CH \cdot CH_2 \cdot CH=CH \cdot CH_2 \cdot CH=CH \cdot CH_2 \cdot CH_2 \cdot CH_2 \cdot CH_2 \cdot CH_2 \cdot CH_2 \cdot COOH$$
linolenic acid ($18:3\omega3$)

$$CH_3 \cdot CH_2 \cdot CH_2 \cdot CH_2 \cdot CH_2 \cdot CH=CH \cdot CH_2 \cdot CH=CH \cdot CH_2 \cdot CH=CH \cdot CH_2 \cdot CH=CH \cdot CH_2 \cdot CH_2 \cdot CH_2 \cdot COOH$$
arachidonic acid ($20:4\omega6$)

$$CH_3 \cdot CH_2 \cdot CH=CH \cdot CH_2 \cdot CH=CH \cdot CH_2 \cdot CH=CH \cdot CH_2 \cdot CH=CH \cdot CH_2 \cdot CH=CH \cdot CH_2 \cdot CH_2 \cdot CH_2 \cdot COOH$$
icosapentaenoic acid ($20:5\omega3$)

$$CH_3 \cdot CH_2 \cdot CH=CH \cdot CH_2 \cdot CH=CH \cdot CH_2 \cdot CH=CH \cdot CH_2 \cdot CH=CH \cdot CH_2 \cdot CH=CH \cdot CH_2 \cdot CH=CH \cdot CH_2 \cdot CH_2$$
docosahexaenoic acid ($22:6\omega3$) |
                                                                                      COOH

However, it is very important to bear in mind that the $\omega3$ HUFA,
which are essential dietary constituents for marine fish, are also
effective for most freshwater fishes, except *Tilapia* which require
$18:2\omega6$ or $20:4\omega6$ as EFA for maximal growth. Furthermore, in
salmonids such as the rainbow trout, the yamame (*Oncorhynchus
masou*) and whitefish (*Coregonus* spp.), $\omega3$ HUFA (e.g. $20:5\omega3$
and $22:6\omega3$) have a biological or EFA efficiency which is two-fold
higher than that of $18:3\omega3$. A similar situation pertains in mammals
where $20:4\omega6$ lipids are assigned a higher EFA efficiency than the
$18:2\omega6$ series. In the carp, eel, chum salmon and coho salmon it
has been found that the supplemental growth effect of 0.5% $20:5\omega3$
or 0.5% $\omega3$ HUFA mixtures ($20:5\omega3:22:6\omega3=1:1$) exceed that of
higher levels (1%) of $18:3\omega3$. Thus, for a range of fish species,
diets containing pollock liver oil produce the best weight gains.
Generally, the commercial diets used for rainbow trout, carp, eel,
ayu, chum salmon and red seabream in Japan contain lipids at
levels of 4–6%, with the lipid being derived from fish meal where

Table 5.6 Essential fatty acid requirements of fish

Fish species	Requirement	Reference*
Rainbow trout	18:3ω3 1%	Castell *et al.* (1972)
	18:3ω3 0.8%	Watanabe *et al.* (1974)
	18:3ω3 20% of lipid	Takeuchi and Watanabe
	ω3 HUFA 10% of lipid	(1977)
Carp	18:2ω6 1% and 18:3ω3 1%	Watanabe *et al.* (1974)
		Takeuchi and Watanabe
		(1977)
Eel	18:2ω6 0.5% and	Takeuchi *et al.* (1980)
	18:3ω3 0.5%	
Chum salmon	18:2ω6 1% and 18:3ω3 1%	Takeuchi *et al.* (1979)
	ω3 HUFA 0.5%	Takeuchi *et al.* (1980)
Coho salmon	Tri-18:3ω3 1−2.5%	Yu and Sinnhuber
		(1979)
Ayu	18:3ω3 1% or 20:5ω3 1%	Kanazawa *et al.* (1982)
Tilapia zillii	18:2ω6 1% or 20:4ω6 1%	Kanazawa *et al.* (1980)
Tilapia nilotica	18:2ω6 0.5%	Takeuchi *et al.* (1983)
Red seabream	ω3 HUFA 0.5% or	Yone *et al.* (1978)
	20:5ω3 0.5%	
Turbot	ω3 HUFA 0.8%	Gatesoupe *et al.* (1977)
Yellowtail	ω3 HUFA 2%	Deshimaru *et al.* (1984)
Yamame	18:3ω3 1%	Watanabe *et al.* (1986)
(*Oncorhynchus*	ω3 HUFA 0.5%	
masou)		
Coregonus	18:3ω3 1%	Watanabe *et al.* (1986)
lavaretus	ω3 HUFA 0.5%	
maraena		

*All references cited in Watanabe (1982), except for Watanabe *et al.* (1986).

the content of ω3 fatty acids (mainly consisting of 20:5ω3 and 22:6ω3) comprises 0.4−0.6% of the diets. Such practical diets, however, always satisfy the EFA requirement of these fish species.

Thus, when fish feeds contain fish oil rich in ω3 HUFA, there will be no EFA deficiency for any cultured species except *Tilapia*. Vegetable oils rich in 18:2ω6 should be included in the diets of *Tilapia* to obtain maximal growth. The fatty acid composition of some lipid sources is shown in Table 5.7.

Lipids from marine fish, high in polyunsaturated fatty acids, especially the ω3 HUFA series, in addition to providing a complete EFA specification and hence good growth and feed conversion, also have a protein-sparing action in many fish species. However, these lipids are susceptible to auto-oxidation or rancidity when

Table 5.7 The concentration of ω6 and ω3 fatty acids in some lipids (percentages)

Lipid source	ω6	ω3	
		18:3ω3	ω HUFA
Pollock liver oil	2.0−3.5	0.2−2.0	12−20
Cod liver oil	2.0−3.5	1.0−1.5	20−25
Squid liver oil	2.0−4.0	1.0−1.5	25−30
Herring oil	1.5−2.5	0.5−1.0	11−15
Sardine oil	2.0−4.0	1.0−2.0	20−25
Bonito oil	2.5−4.5	1.0−2.0	20−30
Soybean oil	49−52	1.5−11.0	—
Corn oil	34−62	0−3.0	—
Cotton seed oil	34−55	—	—
Olive oil	5−8	0.5−1.5	—
Safflower oil	39−79	trace−6.0	—

exposed to atmospheric oxygen. Consequently, the positive nutritional value of ω3 fatty acids in fish oils will become a disadvantage if adequate care is not taken in the preparation and storage of diets.

Hydrogenated fish oils or animal fats like lard and tallow are especially suitable as energy sources in fish diets because of their greater resistance to auto-oxidation. In addition, these lipids are widely used as energy sources in animal feeds and hence are readily available. Animal fats or hydrogenated oils with high melting points (mp) have been reported to be less easily digested by coldwater fish than lipids with lower mp and, consequently, are seldom used in fish rations. However, more recent studies on the energy requirements of fish have indicated that animal fats, such as lard and tallow, or hydrogenated beef tallow, together with other lipids, can be used as an energy source without any adverse effects on feed efficiency, fish growth or survivial. In both the carp and rainbow trout, the digestibility of hydrogenated fish oils was found to be affected by their mp and increased as mp decreased (Figs 5.4 and 5.5). Hydrogenated oils with mp of 53°C were poorly digested in both species, especially in fish weighing less than 10 g. On the other hand, a beef tallow and hydrogenated fish oil of mp 38°C was found to be effectively utilised, with digestibilities of more than 70% regardless of fish size and water temperature. This indicates that such lipids are freely available to the fish as a dietary energy source when they are used with appropriate amounts of marine

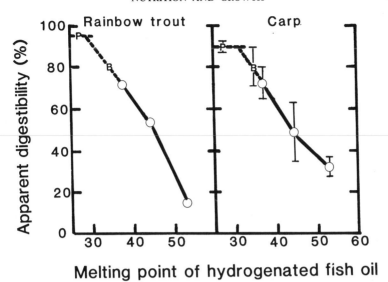

Fig. 5.4 Relationship between melting points of hydrogenated fish oils and their apparent digestibility in rainbow trout and carp. The signs P and B in the figures indicate those obtained from pollock liver oil and beef tallow respectively. Mean ± SD in the right figure was calculated from the values determined at different water temperatures.

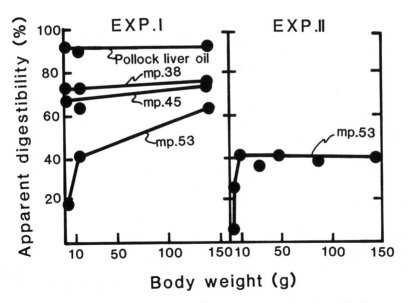

Fig. 5.5 Relationship between fish size and apparent digestibility of hydrogenated fish oils with different melting points (in °C) in carp.

lipids and hence can provide the necessary levels of EFA without any adverse effects on fish.

In intensive aquaculture in Japan, so-called 'feed oil' (pollock liver oil) is usually added to the pellets or moist diets (Table 5.8) before feeding to promote fish growth. However, feeding of high protein–high energy diets frequently leads to lipid accumulation in both the muscle and viscera, changes which result in poor quality fish flesh. Fish feeds should always be formulated with lipids to satisfy EFA requirement and optimise the protein and available energy ratios. They should also take into account the intended feeding rate, the growth stage of the fish and the water temperature.

5.2.3 Carbohydrates

The group of foodstuffs which includes the sugars, starch, cellulose, gums, and related substances is called carbohydrates. They form the largest part of most animals' food supplies and the name is derived from the fact that they contain carbon combined with hydrogen and oxygen, these latter elements usually in the same ratio as in water. Chemically they are polyhydroxy aldehydes and ketones or substances which yield them on hydrolysis. They are classified into four groups: (1) monosaccharides (pentose and hexose types), (2) disaccharides (e.g. sucrose, fructose), (3) trisaccharides (e.g. raffinose), (4) polysaccharides (e.g. dextrin, starch, glycogen, cellulose).

Carbohydrates can act as either an immediate energy source or as a rapidly available energy reserve stored as glycogen in the liver and muscle; alternatively they can act as a long-term energy reserve

Table 5.8 Supplementation of feed oil to practical fish diets

Fish species	Initial fish size in g	Supplementary oil level to diet in %	Water temperature in °C
Rainbow trout	10	5–12	5–20
Carp	40	5–15	10–30
Ayu	3	5–12	10–25
Eel	20	10–20	15–30
Yellowtail	20	10–15	15–30

when converted to fat in the body. Glycogen energy reserves prevent delayed fish mortality after stocking. Carbohydrates also function as pellet binders. Carbohydrates provide the least-cost energy source for fish feeds, although their availability as an energy source is limited in carnivorous fish, such as red seabream, yellowtail and salmonids. Energy requirements of fish are also satisfied by protein and lipids, and fish can survive on carbohydrate-free diets.

Carbohydrates consumed by fish are digested by several enzymes secreted from the pyloric caeca, pancreas, and intestinal mucosa to produce absorbable monosaccharides, which then follow the same Embden−Meyerhof and pentose phosphate shunt pathway of carbohydrate metabolism as in higher vertebrates. The mechanisms of carbohydrate utilisation in fishes vary between carnivorous and omnivorous fish. Omnivorous fish such as carp, catfish and eel can effectively utilise carbohydrates as energy sources. Carp can digest about 85% of ingested α-starch at dietary levels between 19 and 48% and about 55% of β-starch. Thus relatively large amounts of carbohydrates can be included in diets for carp. In contrast, carbohydrate levels of only about 25% are as effective an energy source as lipid for carnivorous fish such as rainbow trout and plaice.

Recent studies on carbohydrate nutrition have demonstrated several reasons for the poor ability of carnivorous fish to utilise carbohydrates, as follows:

1. Several enzymes involved in the digestion and catabolism of carbohydrates are relatively inactive.
2. The glucose tolerance of carnivorous fish is inferior to carp.
3. The low levels of insulin in carnivorous fish mean that their capacity for insulin secretion after glucose administration is poor.
4. Glucose tolerance and hepatic enzyme activities are further aggravated when feeding diets are high in carbohydrates.
5. Cooked starch is more effectively utilised than glucose or dextrin, despite the latter both being absorbed more quickly. Thus dietary inclusion of carboxymethylcellulose enhances the nutritional value of dextrin by delaying its absorption.

The results indicate that carbohydrates can be used to advantage, albeit in limited amounts, even in feeds for carnivorous fish.

5.2.4 Minerals

The essential mineral elements serve the animal body in many different ways, as constituents of the bones and teeth, in organic compounds such as protein and lipids, and also as cofactors for enzymes. Furthermore, they serve a variety of functions as soluble salts in the blood and other body fluids. They are also concerned with the maintenance of osmotic relations and acid−base equilibria and the proper functioning of muscles and nerves. Many of their vital functions are due to ionic interrelationships, for example a balance between calcium, sodium, and potassium in the fluid which bathes the heart muscle is essential for the heart's normal relaxation and contraction while beating.

Our knowledge of the mineral requirements of fish is one of the least advanced areas of fish nutrition. Although many studies have been conducted on osmoregulation, the toxicity of heavy metals and related physiological functions, few are relevant to nutrition. One of the reasons for the paucity of research in this area is the difficulties posed by experimentation. Unlike other nutrients, significant amounts of the total minerals input enter from the surrounding waters, making it difficult to control properly the levels in experimental diets. This is a particular problem for studies of trace metals, such as iron, zinc, manganese, copper and cobalt, and for trials with marine species where the external waters contain significant quantities of most elements. For red seabream only iron, potassium and phosphorus need to be supplied in the diet to meet the mineral requirements of the fish, the remainder coming from the external environment. Interactions between minerals also complicate assessments of dietary requirement.

About twenty inorganic elements are required to meet the structural and metabolic functions of vertebrates. The metabolism of minerals differs from that of most of the other nutrients because they are neither produced nor consumed by the organism. Furthermore, most vertebrates are only able to exercise minimal regulation of the levels of minerals which are absorbed from the food. Nevertheless, most species have the ability to regulate the concentration of ions in the body fluids and so maintain a constant internal milieu. This is achieved mainly by ionic and osmotic regulatory processes in the kidney and gills (see chapter 2).

Calcium, chlorine, magnesium, phosphorus, potassium and sodium are the most important minerals, together with a series of trace elements including copper, iodine, manganese, selenium and zinc. Aluminum, chromium and vanadium may also be required, although the evidence is somewhat equivocal. The mineral requirements of various fish species are summarised in Tables 5.9 and 5.13.

Of all the minerals phosphorus is arguably one of the most important, mainly because of its high requirement in growth and bone mineralisation and also in lipid and carbohydrate metabolism. Certainly the levels of phosphorus required are the highest of all the inorganic ions, with dietary inclusions of 0.4−0.9% of available phosphorus being required for most fish species. Generally, phosphorus requirements are not affected by dietary calcium levels. In controlled experiments the growth of both carp and rainbow trout has been shown to correlate positively with dietary phosphorus levels but not with calcium levels. However, for most fish species it is difficult to study the effects of calcium deficiency because calcium is actively absorbed from the water by the gills. The availability of inorganic phosphorus depends on the solubility of the salt concerned. Thus, phosphorus from tri-calcium phosphate is less available than that from the more soluble mono- and di-calcium phosphates, with tri-calcium phosphate having an especially low availability, particularly in stomachless fish like the carp (Table 5.10). Phosphorus

Table 5.9 Mineral requirements of fish

Mineral	Carp	Rainbow trout	Other species
P	0.6−0.7%	0.7−0.8%	Red seabream 0.68% Channel catfish 0.42% Chum salmon 0.5−0.6% *Tilapia* 0.9% Eel 0.58%
Mg	0.04−0.05%	0.06−0.07%	Eel 0.04%
Zn	15−30 ppm	15−30 ppm	
Mn	13 ppm	13 ppm	
Cu	3 ppm	3 ppm	
Co	0.1 ppm	0.1 ppm	
Fe	150 ppm		Eel 170 ppm Red seabream 150 ppm
Ca	Deficiency was not detected at 0.028%	Deficiency was not detected at 0.03%	Eel 0.27%
Se			Salmo salar 0.03−0.04 ppm (with VE)
I			Chinook salmon (0.5−8.5 g) 0.6 ppm Chinook salmon (8.5−50 g) 1.1 ppm

this functional role of vitamins has been used to advantage as a means of assessing the nutritional status of an animal with respect to that vitamin.

None of the fat-soluble vitamins is known to function as a coenzyme. Vitamin A is involved in the metabolism of mucopoly-saccharides and visual pigments and in the general maintenance of epithelial tissues. Vitamin D functions in calcium homeostasis, possibly by induction of calcium-binding proteins. Vitamin E is a lipid-soluble antioxidant and may terminate peroxidative chain reactions among the highly unsaturated fatty acids of biomembranes. Vitamin K is involved in electron transport and oxidative phos-phorelation; it is also a cofactor in the blood coagulation process.

The requirement for a vitamin may be affected by various factors, such as the size, age and growth rate of fish, also water temperature and the composition of diets. For instance, the requirement for vitamin E may increase as the polyunsaturated fatty acid level in the diet increases. In extensive culture and semi-intensive culture in ponds or lakes, natural live foods are often sufficiently abundant to provide for the essential vitamins. However, in intensive high-density culture, such as in heavily stocked ponds, or in cages and raceways, natural foods are absent or very limited, hence vitamins must be supplied in the diet to achieve normal growth.

Recommended levels of vitamins in salmonid diets are sum-marised in Table 5.13. The inclusion levels of each vitamin in fish feeds are always higher than the required levels to provide a safety margin. Certain vitamins may be destroyed during feed manufacture by heat, moisture, alterations in pH, the presence of some metals, lipid oxidation, etc. Destruction of vitamin C (ascorbic acid) due to oxidation is one of the most significant problems in feed manu-facture. Some vitamins are also lost during storage, thus feed should be used soon after pelleting. In the case of crustacean feeds, which are usually kept in a tank or pond for a long time, allowance should be made for the leaching of vitamins from pellets, although microencapsulation techniques are now being adopted to ameliorate this problem.

5.3 LARVAL DIETS

5.3.1 Live foods

Formulated starter diets can be used as the first-feeding diet for the fry of many cultured freshwater fish. In contrast marine fish

Table 5.13 Recommended nutrient levels in salmonid diets

		Recommended level	Deficiency symptom codes*
Vitamins[†]			
Fat-soluble:			
Vitamin A	(IU/kg feed)	3500	03, 18, 27, 28, 40, 42, 44, 47, 55
Vitamin D$_3$	(IU/kg feed)	3000	62, 67
Vitamin E	(IU/kg feed)	100	01, 03, 10, 24, 25, 28, 30, 32, 34
Vitamin K	(mg/kg feed)	10	01, 13
Water-soluble:	(mg/kg feed)		
Ascorbic acid		300	01, 03, 08, 22, 28, 43, 49, 51, 60
B$_{12}$		0.02	01, 34, 39
Biotin		0.4	06, 14, 15, 19, 30, 32, 34, 46, 64
Choline		3000	30, 42
Folic acid		5	01, 14, 33, 49
Inositol		400	01, 23, 30, 48
Niacin		150	01, 25, 44, 46, 48, 49, 57, 65, 67
Pantothenic acid		60	04, 05, 12, 20, 29, 48, 53, 60, 66
Pyridoxine		10	04, 14, 15, 26, 35, 54, 61, 64, 66
Riboflavin		20	09, 11, 14, 27, 40, 47, 57, 59, 68
Thiamine		10	06, 09, 15, 16, 55, 26, 45, 49, 54
Minerals**			
Calcium	(g/kg feed)	0.2–0.3	31, 37
Phosphorus,			
inorganic	(g/kg feed)	7–8	17, 62
Magnesium	(g/kg feed)	0.5–0.7	02, 07, 37
Copper	(mg/kg feed)	3	37
Manganese	(mg/kg feed)	12–13	37
Selenium	(mg/kg feed)	0.1–0.4	21, 24
Zinc	(mg/kg feed)	15–30	09, 11, 27, 47
Iodine	(µg/kg feed)	0.6–1.1	36
Iron	(µg/kg feed)	Required	01

*Refer to Table 5.12.
[†]Based on NRC (1981) and practice in Fish Nutritional Laboratory, Ontario Ministry of Natural Resources.
**Lall (1979).

larvae require live foods, such as rotifers, *Brachionus plicatilis*, and marine copepods as their initial feeds. No artificial diet is currently available which is able to meet fully all the requirements of marine larvae as regards size and formulation. At present rotifers are widely used as the initial diet for many cultured fish larvae; they are usually mass cultured using marine *Chlorella* as a food organism. More recently baker's yeast has also been found to be suitable as a partial food source for rotifers. However, when rotifers which have been exclusively cultured on yeast are fed to larval fish, sudden heavy losses of the larvae frequently occur. These unpredictable mortalities can be avoided by culturing the rotifers with a mixture of both yeast and *Chlorella*, or by firstly culturing the rotifers on yeast and then substituting with marine *Chlorella* a short period before the rotifers are fed to the fish.

Recent investigations on the relationship between the nutritional quality of live foods and their food organisms have demonstrated that the content of EFA in the live foods is the principle determinant of their dietary value. The most striking difference in fatty acid profile between rotifers cultured with baker's yeast and those maintained on *Chlorella* is their content of EFA (Table 5.14). The rotifers cultured with yeast are low in ω3 HUFA, while those cultured with *Chlorella* contain high levels of 20:5ω3, which is the principal EFA for marine fish. Rotifers fed on both yeast and *Chlorella* show intermediate values. This difference is probably the reason why rotifers cultured with yeast are always inferior to those grown with *Chlorella* as far as their nutritional quality as a live food for fish is concerned. When marine *Chlorella* is used as the culture organism, the low levels of ω3 HUFA food in yeast-fed rotifers increase in proportion to the length of time the rotifers are fed on *Chlorella*. The dietary value of the rotifers for various fish larvae was also found to be significantly improved by secondary culture for more than 6 hours with marine *Chlorella*, but not with freshwater *Chlorella*.

Based upon these findings, a new kind of yeast has been developed as a culture organism for rotifers so as to improve their

Table 5.14 Concentration of ω3 fatty acids in total lipids from rotifers cultured with yeast and marine *Chlorella* at Nagasaki Prefectural Institute of Fisheries in the period 1975−7

	Rotifers cultured with		
Fatty acid	Yeast	Yeast + *Chlorella*	*Chlorella*
20:4ω3	0.4−0.5	0.4−0.6	tr−0.2
20:5ω3	1.0−1.9	8.1−11.8	22.8−27.7
22:5ω3	tr−0.3	1.7−2.9	3.0−3.8
22:6ω3	tr−0.5	tr−0.9	tr−0.5
ω3 HUFA	2.2−3.1	11.3−14.7	26.2−30.9
Lipid (%)	1.4−2.3	2.2−2.8	3.7−4.2

nutritional value as a food for fish larvae. This new type of yeast (designated ω-yeast) can be produced by adding a fish oil rich in ω3 HUFA to the culture medium for baker's yeast; this results in a high content of lipid and ω3 HUFA in the yeast. Using this modified yeast has significantly improved the dietary value of rotifers to fish larvae when compared to that of rotifers cultured on marine *Chlorella*.

A more direct method of preparation of appropriate larval fish diets has also been developed. Lipids containing ω3 HUFA are homogenised with a small amount of raw egg yolk and water and the resulting emulsion included in the culture medium for rotifers together with baker's yeast. The rotifer readily takes up the lipids, and the concentration of ω3 HUFA reaches a maximum 6–12 hours later. These indirect and direct methods of HUFA application are also effective in improving the dietary value of other living organisms. Using either method, it is also possible to improve further the dietary value of live foods by allowing them to take up fat-soluble vitamins in addition to the ω3 HUFA from the culture medium.

Nauplii of *Artemia* have been widely used as a food in the production of juvenile fish. However, feeding with such diets sometimes results in high mortalities amongst larval marine fish. In part the extent of the mortality would appear to be determined by the geographical source of the *Artemia* and its variable fatty acid composition. There are two major types of *Artemia*: one (the freshwater type) contains a high concentration of 18:3ω3, which is the EFA for freshwater fish, and the other (the marine type) is high in 20:5ω3, which is the EFA for marine fish. The dietary value of the freshwater type of *Artemia* can be improved by feeding with marine *Chlorella* and ω-yeast, both of which contain substantial amounts of the EFA required by marine fish. These results suggest that the class of EFA contained in *Artemia* is the principal determinant of its value as a source of nutrient for cultured fish larvae. The food value of *Artemia* is also considerably improved when it is fed to fish together with *Tigriopus* or *Acartia*, both rich in 20:5ω3 and 22:6ω3. *Artemia* nauplii are also found to take up lipids readily by both the direct and indirect methods of application, the concentration of ω3 HUFA reaching a maximum 12 hours after feeding. The dietary value of the nauplii to fish larvae can also be improved by incorporating ω3 HUFA from emulsified lipids or ω-yeast. Because of the variation in fatty acid composition of *Artemia*, it is essential to analyse all *Artemia* supplies before they are fed to fish larvae. If the fatty acid composition is not known, then the *Artemia* should be mixed with other marine copepods or alternatively prefed on lipids containing ω3 HUFA.

5.3.2 Formulated larval diets

The mass culture of rotifers, *Chlorella*, or any other living organism not only requires considerable manpower and costly equipment, but also depends upon the maintenance of appropriate environmental conditions. For example, the mass propagation of one million fry of red seabream requires 5×10^9 rotifers per day, a production which depends on the provision of four 200 tonne tanks. In addition, other tanks are required to culture the large volumes of marine *Chlorella* which serves as the food for the rotifers, as described above. The success of mass culture of *Chlorella* is also greatly affected by environmental conditons.

Because of these requirements there has been much interest in the rearing of larval marine fish solely on artificial microdiets. In Japan, considerable efforts have been made to meet this challenge. To be successful artificial microdiets must satisfy the following conditions. They must:

1. float in the water for more than 30 minutes without any leaching of nutrients;
2. be palatable;
3. contain all the nutrients required by the larval fish;
4. be digested and absorbed by the larval fish;
5. be almost the same size as rotifers after absorption of water;
6. not cause any pathological change.

Three basic types of microdiets appear to satisfy these conditions. They are as follows:

1. micro-encapsulated diets prepared by encapsulating a solution, colloid, or suspension of dietary ingredients with a membrane;
2. micro-bound diets prepared by combining dietary ingredients with binder;
3. micro-coated diets prepared by coating a micro-bound diet with a water-insoluble material such as zein or a cholesterol–lecithin mixture.

The dietary ingredients include egg yolk, short-necked clam extract, bonito extract, yeast, skimmed milk, fish meal, squid meal, krill meal, beef liver, lecithin and squid liver oil. At present such microdiets have only been used to partially replace rotifers in some marine and freshwater larval rearing programmes and it will still take many years before such techniques are able to replace live foods completely.

5.4 BROODSTOCK DIETS

The ever-growing demand of fish breeders for artificially produced seed ideally requires a year-round (rather than seasonal) supply of fertile eggs of high quality with a high survival and growth rate to match those occurring naturally. The requirement for year-round production has been achieved with some species by adjustment of the endocrine balance through variation in the photoperiod or temperature regimes (chapter 4; section 4.3.3).

Nutrition is known to have a considerable effect upon gonadal growth and fecundity, although the few published papers on the subject have given equivocal results. Consequently, precise information on the nutritional requirements for gonadal maturation in broodstock is lacking.

Recently, several nutritional experiments have been carried out with broodstock of both rainbow trout and red seabream. In an experiment involving rainbow trout, fingerlings (3.5 g live weight) were grown at natural water temperatures (5–20°C) for 3 years and given either a commercial diet (43–47% crude protein) or one of three experimental diets (two of these were low protein/high energy diets, 33–35% crude protein and 390 kcal/100 g, and the other contained no trace element supplement).

The average gain over the experimental period was 1.5 kg, except for the trout given the diet lacking a trace metal supplement which gained 1 kg on average (Fig. 5.6). During this time, the fish spawned twice but the quality of eggs was examined only after the second spawning as eggs produced during the first spawning may be poorer in quality, even in well-fed stocks. For the second spawning, there were no differences in egg production per female, egg diameter, eyeing rate and hatchability between treatments, other than for trout given the diet lacking supplementary trace

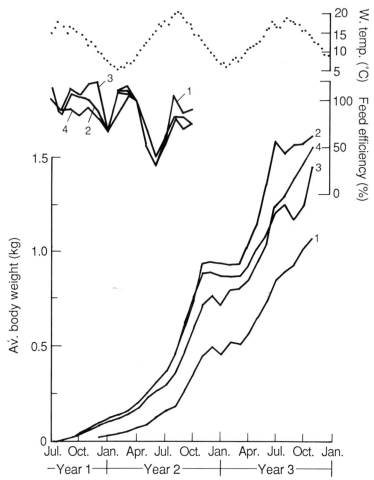

Fig. 5.6 Results of long-term feeding experiments in rainbow trout. Curve 1, fish fed the diet without supplement of trace element and curves 2 and 3, those fed white fish meal diets containing respectively 5% or 7% beef tallow as an energy source. Curve 4, fish fed the control commercial diet.

minerals (Table 5.15) where the eyeing and hatch rates were significantly reduced (3.7% and 0.4% respectively).

These results demonstrate that a diet containing a lower protein content than that normally employed, but with a high energy level, is as effective for both fingerlings and broodstock rainbow trout as more conventional diets. In contrast, a trace metal supplement added to the diet has been shown to be indispensable for reproduction of rainbow trout. Of the minerals analysed, the most striking change was in manganese concentration, which fell from 4.1 ± 0.7 $\mu g/g$

Table 5.15 Effect of low-protein diets with a high-energy content and no supplemental trace metals on the spawning and the quality of eggs of rainbow trout when compared with a commercial high protein diet

	Number of eggs produced per female	Egg size in mm	Eyeing rate in %	Total hatch in %
Commercial high protein diet (43–47%)	3631 ± 836*	5.3 ± 0.25	89.0 ± 4.3	87.2 ± 3.7
Low-protein, high energy diet (33–35%, beef tallow 5%)	3954 ± 1006	5.2 ± 0.29	90.5 ± 4.9	87.4 ± 1.0
Low-protein, high energy diet (33–35%, beef tallow 7%)	3050 ± 1156	5.1 ± 0.12	86.3 ± 6.3	86.3 ± 6.3
Diet without supplemental trace metals	1975 ± 342	5.1 ± 0.00	3.7 ± 5.0	0.4 ± 0.9

*Mean ± SD ($n = 3–12$).

eggs in females given a commercial diet to 1.6 ± 0.1 $\mu g/g$ in females given the experimental diet lacking a trace metal supplement.

In a second experiment with rainbow trout, broodstock were given experimental diets for only 3 months prior to spawning. Four of the treatments were used to examine further the effects of protein and energy balance on egg quantity and quality. It was confirmed that broodstock given a diet with 36% crude protein and 18% lipid performed as well as those given a diet with 46% crude protein and 15% lipid. It was also shown that beef tallow used at a level of 7% as an energy source had no adverse effect on the reproduction of rainbow trout.

As anticipated, the use of partially-defined diets deficient in EFA led to the lowest values for total egg production, percentage of eyed eggs produced, and total hatch. It is of particular interest to note that addition of linoleic acid (18:2ω6) to the EFA-deficient broodstock diet led to a marked improvement in percentage of fertilisation, percentage of eyed eggs, and total hatch, compared with broodstock given diets lacking EFA. Linoleic acid is known to be inferior to linolenic acid (18:3ω3) as an EFA for rainbow trout fingerlings but the situation seems quite different with broodstock.

In this context, it is of considerable interest to note that rainbow trout have been grown through a complete generation using a semi-purified test diet containing 1% linolenic acid as the sole dietary EFA. The eggs produced, however, contained a small amount of arachidonic acid (20:4ω6), showing that the trout broodstock had tenaciously retained the small amounts of ω6 fatty acids present in incompletely extracted diet ingredients, such as casein, dextrin, and gelatin. It is possible, therefore, that there is a small but absolute requirement for ω6 fatty acids in rainbow trout.

In red seabream, it is well known that acceptable diets are actively eaten by the broodstock even during spawning and that feeding broodstock with krill, Mysis, shrimp, and crab wastes results in pigmentation of the eggs within a matter of hours. This suggests that the nutritional value of the diet given to broodstock shortly before spawning may affect the results of spawning. It also indicates that pigments and other fat-soluble materials in diets are easily incorporated into eggs and the quality of eggs may therefore be improved by feeding broodstock with some fat-soluble nutrients, such as EFA and vitamins. This has been studied further using broodstock that had been reared on five experimental diets for 5 months in floating net cages. They were then transferred to spawning tanks in the aquarium for the investigation of spawning and egg quality. The results of natural spawning by red seabream broodstock fed diets of different nutritional quality are shown in Table 5.16. In the case of red seabream the percentage of buoyant eggs, floating on the water surface, is very important for evaluation of egg quality. Buoyant eggs generally have high rates of hatching and normal development. On the other hand, eggs which sink to the bottom of tanks comprise mainly unfertilised or dead eggs. As shown in Table 5.16, for the broodstock fed on the control diet (diet 1: white fish meal diet with 45% crude protein) the average percentage of buoyant eggs during about 60 days of spawning was 80.9%. For the group fed a low-protein diet (diet 2), or a diet without supplemental phosphorus (diet 3), egg quality was lower than in the control. The proportion of buoyant eggs produced by an EFA-deficient broodstock given the corn oil diet rich in 18:2ω6 was the lowest recorded of all the experimental groups, more than 75% of eggs produced being inviable, in marked contrast to the rainbow trout broodstock mentioned above. The value was highest in the group receiving the diet containing cuttlefish meal as a protein source (88.5%). Percentages of abnormal eggs with more

Table 5.16 Effect on the spawning and egg quality of red seabream of broodstock diets of different nutritional quality*

	Diet number				
	1 Control fish meal diet	2 Low protein fish meal diet	3 Low phosphorus fish meal diet	4 EFA-deficient casein diet	5 Cuttlefish meal diet
Buoyant egg[†] (%)	80.9	54.4	62.1	23.9	88.5
Abnormal egg** (%)	30.7	70.7	67.9	93.7	2.7
Total hatch (%)	69.4	23.6	26.3	0.9	93.9
Normal larvae (%)	62.4	3.8	6.2	—	97.6
Final productivity of fish seed from eggs produced (%)	24.3	0.1	0.3	—	78.9

*Values in the table are all average of about 60 days spawning.
[†]Normal eggs floating on the water surface.
**Eggs with more than two oil globules.

than two oil globules (normal red seabream eggs generally possess one oil globule) amounted to 70.7%, 67.9% and 93.7% in the groups fed the low-protein, the phosphorus-deficient and the EFA-deficient diets respectively, in comparison with 30.7% in the control group. The rate of hatching was also significantly lower in these eggs. The females given the diet containing cuttlefish meal mainly produced normal buoyant eggs with a high rate of hatching, leading to high productivity of viable larvae.

These results demonstrate the close relationship that exists between nutritional quality of broodstock diets and reproduction in red seabream. Later experiments have also shown that nutritional quality of the diets given to broodstock shortly before spawning, or even during spawning, can greatly affect reproduction. Supplementation of diets with carotenoids, such as astaxanthin, phospholipids, and vitamin E was also found to be very effective for improving egg quality.

5.5 PRACTICAL FEED FORMULATION

Formulation of practical fish diets for intensive aquaculture is usually based upon a knowledge of the nutritional requirements of the fish species under cultivation, in particular the optimum protein and energy levels, and the inclusion of essential amino and fatty acids, together with vitamins and minerals.

Formulation must also take into account the market values of the cultured fish (e.g. the protein sources in the diets for seabream, eel and kuruma shrimp are too expensive to be used in diets for carp and *Tilapia*). For example, a practical carp diet must cost less than ¥ 100 ($0.78 or £0.42) per kg, to account for the lower first-sale value of carp, which in 1985 was about ¥ 250–300 ($1.94–2.35 or £1.05–1.27) per kg. Consequently, the protein quality of carp diets, which commonly include whole meal, meat and bone meal, soybean meal, etc., is very low. The net protein utilisation (NPU) of these diets is only 30–35 compared to values of 50–60 for egg yolk, whole egg or white fish meal, and this leads to the excretion of a high nitrogen loading into the surrounding waters. It is very difficult to formulate low price diets which can produce maximum protein retention and growth.

The procedure for diet formulation is outlined in Figure 5.7 and Table 5.17 lists some of the formulations used for grower diets in the commercial culture of rainbow trout, carp, ayu, eels, *Tilapia*,

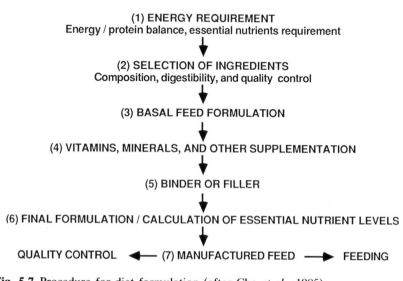

Fig. 5.7 Procedure for diet formulation (after Cho *et al.*, 1985).

Table 5.17 Examples of practical diet formulae used for aquaculture in Japan

	Rainbow trout	Carp	Ayu	Tilapia	Eel*	Red seabream*	Yellowtail*
Fish meal	53	34	52	25	65	43	51
Meat and bone meal	2	4					5
Krill meal						25	
Soybean meal, defatted	5	17	8	18	4.4	2	7
Corn or wheat gluten meal	3	6	3	2	2	3	6
Brewer's yeast					3		
Brewer's grains				6			
Torula yeast	2	2	4				3
Defatted wheat germ		4					
α-Starch					22		
Wheat flour	32.7	28	26.7	28		24.7	25.4
Defatted rice bran		5.7		15.7			
Vitamin mixture	1	1	1	1	1	1	1
Chlorine chloride	0.3	0.3	0.3	0.3	0.3	0.3	0.3
Mineral mixture	1	1	1	1	2.3	1	1
Others		1		3			0.3

*The mash diets for eel, red seabream and yellowtail are mixed with water in a ratio of 50:50 or 40:60, and fed as moist pellets. Feed oil is added to the mixture as needed.

red seabream and yellowtail in Japan. However, some of the ingredients will vary depending on the farm and its location. For intensive freshwater fish culture in most countries, diets are made into a dry mash, or more commonly into crumbles or pellets, except for the eel, in which moist (paste) diets are prepared by mixing with lipid and water before feeding to the fish. The size of these prepared dry diets varies with each feed company, but some estimated values of diet size for each stage of fish are shown in Plate 5.1 and listed in Table 5.18. In contrast the Japanese farming of yellowtail and red seabream often utilises minced, chopped or whole locally-available fresh raw fish, although this is gradually being replaced by moist pellets in order to reduce pollution and the risks of introducing diseases. It is also well known that some raw fish contains a thiamine-destroying enzyme (thiaminase) (Table 5.19). Although thiaminase is more commonly found in freshwater fish than seawater fish, feeding raw fish such as anchovy and

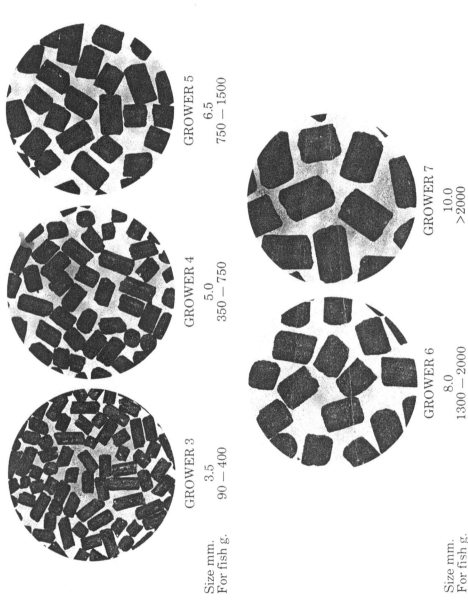

GROWER 3
3.5
90 — 400

GROWER 4
5.0
350 — 750

GROWER 5
6.5
750 — 1500

GROWER 6
8.0
1300 — 2000

GROWER 7
10.0
>2000

Size mm.
For fish g.

Size mm.
For fish g.

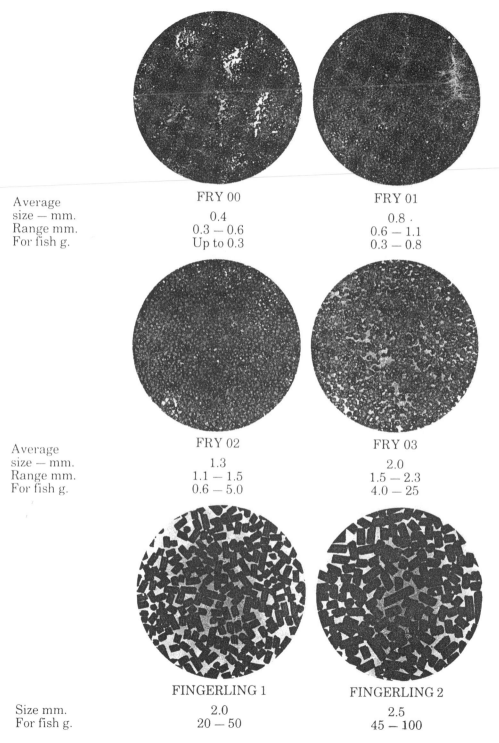

	FRY 00	FRY 01
Average size — mm.	0.4	0.8 ·
Range mm.	0.3 — 0.6	0.6 — 1.1
For fish g.	Up to 0.3	0.3 — 0.8

	FRY 02	FRY 03
Average size — mm.	1.3	2.0
Range mm.	1.1 — 1.5	1.5 — 2.3
For fish g.	0.6 — 5.0	4.0 — 25

	FINGERLING 1	FINGERLING 2
Size mm.	2.0	2.5
For fish g.	20 — 50	45 — 100

Plate 5.1 Range of diet sizes used for rearing salmon from fry to marketable fish (actual size).

Table 5.18 Estimated values of diet sizes for each stage of rainbow trout

Diet size number	Average diet size in mm diameter	Range of diameter in mm	Suitable size of fish in g
1C	0.47	0.32−0.62	2−4
2C	0.78	0.62−1.00	4−7
3C	1.18	1.00−1.60	7−14
4C	1.95	1.60−2.20	14−40
5C	2.40	2.20−2.41	40−200
6C	3.10	2.41−3.65	200−500
7C	4.50	4.10−5.45	500−
8C	6.20	5.45−7.20	500−

NB: Size number and diet size vary according to fish species and feed company.

Table 5.19 Distribution of thiaminase in marine fish species

	Thiaminase activity in tissue (ppm)					
	Flesh	Viscera	Liver	Intestine	Pyloric caeca	Kidney
Sardine	0	14	—	—	—	—
Anchovy	25	55	94	80	83	272
Mackerel	19	172	272	210	—	494
Cornetfish	145	455	—	—	—	—

Source: Ogino (1985).

mackerel with strong thiaminase activity may result in thiamine deficiency in maricultured species. This is also true for moist pellets containing raw fish. The thiamine level of prepared moist pellets may also be reduced by contact with the thiaminases found in either raw fish or other feed ingredients. The amount of thiamine needed for fish can be met by alternately feeding moist pellets containing raw fish and a second diet high in thiamine and having no thiaminase-containing raw fish; otherwise thiaminase should be inactivated by heat before use.

Formulated mash diets are usually mixed with minced trash fish in a ratio of either 50:50 or 40:60, or alternatively mixed with water alone and formed into moist pellets. These moist pellets containing both mash diet and trash fish have similar qualities to those of fresh raw fish in terms of acceptability or palatability and dietary value (Fig. 5.8). Recently, however, it has been suggested that a proportion of the water-soluble vitamins in moist diets,

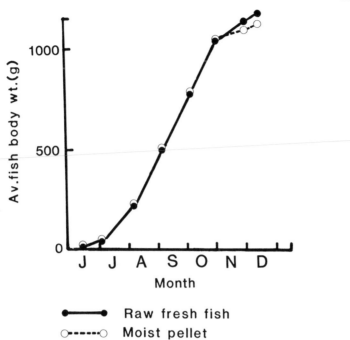

● ● Raw fresh fish
○----○ Moist pellet

Fig. 5.8 Growth of yellowtail fed on trash fish and moist pellets (Yone, 1985).

especially vitamin C, may be destroyed during both processing and storage of diets. Thus it is essential to use only stable forms of vitamin in the preparation of moist diets.

5.6 CONCLUSIONS

Economically-productive aquaculture systems are heavily dependent upon an adequate supply of low-cost feeds with a high nutritional quality which are formulated taking into account the requirements of fish for energy, protein, lipids, vitamins, and minerals. However, fish feeds are not yet sufficiently balanced to meet optimal nutritional requirements, which include the true bioavailability of the ingredients, due to lack of information on fish nutrition. Hence, there is still some wastage of nutrients due to the generous safety margins applied by feed manufacturers. This may not only increase the oxygen requirements of the fish but also contribute to unnecessary discharges of biological wastes to the aquatic environment.

In systems of intensive production food is the most expensive single factor. Thus, for least-cost formulation dietary protein levels must be decreased by the substitution of costly fish meal with other protein ingredients, and non-protein energy levels must be increased. At present the protein source in fish diets is heavily dependent upon fish meal which cannot be expected always to be available in the future. Alternative protein sources should be sought from locally produced inexpensive materials. Further research input is therefore required, especially in the following areas: amino acid and energy requirements; bioavailability of amino acids, together with digestibility of protein and energy in a wide variety of feed ingredients; nutrient interactions; and maximum acceptable limits for each feed ingredient in a diet based both on acceptability by the fish and milling quality. Such information will be needed for each aquacultural species under culture.

For all species with small sized eggs, especially marine fish, feeding strategy is based on live food, with zooplankton being either caught in their natural environment or produced in rearing tanks in a more or less controlled way or produced by monoculture. Thus monoculture of rotifers is the commonest form of larval feed in the world. But mass culture of live food requires costly facilities and may depend upon natural conditions such as the weather. Thus, an artificial starter diet for marine fish is one of the most pressing needs in aquaculture. Considerable research is still needed before artificial microdiets can be widely used by aquaculturists in place of live foods.

In mariculture most fish seed comes from natural sources; these are clearly inadequate to supply the present and future needs of farmers throughout the world. One of the most important priorities for aquaculture is to ensure a year-round (rather than seasonal) supply of fertile eggs of high quality which have high survival and growth rates. Broodstock which produce such eggs must be cultivated to this end. However, useful information on broodstock diets is limited. Whether, and in what way, they should differ from grower diets remains to be researched.

Aquaculture production is increasing annually, but hatchery reared and cultured fish are often lower in quality and thus command lower market prices than wild fish. Although evaluation of aquaculture products is somewhat subjective, farmed fish are often held to be inferior in terms mainly of taste and colouration. Texture, odour and even external shape may also be said to differ. The taste problem is mainly due to a high concentration of body lipids perhaps induced by overfeeding, by a slightly imbalanced diet or by the absence of dietary variety compared with wild fish. Consequently there is a need for research effort to ensure, partly by nutrition, that the cultured product closely resembles its wild counterpart. As regards body lipid level, this may be partly a matter of total energy: protein energy balance in the diet. The problem of pigmentation in the skin of cultured fish appears to have been overcome, at least for Japanese red seabream and yellowtail, by feeding oil rich in astaxanthin or shrimp, although artificially enhancing pigmentation with carotenoids gives a colour tone different from that of wild fish.

Fish cultivation leads to release into the water of inorganic and organic nutrients, both of which can have harmful effects. Inorganic nutrients like phosphate and ammonia can cause eutrophication, whereas organic compounds may lead to oxygen deficit. These side-effects of intensive cultivation originate from the food, hence the type and quality of food given are of great importance. Organic and particulate matter appearing in the water arises either from uneaten food, from leaching of both soluble and particulate material from the food, or from undigested material. Greatest waste and thus highest pollution occur when trash fish are fed, with up to 30% of the feed not being consumed, whereas with moist pellets up to 10% and even with dry feed up to 5% may remain uneaten. Pellet hardness, water stability and sinking speed should also be appropriate both to fish species and farm design.

6 Fish health and disease

6.1 INTRODUCTION

Intensive fish culture has often had the reputation of being a 'high risk' activity. It is true that great care must be taken to prevent system failures (e.g. blocked screens, storm damaged pens) causing asphyxiation and mass mortality or even fish escapement. As described in chapter 2 (section 2.3.2), massive fish kills can bankrupt an uninsured farm within a matter of hours due to pollution of the water supply or to algal toxins. Chapter 3 has also illustrated the need for high quality nutrition to ensure satisfactory survival, growth rates and feed conversion efficiency. With greater emphasis on site selection, farm management and dietary control, the risk profile of fish farming has inproved considerably. However, fish diseases continue to threaten the commercial viability of many intensive systems.

As in man and higher animals, the majority of fish health problems are caused by disease processes involving living agents, such as bacteria, fungi, parasites and viruses. The type of infectious agent influences the resulting clinical signs, which may vary from minor skin irritation in a few fish to a sudden flare-up of disease, with the entire stock showing reduced appetite, lethargy and heavy losses. For example, the scale of mortality in certain bacterial or viral infections may reach 30% or more within several days. Also repeated outbreaks of disease may occur and, although the onset may be less sudden than a fish kill, the financial outcome may be just as severe. The purpose of this chapter is to survey the underlying factors involved in fish diseases in order to offer practical guidance on how best to mitigate the risk of such problems occurring.

Fish have evolved so that they survive and prosper in an aquatic environment which can be subject to changing conditions. Whereas environmental conditions in the middle of the ocean are normally fairly constant, the situation in a fish pond can be very different, with fluctuating water quality as well as the demands placed on fish by being confined, crowded and handled. In short, a fish pond is an unnatural and stressful environment and it is appropriate to start a chapter on health and disease by considering the normal defence mechanisms of fish and then by asking what is stress and how do fish cope with it.

6.2 DEFENCE MECHANISMS

Fish survival depends on the internal environment of the fish being kept constant. This so-called 'homeostasis' is threatened by the surrounding water but, like other vertebrates, fish have various defensive adaptations, starting with a layer of skin. Fish skin acts as a waterproofing barrier to keep the external water out and the tissue fluids in. Its thickness varies greatly from species to species and sometimes even with age and sexual maturity, but it covers the entire fish except for the gills. Numerous mucus-producing goblet cells are present and the resulting film of slimy mucus has anti-infection properties, hinders fish predators, and helps to reduce the frictional resistance of the water when the fish is swimming. Beneath the outer skin layer are the scales which form a tough defensive coat of bony plates in all but a few farmed fish. By contrast, eels do not possess scales but have a thicker skin layer instead, while underneath the skin of all fish lie the main swimming muscles which are attached to the spinal column.

The gills play a key role in homeostasis due not only to their respiratory function but also to their responsibility for regulating salt exchange and for excreting ammonia. The mucous membrane of the gills has a surface area which is therefore larger than the skin and only one layer of cells separates the blood flowing through the gill capillaries from the water passing via the mouth across the outer surface of the gills. As explained in chapter 2 (sections 2.2.4 and 2.2.5), this delicate gill epithelium is the barrier through which oxygen must pass into, and carbon dioxide out of, the fish. The situation is complicated further by the need to eliminate ammonia as the major excretory product of fish and by the large quantities of water needed to dilute it. In a freshwater environment the water which tends to pass into the body through the gills is voided by the kidney producing large amounts of dilute urine, whereas in sea-water the fish avoid dehydration by drinking large amounts of water to compensate for passive water loss from the gills. Species of fish which are able to move frequently between freshwater and seawater environments have adapted so that they can switch their osmoregulatory mechanisms accordingly.

It is clear that severe damage to the waterproofing of the skin (e.g. by a wound) is likely to kill fish by allowing waterlogging or dehydration of their internal tissues, depending on whether the environment is freshwater or seawater. The same can happen if the gills are severely damaged (e.g. by toxins or polluted water). Depending on circumstances, the quality of the environment may be totally unsuitable to support fish life (e.g. insufficient oxygen or excessive temperature) or may be relatively poor, resulting in reduced fish health and performance. But what internal defence mechanisms can fish bring to bear in the face of tissue damage by trauma, toxins or by the threat of microbial invasion?

Like all vertebrates, fish are capable of mounting an inflammatory response following tissue damage provided they survive the initial insult. Inflammation is a generalised protective response which serves to mobilize at the site of tissue damage the cells and tissue fluids needed for repair in order to maintain homeostasis within the fish as a whole. The cardinal signs of inflammation in man have been summarised since ancient times as heat, redness, swelling, pain and loss of function. Not all these signs are evident in cold-blooded fish suffering inflammation, but the principles are very similar. Any escaped blood will clot, while damaged cells release chemicals causing increased local blood flow and capillary per-meability. This allows various blood cells to leave the blood along with proteins which polymerise into fibrous strands. Some of these 'phagocyte' cells actively ingest foreign material and tissue debris. Such phagocytes are widely distributed throughout the body to trap any foreign microbes which gain access. The overall aim of the inflammatory response is to plug any gap which develops in the fishes' integument, remove extraneous matter, and repair damaged tissue by local proliferation. Acute inflammation may resolve com-pletely if the damage is mild and the source immediately removed. However, cells may die if the damage is more severe or the blood supply is cut off, resulting in tissue necrosis. This area of dead tissue may become walled off by new tissue and slough off or become absorbed, unless the lesion is too severe to prevent the fishes' defences becoming overwhelmed. If rapid resolution of acutely inflamed tissue does not occur, then chronic inflammation may supervene where new tissue proliferates adjacent to areas of continuing inflammation. Many fish diseases are characterised by chronic inflammation which often develops as a central zone of necrosis (e.g. containing bacterial or fungal invaders) surrounded by a capsule of cells and fibrous tissue laid down by the fish. Allied with this variable sequence of inflammatory changes are certain other non-specific defensive factors, such as anti-microbial enzymes in fish mucus and various blood serum proteins. For instance, interferon is an anti-viral substance produced by fish (and humans) as a defence against virus infection.

Unlike the defence mechanisms described above, the immune defence system is characterised by being highly specific and able not only to 'remember' a particular microbe but to ensure that a fish which has recovered from infection by that microbe is subsequently immune or more resistant to reinfection. This should not be confused with the fact that many microbes do not cause disease in certain fish which are genetically resistant to infection (e.g. the parasitic fluke *Diplostomum* causes eye lesions in rainbow trout but not in brown trout). By contrast, the basis of this adaptive immunity is concerned with the ability of fish to produce specific antibodies and defence cells in response to foreign molecules.

Vaccines are substances produced from pathogenic microbes which are then administered to fish in order to confer immunity against the microbe. The aim is to treat the organism in order to make it innocuous to fish while retaining or enhancing its immunizing potential. Already vaccines are being widely used against various bacterial diseases of fish and research work is going on to develop vaccines against viral diseases too. They can be administered to large fish by injection for maximum effect, whereas mass vaccination of small fish is achieved by spraying them or by dipping them into a vaccine solution. The in-feed route of vaccine administration would be ideal but has not yet been developed. There are many factors which influence the efficiency of vaccination in fish. For instance, they have to be large enough to have developed 'immunocompetence' and must also be in good physical condition at the time of vaccination. Water temperature has an effect, with the immune response of fish to vaccination being better at higher than at lower water temperatures and even the season of the year may play a part. Certainly, factors such as the concentration of the vaccine and its exposure time to fish are important and the addition of so-called 'adjuvants' to a vaccine often increases its effectiveness. Looking to the future, as these problems are solved it will become possible to vaccinate fish against a number of different diseases at the same time and routine vaccination will become available for an increasing range of different fish species.

6.3 STRESS

Farmed fish are continually subjected to environmental change as well as to husbandry practices, such as grading and transport, which often evoke a panic reaction as the fish try to escape. All these factors can impose considerable stress on the limited homeostatic mechanisms of fish and many species are unsuitable for farming because intensive stocking is too stressful. In such species farming conditions impose stress requiring a homeostatic adjustment which they are unable to make, with ultimately fatal results. Severely stressed fish are nervous, inappetent, often dark-coloured and may be unable to keep their station in the water.

The underlying physiology of stress is fairly similar in most vertebrates and is elicited by a wide range of stimuli, including fright, forced exercise, low oxygen levels and infection. Its initial manifestation is an alarm reaction akin to the 'fight or flight' reaction in man. If the stress persists, a stage of resistance occurs where the fish has adapted to the demands placed upon its metabolism. But if the stress is too severe or persistent, adaptation is lost and a stage of exhaustion is reached before the fish succumbs. It is important to remember the purpose of this adaptation to stress is to bring to bear metabolic adjustments in order to facilitate survival. For instance, blood glucose levels are increased to aid greater muscular activity and corticosteroid hormones are released from the adrenal glands. However, these changes bring penalties, particularly if the fish remains stressed, because it causes reduced inflammatory and immune responses, in turn resulting in lowered resistance to microbial invasion. So whereas increased cortisone output will enable a fish to cope with an initial stress, in the longer term its defence mechanisms may become overwhelmed by a disease organism.

A useful experimental technique for checking if fish are symptomless carriers of the fish disease bacterium *Aeromonas salmonicida* is to inject them with a synthetic corticosteroid such as prednisolone acetate. Apparently healthy fish which have been carrying the organism then die following rapid multiplication of the organism and the ensuing systemic infection. In a similar fashion salmonids become stressed at spawning under the influence of natural corticosteroid activity and frequently succumb to fungal and bacterial infection. This is because the stage of physiological resistance has given way to exhaustion as the fish defences are overwhelmed by the magnitude of the stress involved. The microorganisms which normally inhabit the gut and skin of fish without harmful effect can themselves invade and multiply within fish tissues if this stage of exhaustion is reached. The same phenomenon is well known in man and higher animals with various microorganisms being distributed commonly throughout populations but causing disease only when particular individuals become weakened by some other factor and hence become more susceptible to infection. To summarise, fish disease is generally the end-result of interactions between a stressful environment and fish unable to adjust to the various stresses. More specifically in the case of infectious disease this entails interaction between three components: a susceptible fish host, an infectious agent, and a stressful environment.

Fish susceptibility clearly depends on species, age, genetic strain differences, non-specific resistance to disease and immunocompetence in the presence of particular disease organisms. The virulence of the infectious agent reflects its invasiveness into fish tissue and its ability to form toxins, etc. However, the presence of an infectious agent is not a prerequisite for disease if the environment is sufficiently stressful due to poor water quality or poor standards of fish husbandry and these non-infectious diseases will now be discussed in more detail.

6.4 NON-INFECTIOUS DISEASES

6.4.1 Genetic

An increased proportion of infertile eggs or abnormal fry, such as Siamese twins, is not uncommon. The hatchery manager will often discard the particular batch and may also discard the parent stock. In seeking to establish faster growing strains of fish, he should be aware of the risks of increasing genetic disorders. The greatest

Probably the best known specific problem of poor water quality is gas bubble disease. This is due to supersaturation of the water supply with dissolved gases and is similar to 'divers' bends' or decompression sickness in man, otherwise known as Caisson's disease. If the water is more than about 110% saturated with air or other gases, bubbles will form under the skin and eyes (Plate 6.1) or in the fins and mouth. Air embolism can also develop inside the capillaries of the gills and blood system and severe cases are rapidly fatal. Outbreaks of gas bubble disease in hatcheries are often manifested by fry swimming in an abnormal manner and floating upside down, or on their backs, due to buoyancy of bubbles accumulating in the mouth and yolk sac. Underground water from boreholes and springs frequently has a supersaturation of nitrogen which needs to be equilibrated with the air before it can be safely used. Supersaturation can also occur if air is sucked under pressure into valves and pipelines; the entrained air usually enters via leaks and must be blown off by means of aerators or splashboards upstream of the fish. Less commonly, supersaturation has been observed in fish farms sited immediately below dams, due to air

Plate 6.1 Gas bubble disease causing unilateral pop-eye in young carp viewed from the front.

bubbles forming in pools below the spillway and also in hatcheries receiving very cold water which then warms up rapidly without being given enough time to equilibrate at the higher temperature. Although persistent severe gas bubble disease is fairly easy to identify and remedy, in practice it often occurs intermittently causing occasional mild symptoms and the exact nature of the problem may be more difficult to pinpoint. If the fish farmer suspects gas bubble disease, a simple test is to put his hand in the water and see if bubbles form and cling to the skin.

6.4.4 Miscellaneous

Poor standards of fish husbandry can be responsible for a variety of stress-related health problems. There is often a narrow boundary between inadequate water flow to achieve satisfactory water quality for the biomass of fish in the pond and excessive flow rates leading to exhaustion, especially with young fish. Even if flow rates are suitable for the biomass present, fish may become stressed through overcrowding with excessive competition for automatic feeders and resultant nipping of gill covers and fins. The stress of handling, grading and fish transport is considerable and great care must be taken to avoid scale loss. It may be preferable to anaesthetise fish during loading operations for fish transport purposes and/or to add ice to the water in transport tanks to reduce metabolic activity. Rearing facilities should be designed with the aim of reducing trauma. For example, fish pumps are usually better than nets for transferring fish; also net pens must be anchored to resist distortion and the risk of trapping fish in the resulting folds and corners. It should also be borne in mind that fish are particularly vulnerable to the effects of poor husbandry at certain stages of the life cycle, such as during larval development and later on when undergoing sexual maturation. Finally any chemical treatments, which may be given to reduce fish parasitic infections, are particularly stressful and may easily do more harm than good.

6.5 INFECTIOUS DISEASES

6.5.1 Infectious agents

The description of a disease as 'infectious' presupposes an involvement in the disease process by one or more species of infectious agent (other than the normal invasion of fish tissues by micro-organisms at the time of death). Infectious agents usually belong to one of four different groups: bacteria, fungi, parasites and viruses. They are living organisms and, with the exception of viruses, are made up of individual cells.

Bacteria are usually single-celled organisms which can only be seen under the microscope. They are abundant in soil and water and many live normally inside the gut and on the skin of fish. Various species of bacteria can cause diseases of fish and bacterial disease is usually capable of being treated by drug therapy. Fungi belong to the plant kingdom and infect fish by means of fungal 'spores' in the water. These can enter the fish through damaged skin or via the food before branching out into a mat of fungus often visible to the naked eye. Parasites spend at least part of their lives inhabiting a 'host' animal from which they derive nourishment and shelter. Parasites are very diverse and fish parasites range from microscopic single-celled 'protozoa' to lampreys and leeches which may be several inches long. Certain parasites, such as tape-worms, have complicated life cycles involving several different hosts. Healthy fish in the wild usually have a few parasites, especially on the skin and gills or in the gut, but in the crowded fish farm they can multiply rapidly, particularly if the fish are stressed. Viruses are minute and can only be seen with powerful electron microscopes. Unlike other microorganisms, viruses cannot reproduce independently, but need to invade host cells and insert their genetic material to programme the host into producing more virus. It is not possible to treat virus diseases which should therefore be prevented from entering fish farms by specific precautions.

6.5.2 Epidemiology

Epidemiology is the study of the means by which disease spreads through human populations (sometimes called 'epizootiology' in animals), hence the term epidemic. In order for a fish to become infected with an infectious disease it needs firstly to take in suf-ficient numbers of bacterial cells or virus particles to constitute a so-called 'infectious dose'. This varies with the particular species of microorganism, as does the virulence of the microorganism and therefore the severity of its effect on the fish host. For instance, a cholera epidemic causes many deaths in unvaccinated humans, however healthy, whereas other infectious agents only become pathogenic and cause disease to stressed individuals. So it is in fish, with the existence of 'primary pathogens' able to cause disease to healthy unstressed populations of fish provided they are susceptible, i.e. have had no prior exposure to the particular pathogen or to its vaccine. Such diseases commonly start as an acute flare-up with initial heavy losses giving way later to a more drawn-out phase, although certain pathogens, such as tuberculosis bacteria, cause disease (in man or fish) which is chronic from the outset. Survivors of an epidemic (or 'epizootic') often recover completely, but may continue to harbour the infectious agent and act as 'carriers' capable of infecting other fish. If such carrier fish do not become immune, they may succumb to clinical disease when subsequently stressed.

The process by which an infection proceeds to a clinical outbreak of disease involves a similar interaction of fish host, pathogenic microbe and environment, irrespective of whether it involves a primary pathogen acting upon a healthy host or a 'secondary pathogen' acting as an opportunist invader to cause disease in a weakened host. The three most common routes of entry for a pathogen into the fish host are the skin, gills and gastro-intestinal tract. For instance an infected fish may excrete the pathogen into the water via its faeces to be subsequently ingested by an uninfected fish. This is an example of 'lateral' transmission of infection between adjacent fish in the same fish farm or watershed. However, 'vertical' transmission is also important, particularly for certain viruses, whereby the pathogen spreads from one generation to the next via infected eggs or sperm. Whatever the mode of spread, the existence of carrier fish in fish farms and the frequent escape of carriers into the wild are important in establishing reservoirs of infections, hence the possibility of renewed outbreaks of disease at a later date.

6.5.3 Primary pathogens

Infection of susceptible fish by virulent strains of viruses will invariably result in a high proportion of fish suffering acute disease. Virulence is related to water temperature in most virus diseases and the onset of acute disease may therefore be delayed until a change in water temperature occurs, but the fact that even healthy unstressed fish are affected means that the causative viruses are

'primary pathogens'. Although certain fish bacterial diseases (e.g.
bacterial kidney disease) have somewhat similar characteristics, the
clearest example of a true primary bacterial pathogen in salmonids
is *Aeromonas salmonicida*, the causative agent of furunculosis
disease.

Furunculosis is a severe septicaemic disease of salmonid fish,
although *Aeromonas salmonicida* has also been implicated in an
ulcerative condition of cyprinid fish known as carp erythrodermatitis.
A typical outbreak starts with a short period of reduced appetite
leading to heavy losses of fish showing few, if any, obvious clinical
signs, particularly in the case of salmon and young trout. In older
fish the onset is usually more gradual and accompanied by haemorrhages at the base of the fins, around the gills and in the
abdominal cavity (Plate 6.2). The 'furuncles' which give the disease
its name are dark boil-like swellings containing blood-stained fluid.
These are usually found only on the back and sides of chronically
affected fish, particularly brook trout. Although prompt treatment
with medicated feed containing appropriate antibiotics or sulphonamides will usually save those fish which are still actively
feeding, it is very difficult to eradicate furunculosis on an infected
farm. This is because a few causative bacteria usually persist in
carrier fish between outbreaks. When such carriers are stressed
(e.g. by handling, transport, low oxygen or perhaps just by a rise
in water temperature) the organism will multiply and is then
excreted to reinfect other fish, with a consequent recurrence of

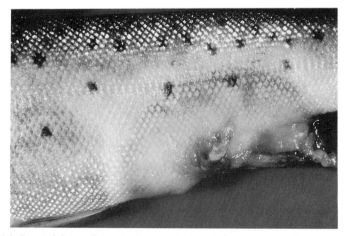

Plate 6.2 Bacterial septicaemia in trout due to *Aeromonas* infection causing
haemorrhage, necrosis and protrusion of the gut at the vent.

disease. In order to prevent furunculosis, fish farms should be sited, whenever possible, away from water courses containing wild salmonids likely to harbour the organism. Great care should also be taken to ensure that any fish introduced onto the farm are not furunculosis carriers.

Salmonids are also susceptible to a number of virus diseases which can cause severe economic loss in infected farms. Table 6.1 summarises the five best known virus diseases of greatest significance to fish farms in terms of host susceptibility and geographical distribution, including the salmonid viruses (IHN, IPN and VHS) channel catfish virus and Spring viraemia of carp. No means of chemotherapy is available for treating these diseases which are caused by true primary pathogens. Control measures are therefore aimed firstly at preventing initial spread of the virus into uninfected regions. Thus legislation is in place in many countries prohibiting the movement of live fish across national boundaries and restricting trade in fish ova to hatcheries which have been tested and shown to be virus-free. Despite these precautions the expansion of fish farming has now resulted in a fairly wide spread of the salmonid virus diseases in particular.

Table 6.1 Characteristics of the important fish virus diseases

Disease	Susceptible fish	Age susceptibility	Usual temperature susceptibility range	Geographical distribution
Channel catfish virus (CCV)	Channel catfish	Fry and finger lings	>25°C	Southern USA
Infectious haematopoietic necrosis (IHN)	Salmonids	Mostly less than 1 year old	<18°C	Japan, N. America
Infectious pancreatic necrosis (IPN)	Salmonids	Fry and up to 20 weeks	any	Worldwide
Spring viraemia of carp (SVC)	Common carp	All but mostly yearlings	<22°C	Europe
Viral haemorrhagic septicaemia (VHS)	Salmonids (grayling, trout, whitefish) and pike	All	<14°C	Europe (mainland)

The clinical picture of acute virus disease in a previously un-infected farm usually involves heavy mortalities and is usually related to the age of the fish and water temperature. For example, IPN virus causes large losses in salmonid fry generally about 6 to 8 weeks after they start feeding. The virus attacks the pancreas and surviving fish are sometimes vulnerable to husbandry stress and can suffer pancreatic necrosis at a later stage after grading or transport. However, adult fish frequently carry the virus within their reproductive organs without any sign of disease. If used for breeding, such IPN carriers may then spread the virus within their ova or milt by vertical transmission to the next generation. IHN has many features in common with IPN although the virus mainly attacks the blood-forming tissue of the salmonid kidney and the disease can strike older fish as well as fry. Clinical IHN disease usually appears at temperatures below 15°C and both vertical and horizontal transmission of the virus takes place, unlike VHS. VHS typically occurs in fingerlings and growers at low water temperatures. A clinical outbreak of VHS usually has three distinct phases and in the acute phase heavy losses occur, while at post-mortem large haemorrhages are seen in the abdominal cavity. This usually gives way to a chronic phase involving dark-coloured fish often with protruding eyes and an anaemic pale appearance of the gills and internal organs due to blood loss. Mortalities cease by the third phase and affected fish have a characteristic slow 'tumbling' motion due to brain damage, together with swollen kidneys. As the water temperature rises to about 10°C in the Spring, clinical signs of VHS start to disappear, only to reappear when the temperature falls during the autumn.

Water temperature changes also play a part in the development of CCV disease of catfish and SVC disease of cyprinids. CCV affects fry and fingerling channel catfish when water temperature rises during the summer above 25°C. Clinical signs of infection include heavy losses with abdominal distension and at post-mortem an enlarged liver and pinpoint haemorrhages are evident. The virus of Spring viraemia typically causes abdominal distension ('dropsy') and mortalities in young carp, particularly in countries where the water temperature rises rapidly above about 13°C during Spring after a long hard Winter. Affected fish often gather at the pond outlet and may become dark-coloured with protruding eyes and haemorrhages on the skin and gills. The stress of handling after overwintering carp at low temperatures appears to exacerbate

SVC and the same virus sometimes attacks the swimbladder causing haemorrhages, with inflammation and similar dropsical symptoms.

6.6   STRESS-MEDIATED INFECTIONS

The immediate environment of any fish is teeming with micro-organisms. A wide variety of bacteria and parasites normally live in harmony with their fish host, inhabiting the skin, gills or the inside of the gastro-intestinal tract. Under normal conditions such 'commensal' organisms do not harm the host and indeed have a vested interest in its continuing existence. However, under the crowded conditions of a fish farm, the fish may become stressed and the normal balance between host and parasite may be upset. A commensal organism on the fish, or other microorganisms in the surrounding water, may then become involved in the resulting disease process as secondary pathogens. The majority of infectious diseases of fish come about due to opportunistic invasion of stressed fish by secondary pathogens and can be avoided or greatly reduced by corrective husbandry measures.

6.6.1   Bacterial septicaemias

Furunculosis has already been described as an acute infection due to a primary bacterial pathogen *Aeromonas salmonicida* spreading rapidly throughout the body. A very similar pattern of acute bacterial septicaemia also occurs in fish due to various secondary pathogens. For example *Vibrio anguillarum* is a common marine bacterium which is often present within the gut of healthy fish. A frequent sequel to handling marine fish (e.g. grading, transport) is an outbreak of vibriosis, especially if water temperature is high and dissolved oxygen levels are low. As with furunculosis, affected fish usually become inappetent before losses commence. When opened at post-mortem fish dying from vibriosis show haemorrhages in the muscle and viscera, sometimes accompanied by skin ulcers. This condition occurs in a wide variety of fish farmed in brackish water or in the sea, including eels where it is sometimes called 'red pest'. So-called 'Hitra disease' of marine farmed salmonids in Northern Europe may involve an unusual *Vibrio* strain able to multiply and cause disease at relatively low water temperatures. The *Vibrio*

organism can only multiply in seawater and the clinical counterpart of vibriosis in freshwater is bacterial septicaemia caused by
Aeromonas or *Pseudomonas* bacteria. This is a particular problem
in carp farming and affected fish are often dark in colour with
swollen fluid-filled abdomens and usually have haemorrhages on
the body surface and at the base of the fins. Heavy losses can occur
in carp due to bacterial septicaemia, particularly if handled after
overwintering and if simultaneously infected by SVC virus. In
salmonids the stress of spawning is a well known predisposing
factor, together with handling, high stocking density and high
levels of suspended solids in the water. *Aeromonas hydrophila* can
survive in fresh and seawater and has caused mortalities in red
seabream (*Pagrus major*) due to progressive infection of pre-
existing skin wounds.

Table 6.2 lists the better known acute bacterial septicaemias of
farmed fish. It is sometimes difficult to differentiate between them
on clinical grounds and isolation of the 'causative' organism is
often necessary to confirm diagnosis. Enteric redmouth (ERM)
has now spread from North America to Europe and causes haemorrhagic septicaemia in trout with the head and gastro-intestinal
tract particularly affected, hence the name. *Edwardsiella* causes
lesions of the skin and internal organs in catfish, carp and eels and
is a particular problem if the water is polluted with organic matter.
In general these bacterial septicaemias are susceptible to chemotherapy
by means of feed medicated with antibiotics or sulphonamides,
although treatment must be initiated quickly while the fish are still
feeding. Clearly it is also important to rectify the stress factors
which predispose to the disease outbreak.

6.6.2   Skin and gill infections

There are a number of important diseases of farmed fish which are
predominately infections of the skin and gills. This is in contrast to
the acute septicaemias, described earlier, which usually affect the
entire system often with accompanying skin lesions. The prevalence
of these external infections is closely linked to husbandry practice,
in particular to water quality, and will be discussed in relation to
the various species of bacteria, parasites and fungi implicated.

The *Cytophaga* group of bacteria are involved in several different
diseases affecting fish skin and gills, including columnaris, cold-

Table 6.2 Acute bacterial septicaemias of farmed fish

Causative agent	Disease	Susceptible fish species
Aeromonas hydrophila	Haemorrhagic septicaemia	Carp, salmonids, seabream
Aeromonas salmonicida	Furunculosis	Salmonids
Edwardsiella tarda	Edwardsiellosis	Catfish, carp, eels
Pseudomonas fluorescens	Haemorrhagic septicaemia	Catfish, carp, eels
Vibrio anguillarum	Vibriosis	Marine fish
Yersinia ruckeri	Enteric Redmouth (ERM)	Salmonids

water disease and bacterial gill disease. Columnaris only occurs at higher water temperatures and causes raised whitish lesions on the head, back and gills which can progress into haemorrhagic ulcers. If carrier fish are present in the water supply, columnaris will frequently break out after crowding or handling provided the temperature is above 15°C. It is caused by the bacterium *Flexibacter columnaris* which infects cyprinids and salmonids and varying success has been achieved in controlling the disease by means of antibiotic treatment via the feed. A related organism *Flexibacter psychrophila* is associated with coldwater disease of salmonids; this is sometimes called peduncle disease as the tail (caudal peduncle) becomes progressively eroded by bacterial infection at temperatures below 10°C (Plate 6.3). Other pathogenic members of the *Cytophaga* group cause an ulcerative disease of farmed yellowtail and also frequently cause secondary infections of skin and fin lesions in marine fish, especially at low water temperature. But the greatest

Plate 6.3 Tail rot in a stressed trout held at low temperature (cf peduncle disease).

economic significance of the group is due to its role in bacterial gill disease. This is an infection of the gills with various species of myxobacteria which occurs worldwide, particularly among young salmonids. Affected fish gather near the water inlet and tend to stack at the surface of the water. In severe infections fish show obvious respiratory distress with gasping and puffed-out gill covers and continuous low-level mortalities. This is due to the mat of microscopic bacteria coating the gills and interfering with gas exchange. These bacteria are good examples of secondary pathogens and it is not possible to reproduce bacterial gill disease by introducing them to healthy fish in clean water. However, high stocking density and increased levels of suspended solids in the water quickly predispose to gill disease. Fingerling rainbow trout appear to become susceptible to the infection when free ammonia levels reach about 0.02 mg/l and hence the only effective long-term remedy is to improve pond hygiene by reduced stocking and improved water flow. Addition to the water of various chemical treatments, such as quaternary ammonium compounds, can help in the short term by removing the mat of mucus and bacteria covering the gills.

Most skin problems in fish are due to external parasitic infection. External parasitic infection of the skin and gills is probably the commonest fish health problem under intensive farming conditions. Parasites which would normally be present at very low levels on wild fish often find the environment of a fish pond, in particular the high density of fish, very conducive to multiplication and spread rapidly among the fish population. They damage fish skin by attaching to the epidermis and many species of parasite will actively feed on the underlying tissues and blood. This can cause intense irritation and affected fish may scrape themselves on the sides and bottom of the pond or 'flash' as they twist over during swimming. Heavy infestations can result in skin lesions, reduced growth and even high mortalities, especially in younger fish. In addition the presence of large parasites visible to the naked eye may downgrade the value of fish and in certain countries make them unsaleable. Parasitic infection of the gills can quickly lead to respiratory distress with affected fish congregating at the inlets or water surface. Such external infections are often due to a combination of different parasites, bacteria and fungi. Figure 6.1 summarises the most important external parasites known to cause specific problems to farmed fish.

The single-celled protozoan parasites are generally only visible

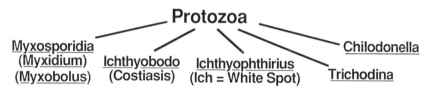

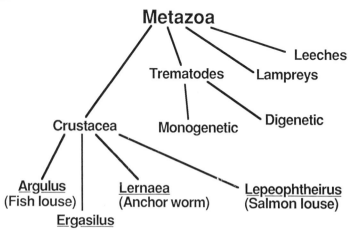

Fig. 6.1 Parasites of the fish integument.

under the microscope with the exception of *Ichthyophthirius*, which is the largest and most commercially significant of the protozoa. This circular parasite can be up to 1 mm in diameter and looks like a white spot, hence the name 'white spot disease'. All fish species are susceptible in freshwater and the organism infects the skin, fins and gills and sometimes the mouth. Water temperature influences the development of white spot and catfish farmers in the USA often recognise an April peak as it approaches the optimum range of 21–24°C, possibly also linked with concurrent handling stresses. The traditional method of control in warm water carp ponds is to use three applications of malachite green sprayed over the entire pond at three-day intervals, but for stubborn outbreaks a mixture of formalin and malachite green has gained acceptance in the USA. Japanese eel farmers cure white spot by pumping saltwater into the eel ponds for several days but this is rarely a practicable solution with other species of the microscopic protozoans.

Costiasis has the greatest commercial significance and is caused by *Ichthyobodo* (formerly classified as *Costia*). Fry are particularly susceptible and heavy gill infestation can sometimes precipitate sudden heavy losses resembling certain virus diseases. In salmonid fry and fingerlings, costiasis is typically a problem associated with overcrowding and high levels of suspended solids in the water. But it is not restricted to coldwater fish or even to fry and fingerlings. For instance, large losses of 500 g fish have been reported from overwintering ponds holding carp, mullet and tilapia. Treatment is usually by formalin and badly affected earth ponds should be emptied and disinfected with lime. *Trichodina* and *Chilodonella* are also found frequently on the skin and gills under conditions of poor water quality. Often they appear together with *Ichthyobodo* or other parasites and attack weakened fish. As with costiasis, fry are particularly susceptible and treatment is with formalin. Myxosporidia are another group of protozoa which can be pathogenic to fish. They include *Myxobolus*, which causes anaemia due to cysts on the gills of carp, mullet, etc., and also *Myxidium*, which causes a progressive white ulcerative condition on the skin of eels. It is difficult to treat myxosporidia and affected fish should be removed and burnt, or placed in a lime pit, before thorough disinfection of the holding tank or pond.

The multicellular fish parasites or metazoans include a diverse array of different organisms, such as various crustaceans, flukes, larval molluscs, leeches and lampreys which infest the skin and gills (together with others which are internal parasites, e.g. nematode worms and tapeworms). Of these the parasitic crustaceans pose the greatest threat to both freshwater and marine fish farms. The fish louse (*Argulus*) occurs widely in freshwater fish farms and is up to 1 cm in length, hence clearly visible to the naked eye (Plate 6.4). It can scuttle crab-wise over the fish's body and sucks blood by means of a proboscis. Heavy infestations cause severe anaemia and can cause mass mortality, particularly in young fish. Fish lice damage often results in open wounds, which may become secondarily infected, and in greatly reduced growth rates. The anchor worm (*Lernaea*) is also up to 1 cm in length and has a rigid anchor-like structure which is embedded deeply into fish skin. Different species of anchor worm can affect all freshwater and marine farmed fish provided the water temperature is above 18°C, with the optimum range being 22–30°C. Young fish can be killed by only a few parasites, whereas older fish may become infested with hundreds,

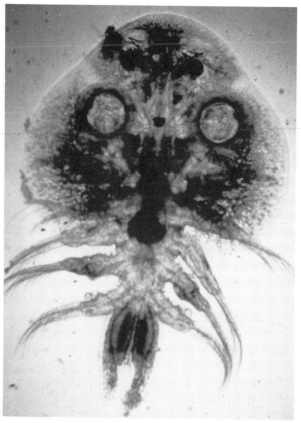

Plate 6.4 Close-up view of the fish louse (*Argulus*) which is a large (approximately 1 cm) fish ectoparasite.

resulting in anaemia, emaciation and open ulcerating wounds. In eel farms, anchor worms frequently attach to the inside of the eels' mouths and heavy infestation may consequently cause the eels to die from starvation. Japanese eel farmers cure anchor worm by pumping seawater into the ponds for a period of 3−4 days and by sprinkling calcium chloride onto the surface of the ponds to kill the egg and larval stages. Potassium permanganate has been used, but the current standard treatment method in Israeli fish ponds infested with either fish louse or anchor worm involves the use of organo-phosphorous compounds by spraying with 0.12−0.15 ppm bromex.

The same method is used to control *Ergasilus* infestation of mullets in brackish water ponds, which if left untreated can cause reduced growth and large losses due to the effects of this crustacean on the gills. Because salmonids are farmed at cooler temperatures, *Argulus* and *Lernaea* are not usually a problem, whereas severe infestation with the salmon louse can cause heavy losses in seawater. The commonest species of salmon louse is *Lepeophtheirus* and wild salmon often have several attached around the vent area. In a crowded salmon farm, large numbers can accumulate and cause severe damage all over the skin surface. Control is most commonly by means of organophosphorous pesticide solutions added to the water (e.g. Neguvon; Dichlorvos), but needs to be repeated as the immature stages of the parasite are not affected. Of the remaining metazoans of significance to the fish farmer, lampreys and leeches rarely reach epidemic proportions unless the fish are severely weakened. However, since leeches can carry blood parasites and hence infect fish when sucking blood (cf malarial mosquitoes), it is important to eradicate them by liming ponds between fish crops.

Of greater importance are the monogenetic and digenetic flukes or trematodes (Plate 6.5). Like the protozoans described earlier, these parasites of skin and gills thrive under conditions of low flow rate and heavy stocking and hence for the salmonid farmer are usually an indicator of poor husbandry practice. But in standing water pond cultivation of carp or catfish, heavy infestations are common, especially as the temperature is rising (e.g. in the period March–May in the south eastern USA). Carp fry are particularly susceptible to gill fluke damage until an average weight of 2–3 g is

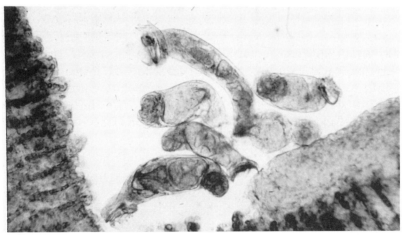

Plate 6.5 The monogenetic fluke *Gyrodactylus* parasitising fish gills.

reached. Control is achieved by using chemicals such a bromex or formalin added to the water.

Any external wound on fish immediately predisposes to secondary infection by fungi which occur so commonly in the aquatic environment. Fungal infections are also a persistent problem in hatcheries as any dead eggs quickly become a focus of fungal proliferation which can then spread to adjacent healthy eggs. The most common species involved is *Saprolegnia*, which will infect fish tissue only in freshwater and at water temperatures usually below about 18°C. Saprolegniasis is a well known problem in Japanese eel farms during Spring and Autumn with patches of white fungal growth spreading from the head over the whole body and up to 70% mortalities of eels being reported. This condition can be partially controlled by the use of antibacterial drugs and so it seems likely that *Saprolegnia*, which does not respond to such drugs, is an opportunist secondary invader. More typically saprolegniasis occurs as a sequel to handling fish, such as mullet or tilapia, during the winter period. Pale coloured tufts of cottonwool-like fungus appear on the skin, at the base of the fins and on the gills. The branching fungus invades the skin and in severe cases will cause death due to the resultant water logging of the fish. In adult salmonids the stress of spawning is an important predisposing factor (Plate 6.6) but *Saprolegnia* in younger fish is usually associated with scale loss following grading or traumatic injury. Malachite green is the most common and cost-effective means of controlling external saprolegniasis. Another fungus (*Branchiomyces*) causes gill rot in warmwater pondfish in Europe and the USA. The fungal strands penetrate the gill capillaries and affected fish succumb rapidly. The disease is not fully understood but appears to be connected with heavy fertilisation of ponds or the presence of decaying vegetation (e.g. after grass cutting). Prevention is by keeping ponds clean and ensuring a continuous inflow of fresh water. Since an outbreak of gill rot can rapidly kill the entire fish stock, if suspected it is necessary to act quickly by adding calcium oxide to the pond.

6.6.3 Chronic diseases

Although the aftermath of an acute septicaemia is usually a stage of chronic inflammation and repair, certain bacterial diseases are

Plate 6.6 Generalised external *Saprolegnia* fungus infection of a mature brown trout.

characterised by chronic inflammation from the outset. Such chronic diseases often develop slowly and the disease process may be far advanced with severe internal pathology before the fish shows any obvious external signs. Table 6.3 summarises the more important chronic bacterial diseases affecting farmed fish. Although fish tuberculosis is the archetype of such diseases, it is rarely encountered nowadays provided that fish viscera are not fed back to growing stock without prior heat treatment. The clinicial picture in this and most other similar infections is of poor growth and emaciation with a trickle of mortalities increasing to a flood if the fish are handled, transported or otherwise stressed. On post-mortem there may be obvious lesions affecting the internal viscera. For instance chronic pasteurellosis of ayu and yellowtail is marked by

Table 6.3 Examples of chronic bacterial diseases of farmed fish

Disease	Causative agent	Principal host
Bacterial kidney disease (BKD)	*Renibacterium salminarum*	Salmonids
Nocardiosis	*Nocardia* spp.	Yellowtail, salmonids
Pasteurellosis	*Pasteurella* spp.	Ayu, yellowtail
Tuberculosis	*Mycobacterium* spp.	Various

nodular lesions in the liver, spleen and kidney. Yellowtail also suffer heavy mortalities due to nocardiosis which is manifested by granulomatous lumps developing throughout the muscle, viscera and in the skin.

The most commercially significant chronic bacterial infection of salmonids is bacterial kidney disease (BKD) due to *Renibacterium salminarum*. BKD developes slowly and affected growers may appear normal but become prone to heavy mortalities, for example if transferred as smolts to seawater or if subjected to a diet change. Internal pathology varies from small white nodules to areas of frank necrosis in the kidney and other viscera, sometimes accompanied by a white membrane in the abdominal cavity. As with other chronic bacterial diseases, treatment is rarely possible since BKD may well be generalised before its presence is suspected. Even if treated early with appropriate antimicrobial drugs, it is difficult to eradicate the causative bacteria which become surrounded by a fibrous capsule laid down as a result of pathological changes within the fish. Prevention is therefore all-important and needs to take account of the possible risks arising from both horizontal transmission between fish and vertical transmission via infected eggs. Although the scale of losses from BKD can be reduced by conservative husbandry, the organism behaves like a primary pathogen and is the subject of legislative control to prevent further spread in certain countries.

Among the internal parasites associated with chronic disease, the multicellular metazoans include nematode roundworms and cestode tapeworms. These can occur in large numbers within the alimentary canal and surrounding viscera and may reduce growth rate. Although not usually a potent health hazard, they look unsightly and may render fish unsaleable in the market place. A particular problem in carp and rainbow trout farming is the eye fluke (*Diplostomum spathaceum*). The life cycle of this parasite involves different stages occurring in snails and then fish, with fish-eating birds being the final host. If present in large numbers, the infective stage for fish will cause cataract and blindness. Prevention of eye fluke infestation is achieved by breaking the parasitic life cycle (e.g. by use of bird netting).

By comparison with the metazoans, the various unicellular protozoans have somewhat greater commercial importance as endoparasites and are summarised in Table 6.4.

Hexamita and *Eimeria* are gut parasites of salmonids and cyprinids

Table 6.4 Commercially important protozan internal parasites of farmed fish

Causative organism	Protozoal type	Principal host/disease
Ceratomyxa shasta	Myxosporidia	Salmonids
Eimeria spp.	Coccidia	Carp coccidiosis
Hexamita (=*Octomitus*)	Flagellate	Salmonids
Myxosoma cerebralis	Myxosporidia	Salmonid whirling disease
Pleistophora spp.	Microsporidia	Cripple body disease of eels
PKD agent	?	Proliferative kidney disease of salmonids (PKD)

respectively. *Hexamita* (*Octomitus*) is commonly found in salmonid fingerlings showing ill-thrift, but it is often difficult to tell whether this is cause or effect. Various species of *Eimeria* cause nodular coccidiosis and coccidial enteritis of carp during late Winter and the resulting intestinal inflammation leads to emaciation and is often fatal. *Ceratomyxa shasta* and *Myxosoma cerebralis* are both myxosporidian parasites of salmonids. *Ceratomyxa* is restricted to the western USA and can cause large-scale losses on trout farms with fish showing a fluid-filled abdominal cavity and nodules developing throughout the muscles and viscera. *Myxosoma cerebralis* is the causative agent of whirling disease, so called because the parasite damages the balance organs within the head of affected fish which then swim in a characteristic whirling manner. It now seems that *Myxosoma* spores need to undergo some form of maturation stage within worms on the mud bottom of the pond before they can become infective to fish. Control is therefore possible on infected farms by rearing fish within concrete or fibreglass tanks and away from earth ponds (at least until after they reach about 7 cm in length, at which stage the cartilage in the head becomes bone and the fish cannot then go down with the disease even if they are carriers). Because high mortalities and severe deformity can occur in clinical outbreaks, whirling disease is the subject of legal restrictions in many countries to prevent spread. *Pleistophora* attacks the muscles of eels which become crippled and misshapen, hence the name cripple body disease. Proliferative kidney disease resembles BKD with gross changes occurring in affected salmonids which can suffer heavy losses if handled or otherwise stressed. Treatment is not possible and it seems likely that this condition is parasitic in origin.

Although most fungal diseases of fish involve the skin and gills as described earlier, various fungi are associated with chronic internal infection. The fungus usually enters fish via fish feed in the form of fungal spores. This can come about if fish feed is made from marine fish suffering from certain fungal diseases (notably *Ichthyophonus*) or if it becomes contaminated by moulds or fungi present in the soil, etc. The resulting fungal infection usually starts in the gut and then spreads progressively throughout the rest of the body. Affected fish may have enlarged abdomens and appear emaciated with mounting losses. At post-mortem large granulomatous growths containing the fungus are present in the viscera resembling chronic bacterial infections such as BKD or tuberculosis. Treatment is impossible and prevention is by careful manufacture and storage of fish feed prior to use.

6.7 DISEASE PREVENTION AND CONTROL

Why does infectious dropsy occur so rarely in Israeli carp ponds when in continental Europe it regularly causes widespread economic loss? The virus is almost certainly present, but Israel has a mild brief winter and fish are kept on a high plane of nutrition for over nine months of the year. The immune status of carp in such a climate is also superior and the resulting balance between the fish, its environment, and the infectious agent means that disease does not ensue. On the other hand, the use of standing water ponds means that carp reared in such systems are more vulnerable to attacks from external parasites than is the case with carp reared at even higher stocking densities in cages moored in running water streams (e.g. Indonesia) because of the favourable conditions for parasite multiplication and transmission under conditions of static water. Clearly if a particular disease organism is excluded from the fish farm environment, health management of the stock is made easier. But the above examples show that ecological factors and husbandry practice can modify the effect of a particular pathogen on farmed fish, whereas these effects will, of course, vary greatly between different species of pathogen and fish. In the following section the principles of preventive medicine will be described in relation to avoiding fish disease and mitigating its effects, together with a brief survey of different methods of treating such diseases.

6.7.1 Controlling the movement of fish

Statutory controls on the movement of live fish or ova are the first
defence against the spread of fish disease organisms into countries
or watersheds where they are not present. Primary pathogens, such
as fish viruses, can even be introduced along with imports of fish
carcases for human consumption or as fishmeal. However, in practice the greatest risk arises from international trade in live fish
stock infected with pathogens. The problem is complicated by the
possibility of vertically-transmitted diseases (e.g. IPN and IHN
virus) being transferred via egg importations. Pathogens may also
be introduced by carrier fish destined for other markets involving
different species of fish. For example, white spot disease was
introduced into Israeli fish farms and spread rapidly following the
disposal of infected ornamental fish into canals.

In an attempt to limit the spread of important diseases, certain
countries have therefore banned the importation of all live fish
belonging to the farmed groups most at risk. In addition the
importation of fish ova is frequently also banned. Exceptions may
be permitted if the country of origin is known to be free of the
pathogens causing concern, provided the fish disease authorities
are deemed competent to judge. For example, the UK authorities
prohibit the importation of live fish of the salmonid family and are
only prepared to grant individual import licences for batches of
salmonid eggs provided they are thoroughly disinfected and come
from an overseas hatchery source which is routinely inspected
and found free of specific disease organisms by approved testing
methods. Similarly United States regulations effective 1 July 1968
(Code of Federal Regulations, Title 50, Section 13.7) require that
salmonids and salmonid eggs be certified free from the infective
organisms of viral hemorrhagic septicaemia and whirling disease
before they may be imported into the United States.

Clearly it is important to identify outbreaks of the most commerically important fish diseases which could threaten the livelihood
of fish farms or even harm stocks of wild fish if allowed to spread
in an uncontrolled fashion. This is particularly true for exotic
pathogens not known to be present in the country, but is also
important in the case of primary pathogens which are already
present but not widely spread throughout different watersheds. For
the most severe diseases the ideal arrangement would be a system
of compulsory notification to the authorities on suspicion of the

disease, with an immediate ban on any fish movements out of the area until such time as appropriate testing had shown the causative organism to be absent; whereas proof of the disease would result in slaughter and hygienic disposal of all stocks followed by rigorous disinfection before restocking, with full compensation for any financial loss incurred. This is the situation in many countries where disease epidemics in conventional livestock are concerned (e.g. foot and mouth disease of cattle), but is rarely adopted in fish diseases, perhaps because of insufficient pathological expertise or because the aquaculture industry was formerly not judged of sufficient economic value to merit the necessary government resource.

It is important to take account of the varying commercial significance posed by different pathogens within a particular country and modify control measures accordingly. For example, in Great Britain, eight fish diseases are currently classed as 'notifiable' (Table 6.5), which means that under law any fish farmers suspecting the presence of any of these diseases should immediately report the matter to the relevant authorities. However, the authorities react to such notification in different ways depending on the particular organism, with the exotic viruses (such as IHN and VHS) likely to provoke the strongest reaction to prevent further spread and possibly eradicate the organism from the nation. At the other extreme *Aeromonas salmonicida* is very common in wild and farmed salmonids and is deemed to be more commercially significant to salmon farming than trout farming, hence furunculosis is notifiable only within the salmon farming region of the UK (i.e. Scotland). In between these two extremes (e.g. bacterial kidney disease) the policy is flexible and pragmatic with movement restrictions usually applied to infected hatcheries trading in live fish but not applied to farms producing dead sale fish for the table market.

Table 6.5 List of notifiable fish diseases in Great Britain (1987)

Believed to be absent	Known to be present
1. Infectious haematopoietic necrosis (IHN)	1. Bacterial kidney disease
2. Spring viraemia of carp (SVC) has occurred previously	2. Enteric redmouth (ERM): Scotland
3. Viral haemorrhagic septicaemia (VHS)	3. Furunculosis of salmon
	4. Infectious pancreatic necrosis (IPN)
	5. Whirling disease

6.7.2 Preventive medicine

The choice of site has immediate implications for fish health, most obviously as regards water source, and compromises often need to be made. For example, in order to gain access to sufficient water of suitable quality it may be necessary to put up with the presence of wild fish which may act as a potential disease source. Also, there may be supplies of underground water available which will be free of wild fish or infectious agents, but probably need artificial aeration. Pumps may be installed to elevate groundwater or to increase the available flow of surface water, but this introduces the possibility of gas bubble disease, in addition to the obvious risks of pump failure.

The exploitation of a suitable site should be carefully planned in order to facilitate disease prevention and control. Whenever possible, the hatchery should have an independent water supply or, in the case of a running water system, it should be sited upstream of the main production ponds. The provision of gravel filters on the hatchery inlet can assist in removing particulate matter, fungal spores, and parasites, to ensure clean water supplies during the crucial stages of egg incubation and larval rearing. Any bought-in ova must be of known health status and disinfected in an approved manner prior to arrival. The overall layout of the farm should accommodate the changing water requirements of growing stock over their production cycle, in particular satisfactory dissolved oxygen and removal of metabolic wastes. At the same time different age classes should not be mixed, hence the need for separate rearing facilities for fry, growers, adult fish and broodstock, together with quarantine ponds isolated from the main farm if fish need to be introduced from elsewhere. It may be difficult to avoid the possibility of cross-infection between different age classes in the case of floating cage units moored on the same body of water, but for ponds or raceways it is usually practicable, provided the individual units are arranged in parallel with regard to the water supply rather than in series. Farm design should enable adequate observation of stocks as well as ease of access for feeding, regular removal of any dead fish and handling with minimum stress. Using fish pumps for transfer of live fish between ponds, coupled where necessary with the use of anaesthetics and artificial aeration, can do much to reduce scale loss and subsequent stress-related health problems, particularly under conditions of less than ideal water temperature. As a general rule, the less fish are handled the better and they should always be starved beforehand. It is important to exclude predators, such as fish-eating birds, which not only kill fish and cause unsightly or harmful stab wounds, but also may act as hosts in parasitic life cycles involving fish (e.g. eye fluke).

Another important health factor in the choice of husbandry system is ease of disinfection. It should only be possible to enter the hatchery by means of a footbath containing an appropriate germicidal disinfectant such as an iodophor and the same should apply for the quarantine pond area. Fish viruses can enter on the tyres of lorries transporting fish feed which should also be disinfected on arrival. Concrete raceways or fibre-glass tanks should be scrubbed down and sterilised after emptying with an iodophor or with 200 ppm chlorine solution, but this should be thoroughly washed out prior to restocking since these chemicals are highly poisonous to fish. Earth ponds should be drained and left to dry out in the sun (a good disinfectant) before reworking at least annually and may be limed in order to remove fish parasites present. It may be difficult to disinfect bag nets but, when fouled, they should be hung to dry out before being cleaned with a pressure hose. Equipment such as grading tables and dip nets should be dipped in chlorine solution, then thoroughly rinsed before transfer from one pond to another. Any dead or dying fish should be removed daily and burnt or placed in a pit with quick-lime. Not only do such fish act as a focus of bacterial and fungal multiplication, but if left to decompose on the bottom of earthen ponds they can produce botulinum toxin and precipitate an outbreak of acute botulism among cannibal fish. Food storage and distribution should ensure that food is kept dry, free from mould contamination and fed soon after manufacture to avoid problems of rancid pellets (or decomposition of waste fish feed). Use of mouldy pellets can cause an acute haemorrhagic syndrome due to mould toxins, whereas feeding decomposing waste fish can cause mass mortality due to botulism (i.e. a fish kill; chapter 2; section 2.3.2). Overall preventive medicine at farm level means avoiding the introduction of pathogens to the farm and reducing husbandry stresses on the farm. Perhaps the ideal situation is where a farm is self-sufficient in terms of replacement stock, is under no threat from an adjacent wild fish population and has surplus water available throughout the year ensuring optimum water quality.

Fish vaccination is one aspect of preventive medicine already described as holding considerable promise (section 6.2). Commercial vaccines are now available for a number of bacterial diseases, particularly those affecting salmonids. Most success to date has been achieved with vibriosis and enteric redmouth vaccines and injection into the abdominal cavity elicits the strongest immunity. However, injection is usually impractical and preferred methods of delivery include spraying fish with the vaccine, oral administration via the diet and also dipping the fish into a solution of the vaccine which is then absorbed across their gills and epithelium. This method is often used while fish are crowded together prior to transport. It is claimed that vaccinating salmonid fingerlings in this way will confer protection for 18 months depending on circumstances, whereas less success has been obtained with various *Aeromonas* vaccines and with viral vaccines. Part of the problem in developing fish vaccines is the need to protect against different strains of the same disease organism. However, the range and reliability of vaccines will undoubtedly increase and in time should revolutionise certain aspects of fish disease control. Not only do they offer real scope in preventing virus diseases for which treatment is not possible, but they should enable the focus to switch more generally from curative to preventative measures. This should in turn allow the fish farmer to avoid an over-reliance on current treatment methods which are often short-term, limited in efficacy, expensive, and mean that a drug withdrawal period is necessary prior to slaughter to avoid the risk of residue problems in the final product.

6.7.3   Disease treatment

Good management is impossible without keeping up-to-date accurate records on factors such as water flow, water temperature, fish weights, feed consumption, mortalities, etc. This can often help to give early warning of disease problems, particularly if supplemented by routine health monitoring. For example, daily microscopic examination of gills and skin from a representative sample of fry may show a build-up of external parasites, indicating a need to increase water flow rate or to initiate treatment. Whenever possible a firm diagnosis should always be made by identifying the causative organism prior to treatment so that the most appropriate treatment regime may be implemented. Detailed records of all such treatments should also be maintained. Fish are generally treated by adding chemicals either to the water or to the feed, but also occasionally by injection and the various methods will now be discussed in more detail.

External treatment using chemicals added to water is the usual means of controlling infections of the skin and gills caused by external parasitism, external bacterial infections (e.g. bacterial gill disease) and external fungal infections (e.g. *Saprolegnia*). Adding chemicals to water is also used to control toxic blooms of algae which might otherwise cause fish kills. Effective control typically involves achieving a concentration of the chemical in the water sufficient to kill the infectious agent without actually harming the fish. As the margin of safety is often low and treatment is invariably stressful in itself, it is vital to treat only when necessary and to do so with care. Prior starvation for about 24 hours reduces the stress of treatment and it is always preferable to undertake an initial trial treatment with a few fish to check their response prior to implementing the main treatment. Calculation of the required quantity of chemical must take into account water flow rates and the usable volume of the pond and should be independently cross-checked for accuracy. Treatment should be undertaken at the time of day when water temperature is at its lowest and adequate oxygen levels must be maintained throughout. It is important to monitor fish behaviour continuously during treatment and the fish farmer must be ready to switch on aerators or to flush through with freshwater if the fish become distressed.

The practical implementation of external chemical treatment obviously relates to the type of husbandry system. For example, floating cages are usually treated by inserting an underwater polythene sheet or tarpaulin to enclose the fish. The chemical is then added as a bath treatment while oxygen or compressed air is bubbled into the water cage for the duration of the treatment. By contrast, a large number of tanks or ponds with a common water supply can be treated simultaneously by adding a constant volume of chemical to the water supply over a given period to achieve the necessary concentration. This flowing treatment method is undertaken using either a constant-head siphon or a constant-volume delivery pump. For treating standing water ponds it is necessary to spray a very dilute solution of the treatment chemical over the entire pond for a prolonged period.

Table 6.6 illustrates the more common bath treatments used for salmonids and it is important to bear in mind that the target

Table 6.6 Common types of bath treatment for salmonids

Treatment chemical	Bath concentration in mg/l	Duration of bath in hours	Disease indications
Chloramine-T	10	24	External parasitism
Dichlorvos (e.g. Nuvan 50EC)	1	1	Salmon louse
Formalin 40%	167−250	1	External parasitism
Hyamine 3500	2−4	1	Myxobacterial gill disease
Malachite green (zinc-free)	1−2	1	*Saprolegnia* fungus
Proflavine hemisulphate	20	0.5	Fin and tail rot
Sodium chloride (iodine-free salt)	10 000	1	External parasitism

concentration may vary with the age of fish and with water quality. Also, some chemicals remove oxygen from the water or increase the fish demand for oxygen. For example, formalin and the quaternary ammonium compounds (e.g. Hyamine 3500) are more toxic under conditions of soft unbuffered water and the lower range of concentration should be used if water hardness is less than 100 mg/l of calcium carbonate. Also it may on occasion be more practicable to use a higher concentration for a much shorter period and to dip the fish into the chemical or to flush a slug of the chemical rapidly through a tank or raceway of fish. Permanent damage can be done to salmonid gills by frequent treatment with formalin, hence treatment should only be repeated if absolutely necessary and not within 1−2 days of the first treatment.

Table 6.7 illustrates the much lower concentrations of chemicals used in Israel to treat external parasitic and fungal infections by spraying onto fish ponds. A single application of the organophosphorous compound bromex kills most external parasites of carp and other pond fish within six hours and is cheap and relatively safe to fish, whereas malachite green is effective against white spot and fungus, provided treatment is repeated every third day.

The usual means of treating systemic bacterial diseases or intestinal parasites is by the incorporation of appropriate drugs into the fish feed. Such antimicrobial drugs include antibiotics, nitrofurans and

Table 6.7 Concentration of chemicals for spraying Israeli fish ponds (after Sarig, 1971)

Infectious agent	Treatment chemical	Required concentration in mg/l	Safety index
Argulus (fish louse)	Bromex	0.12	32
Lernaea (anchor worm)	Bromex	0.12	32
Dactylogyrus vastator	Bromex	0.12	32
Dactylogyrus extensus	Bromex	0.15	30
Ergasilus	Bromex	0.15	30
Ichthyophthirius (white spot)	Malachite green	0.15	7
Chilodonella	Formalin	40	
Saprolegnia fungus	Malachite green	0.15	7

NB 1. The concentration given is the quantity of active ingredient needed to kill the parasite within 6 hours.
2. The safety index is the ratio of minimum lethal fish dose to the concentration which kills the parasite.
3. Bromex is dimethyl 1,2-dibromo-2,2-dichlorethyl phosphate.

sulphonamides and whenever possible the infectious agent should be isolated and its drug sensitivity pattern determined in the laboratory in order to decide which drug should be used.

In many countries antimicrobial drugs are closely controlled by the authorities and can only be prescribed by a veterinarian who has satisfied himself that their use is justified. In the UK, for example, the tetracyclines are classed as 'prescription-only-medicines' (POM) and cannot be purchased without a prescription signed by a veterinarian (or by a medical doctor). Moreover, the range of drugs which may be used in animal production for human food also varies between different countries. For example, the use of chloramphenicol in food animals, including farmed fish, is now banned in the USA and parts of Europe in order not to jeopardise its value in human medicine. This is related to fears that indiscriminate use will quickly lead to drug resistance occurring and that resistance factors could be then passed from fish bacteria to human bacteria. Prolonged use of many fish antimicrobials usually does produce drug resistant strains of bacteria, although infectious drug resistance between different species of fish bacteria has not been a significant problem to date. Insofar as systemic treatment will usually lead to residues of the drug accumulating in fish tissues, it is important to observe a drug withdrawal period prior to slaughter. This is normally from 1 to 4 weeks depending on the particular drug and water temperature but may be longer at very low temperatures. A withdrawal period allows time for the drug to be excreted from the fish and ensures the fish farmer's customers do not consume drug residues.

The main practical problem with using an in-feed medication route is that it relies on fish eating the feed, whereas appetite is often much reduced, particulary in acute bacterial diseases. For each antimicrobial drug the recommended dose level is normally given in terms of unit body weight over a prescribed treatment period. After diagnosing the disease and choosing the most appropriate drug, it is therefore vital to know the fishes' feeding rate since this will determine the drug dosage per tonne of feed. Obviously there is a ten-fold difference in the quantity of drug to be added per tonne of feed where the daily feeding rates vary between 0.5% and 5% of body weight, so the calculation is critical. If in doubt, use a feeding rate for treatment which is 0.5% lower than that used immediately prior to treatment to compensate for reduced appetite due to disease and also to possible palatability problems with certain drugs (e.g. sulphamerazine). The physical incorporation of drug into pelleted feed may also reduce the potency of active ingredient present due to inadequate mixing or to the high temperatures reached in the milling process, whereas the high moisture levels within waste fish feed also degrade some antibiotics which must therefore be added to such feed immediately before use. In any event, the difficulty of ensuring that each fish receives a therapeutic dose of the drug means that, at best, only those fish which are eating keenly will obtain proper treatment and those already inappetent will probably succumb to the disease. On occasion, it may be permissible to adopt a prophylactic course of medicated feed in advance of an event such as transfer of salmon smolts to seawater or the temperature rise during Spring which predisposes to losses on certain farms (e.g. with mixed aeromonad and viral infections). However, this should only be done if infection is already present and experience shows that an epidemic will otherwise inevitably occur.

Table 6.8 lists the therapeutic drugs commonly added to fish feed, together with recommended dose levels and treatment period for different conditions. It is important to follow the dosage recommendations and therapeutics should never be used at a lower level or over a longer period in order to improve growth rate and survival. There are instead various antimicrobial growth promoters

Table 6.8 Drugs commonly used for fish chemotherapy by addition to feed

Drug	Recommended daily dose of active ingredient in mg/kg body weight	Recommended treatment period in days	Main disease indication
Di-n-butyl tin oxide	250	5	Tapeworms
Furazolidone	110	5−10	*Hexamita* (*Octomitus*)
Nifurpirinol	0.8−3.5	5−10	Bacterial septicaemias
Oxolinic acid	10	10	Bacterial septicaemias
Oxytetracycline	75	5−10	Bacterial septicaemias
Sulphamerazine	110−220	10	Bacterial septicaemias
Trimethoprim/ Sulphadiazine	30	5−7	Bacterial septicaemias

(e.g. virginiamycin) which act only within the gut and are not used in animal or human medicine, hence may hold some promise as performance improvers in fish. Although the choice of treatment for intestinal parasites is limited, in treating bacterial septicaemias the infectious agent is often sensitive to a variety of different antimicrobials and the choice of drug is then dependent on the economics of treatment. The sulphonamides and nitrofurans are generally the cheapest groups, although both can pose palatability and resistance problems. The use of potentiated sulphonamides (e.g. trimethoprim combined with sulphadiazine) enables superior efficacy at a lowered dose for each ingredient. Of the antibiotics, oxytetracycline is probably the most widely used bacterial disease treatment, while resistant strains are frequently treated instead with synthetic antimicrobials, such as oxolinic acid.

Injection has a limited role in fish treatment due to the need for prior anaesthesia and the stress of handling. Economic considerations have usually restricted its use to valuable individual fish such as broodstock. However, the use of automatic multi-dose syringe techniques has enabled mass injection of carp to become widely adopted in Eastern Europe. Antibiotic injection is usually performed at the end of Winter on farms suffering from SVC virus to prevent secondary outbreaks of *Aeromonas* infection. In a similar fashion, mass injection of salmon smolts immediately prior to seawater transfer is occasionally undertaken in an attempt to remove furunculosis carriers. Prior sedation (chapter 4; section 4.3.1) is important to prevent fish struggling and losing scales and oxytetracycline is usually given by injection into the abdominal cavity in such cases.

6.8   RECOMMENDATIONS FOR FARMING PRACTICE

The majority of fish health problems are related to environmental stress, such as the effect of handling or poor water quality. Where an infectious disease is involved, this is the end-result of interaction between the pathogenic organism, a susceptible fish host and environmental stress. Good husbandry practice should aim to minimise the occurrence of fish diseases and to mitigate their effects. Table 6.9 lists the fish diseases covered in this chapter and some of the key points can be summarised as follows:

1. Certain fish pathogens may be primary agents of disease to susceptible fish and every effort should be made to restrict their spread by controlling movements of fish stock, including ova.
2. Fish farms should only purchase stock from hatcheries certified as being free of those primary fish pathogens of commercial significance and ova should be properly disinfected before arrival. All introduced stock should then be held in quarantine ponds until they have been retested and shown to be free of the infectious agents in question.
3. The water supply should be kept free of wild fish and other animals, such as fish-eating birds, which may harbour disease organisms or enable parasitic life cycles to be completed.
4. The hatchery should be sited independently or upstream of the main farm and isolated by means of disinfectant foot baths, etc. Different age classes of fish should be kept apart and fish farming equipment should be disinfected before use on different groups of fish.
5. All ponds should be thoroughly dried out and disinfected whenever they are emptied, or at least annually.

Table 6.9 Fish diseases covered in this chapter

Disease (common name)	Infectious agent (where appropriate)
Anchor worm	*Levnaea* spp.
Bacterial gill disease	*Myxobacteria* spp.
Bacterial kidney disease (BKD)	*Renibacterium salminarum*
Blue sac disease	
Botulism	*Clostridium botulinum*
Brown blood disease/ methaemoglobinaemia	
Channel catfish virus disease (CCVD)	(virus)
Coccidial enteritis	*Eimeria* spp.
Coccidiosis	*Eimeria* spp.
Coldwater disease (= peduncle disease)	*Myxobacteria* spp.
Columnaris	*Flexibacter columnaris*
Costiasis	*Ichthyobodo* spp.
Cripple body disease	*Pleistophora* spp.
Edwardsiellosis	*Edwarsiella* spp.
Enteric redmouth (ERM)	*Yersinia ruckeri*
Exocrine pancreas disease (EPD)	
Eye fluke	*Diplostomum spathaceum*
Fish louse	*Argulus* spp.
Fungus	*Saprolegnia* spp.
Furunculosis	*Aeromonas salmonicida*
Gas bubble disease	
Gill rot	*Branchiomyces* spp.
Haemorrhagic septicaemia	*Aeromonas* and *Pseudomonas* spp.
Hexamitiasis	*Hexamita* spp. (≡*Octomitus*)
Hitra disease	*Vibrio* spp.
Ich (= white spot)	*Ichthyophthirius* spp.
Infectious dropsy	(virus)
Infectious haematopoietic necrosis (IHN)	(virus)
Infectious pancreatic necrosis (IPN)	(virus)
Lamprey	(jawless fish)
Leech	(annelid−worm)
Nocardiosis	*Nocardia* spp.
Pasteurellosis	*Pasteurella piscicida*
Peduncle disease (= coldwater disease)	*Myxobacteria* spp.
Proliferative kidney disease (PKD)	(protozoan−myxosporidean)
Red pest	*Vibrio* spp.
Roundworm	(nematode)
Salmon louse	*Lepeophtheirus salmonis*
Saprolegniasis	*Saprolegnia* spp.
Spring viraemia of carp (SVC)	(virus)
Tapeworm	(cestode)
Tuberculosis	*Myxobacteria* spp.
Vibriosis	*Vibrio* spp.
Viral haemorrhagic septicaemia (VHS)	(virus)
Whirling disease	*Myxosoma cerebralis*
White spot (= Ich)	*Ichthyophthirius* spp.

6. Comprehensive records should be kept detailing water temperature, flow rate, dissolved oxygen level, fish weights, feed consumption, daily mortalities, etc. Details of any fish treatments should also be recorded and simple microscopic examination of skin and gills can be a valuable aid in routine health monitoring.

7. Water flow rates and stocking densities should be monitored closely to ensure that water quality is kept within the parameters needed for good performance and freedom from external diseases of skin and gills.

8. Fish handling should be kept to a minimum and undertaken only after prior starvation and at the time of day when water temperature is at its lowest. Grading equipment should be designed to avoid scale loss and the use of fish anaesthetics can help to reduce the stress of handling.

9. Fish feed should be scientifically formulated and the use of pelleted diets has a number of health advantages over waste fish feeding. Fish feed should be stored properly, kept dry and used soon after manufacture.

10. The emphasis should be on prevention rather than treatment of disease and fish vaccination is now playing an increasing role in this.

11. Whenever possible fish treatment should only be undertaken after a definite diagnosis has been established enabling the appropriate drug to be chosen. External infections are usually treated by means of chemicals added to the water and systemic infections are usually treated by antimicrobial drugs added to the feed.

12. Accurate calculations of the quantity of drug to be added to the water are essential and a preliminary trial treatment is prudent. In-feed medication must take account of variable appetite and appropriate withdrawal periods must be observed in order to avoid drug residues occurring in fish at slaughter.

7 Intensive marine farming in Japan

7.1 INTRODUCTION

Japanese consumption of fishery products, as far as the edible portion is concerned, represents about 35 kg per capita per year, two to four times higher than that eaten by Europeans and Americans and constituting about 25% of the Japanese intake of total protein. Hence, fishery products are a major protein source for the Japanese.

In 1984, the total supply of fish and shellfish was about 13.8 million tonnes. Of this, the domestic production was about 12.8 million metric tonnes, 54.3% of which came from offshore fisheries, 17.8% from distant harvests, 17.7% from coastal waters and 10.8% from inland fisheries and aquaculture.

Of the total supply, as much as about 8.8 million tonnes (i.e. 70%) comprises low value species, such as sardine or anchovy, pollock, mackerel, sand eel and others. This includes about 1.8 million tonnes which are used directly as feed for mariculture species and 1.9 million tonnes which are further processed for the production of fish meal. In all, the use of fishery products in fish feed represents about 27% of total finfish supplies. Thus, the effectiveness of aquaculture in Japan is very dependent on the continued supplies of large amounts of low valued finfish caught by the off-shore seine net fishery for sardine and mackerel. Catches of high market value species, such as tuna (0.36 million tonnes), yellowtail (38 thousand tonnes), etc., in contrast are relatively low and, because of this limited production and the insatiable Japanese demand for high value fish, about 1 million tonnes are imported each year from various countries to counterbalance supply and demand.

Japanese aquaculture, especially mariculture, has developed very rapidly during the last 10 to 15 years largely in response to the enactment of 200-mile exclusive fishery zones by more than 88

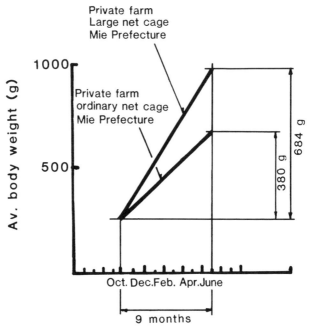

Private farm
Large net cage
Mie Prefecture

Private farm
ordinary net cage
Mie Prefecture

Fig. 7.2 Comparison of growth of yellowtail cultured in conventional and large-scale net cages.

the maintenance of broodstock and the development of artificial spawning techniques.

Rearing techniques for various kinds of fish and methods for the mass production of live foods for fish larvae have advanced markedly in recent years. At present more than 20 fish species are mass-produced in governmental and private hatcheries throughout the prefectures of Japan. These are produced for release into coastal waters and for commercial on-growing (Table 7.2). The number of species successfully cultured exceeds 60 if those which have been reared on an experimental scale are included. There are, however, some problems with hatchery-reared juveniles, primarily with regard to the incidence of deformities, such as short-tail, pughead, curvature of the skeleton, etc. Among these abnormalities, skeletal deformity in red seabream and abnormal coloration in Japanese flounder have recently been of significant commercial concern because they affect consumer acceptance of the product.

A further problem with some hatchery-reared and cultured fish is that, in general, they are poorer in quality than wild-caught fish

Table 7.2 Hatchery-produced seedlings of main finfish species (1985)

Common name	Scientific name	Number of seedlings (× 1000)		
		Release	Culture	Total
Red seabream	*Pagrus major*	21 388	24 866	46 410
Black seabream	*Acanthopagrus schlegeli*	11 033	836	11 869
Striped knifejaw*	*Oplegnathus fasciatus*	880	542	1 422
Flounder	*Paralichthys olivaceus*	7 949	5 310	13 300
Striped jack	*Caranx delicatissimus*	133	308	441
Yellowtail	*Seriola quinqueradiata*	1 253	0	1 253

*Data for 1984.

as far as taste (texture) and coloration are concerned, as described in chapter 5 (section 5.6). Captive marine fish are generally forced to eat diets in amounts far exceeding what is required for optimum growth, mainly to produce appropriately sized fish at the correct time of the year for the market. Cultured yellowtail, red seabream, coho salmon (*Oncorhynchus kisutch*) and ayu (*Plecoglossus altivelis*) in particular suffer from excess lipid accumulation in muscle and viscera.

7.3 GENERAL ASPECTS OF MASS PROPAGATION OF MARINE SPECIES

Mariculture operates successfully by producing market-size fish of high commercial value using juvenile fish to stock the farms, the main form of culture being the production of large numbers of fish fry for net-cage culture. Previously, most fish seed (eggs or fry) came from natural sources. However, these supplies are unable to meet the current needs of farmers and the continued expansion of productive aquaculture requires significant improvements to be made in the supplies of seed to stock the net cages. Improvement in mariculture must therefore take account of the maintenance of broodstock, spawn-taking, larval rearing and the mass culture of live foods, as well as the growth (and diseases) of the farmed stock.

7.3.1 Maintenance of broodstock

The broodstock fish used for egg collection are usually from three different sources: firstly, broodstock which have matured in the wild but are kept in captivity until spawning; secondly, wild juvenile fish reared to maturity in culture and, lastly, the culture of brood-stock from eggs. Fish farmers were almost completely dependent on wild broodstock until technical advances allowed spawn to be taken from cultured fish. The further development of such techniques is essential for management and mass propagation of fish seed. At present, seed has been artificially produced for more than 20 fish species using combinations of semi-natural spawning and/or hormone treatment.

Cultivation of broodstock generally takes place in floating net cages in the sea or in concrete tanks on land. Net-cage culture of broodstock has various advantages in terms of the cheaper costs of maintenance, the quality of the water and the ability to stock at higher densities, when compared with rearing in concrete tanks. Management of fish is, however, easier in the land-based tanks. The most popular size of net cage for broodstock is $5 \times 5 \times 5$ m^3 with stocking densities of $4-8$ kg/m^3 of water. Generally, the fish maintained in net cages are fed minced fish or formulated feeds. When the spawning season approaches, the broodstock are transferred to spawning tanks on land for egg collection and are only returned to the net cages after spawning.

7.3.2 Spawning (egg collection)

Maturing broodstock are transferred to spawning tanks on land from floating net cages in the sea about 3 to 6 weeks before the initiation of final maturation and spawning. Broodstock are usually stocked in the spawning tanks at densities of about 1 kg/m^3 of water with a sex ratio of 1:1. Fertilised eggs are obtained by natural spawning or semi-naturally by stripping broodfish after hormone injection. However, fertilised eggs obtained by natural spawning are of superior quality to those induced by hormone injection and stripping. In the case of semi-natural spawning, acetone-dried pituitary glands of grass carp (*Hypophthalmichthys molitrix*) are frequently utilised to induce gonad maturation in a number of cultured species (chapter 4; section 4.3.1). The fertilised eggs which float on the surface of the water are collected in fine-meshed nets stretched across the overflow water from the spawning tanks. These eggs are then washed thoroughly with water and transferred to indoor incubation tanks or larval rearing tanks whose capacities range from 6 to 100 m³. The rate of water supply, aeration and light intensity in the tanks are carefully controlled during egg incubation, although the levels vary with fish species. The number of eggs spawned and those present in the incubation tank are usually estimated by the weight or volume method. In the weight method, the number of eggs contained in 1 g of eggs is counted and the total obtained by multiplying the number by the total weight of eggs collected. In the volume method, the volume of a small number of eggs is measured and the total number of eggs calculated from the total egg volume. The total number of eggs can also be estimated from counts of the number of eggs in an aliquot of water by knowing the total volume of water in the tank. The estimated number of eggs per gram of eggs is 1200 for yellowtail, 1300 for redeye mullet, 2000 for red seabream, 2800 for striped knife-jaw, and 4500 for Japanese flounder.

Cultivated species of seabream and striped knife-jaw are reported to produce better quality eggs in terms of eyeing rates and hatchability than those from natural stocks, whereas eggs from cultured yellowtail, Japanese flounder and seabass are similar to those of wild broodstock. The spawning seasons of the various species are different (Table 7.3), partly because water temperature differentially affects the rates of gonad maturation. It is also considered that light intensity, oxygen content and salinity are important environmental determinants of the timing of maturation.

Table 7.3 Spawning seasons and optimum water temperatures of the main maricultured species

Fish	Species	Season	Optimum temperature in °C
Striped jack	*Caranx delicatissimus*	Dec−Jan	15−19
Yellowtail	*Seriola quinqueradiata*	Mar−Jun	18−20
Red seabream	*Pagrus major*	Mar−Jun	17−21
Black seabream	*Acanthopagrus schlegeli*	Mar−Jun	17−21
Japanese flounder	*Paralichthys olivaceus*	Feb−May	14−16
Tiger puffer	*Takifugu rubripes*	Apr−May	17−20

7.3.3 Larval rearing

Newly-hatched larvae are either kept in the same egg incubation tanks or they are transferred to larval rearing tanks, indoor or outdoor, each with a capacity of 1–200 m^3 at stocking densities of 10 000 to 50 000 larvae per m^3, depending on the species and tank volume. Tanks of many different designs and fabrication are used, including square and angled, concrete or fibre-glass. In general, these are similar to those used for the larval rearing and on-growing of salmonids (see chapter 3).

Larval rearing tanks are equipped with water and air supply systems. Sea water is supplied to individual tanks from a reservoir, which receives pumped filtered sea water. Water is exchanged as needed during the first two to three weeks, the rate being gradually increased with the growth of the larvae. For both outdoor and indoor tanks, light intensity is usually adjusted to about 5000 lux at the water surface by covering the roof or water surface with dark curtains. Marine *Chlorella* or so-called green water are commonly added to the tanks to stabilise the water quality and food supply for the rotifers, the concentration of *Chlorella* being about 300–400 cells/ml of water. Optimum water temperature and salinity for normal growth of larvae vary with species but greatly affect the success of larval rearing.

The management of water quality in the larval tanks is also very important (Plate 7.3). Tank bottoms must be cleaned by removing unconsumed feed, dead fish and debris, usually with the aid of a siphon or automatic vacuum cleaner. This cleaning is especially important during and after the initiation of feeding with artificial microdiets. Debris should also be removed from the surface to allow for air-bladder inflation of larvae; this in turn enables normal development of the skeleton. Recent studies have demonstrated that lordosis (i.e. spine curvature) occurs in fish with deflated swim-bladders due to their failure to gulp air at the water surface when they reach a body length of 4–4.5 mm (i.e. approximately 10–15 days after hatching).

Once fish larvae have grown to 10–20 mm in length, they are generally transferred to floating net cages in the sea in order to prevent deterioration in water quality, due to the intensive feeding with trash or moist pellets. The net cages measure 2 × 2 × 2, 3 × 3 × 3 or 5 × 5 × 5 m^3, depending upon fish species, fish size and location. The mesh size of the nets and stocking densities

Plate 7.3 Larval rearing tanks with computer monitoring system.

of the fish also vary with species, although normally the nets are 2–6 mm in gauge and the fish stocked at 1000–4000 per m³ of water.

7.3.4 Feeding schedule

As described in chapter 5 (section 5.3.1), almost all larvae of marine species are initially fed on live foods, such as the rotifer *Brachionus plicatilis* (Plate 4.7) and marine copepods. The live food chosen depends on the size of the larvae and rotifers are mass cultured in indoor tanks of 40 m³ (Plate 7.4). Hatched larvae with body lengths greater than 2–3 mm are given rotifers as the initial diet and this is continued for about 30 days after hatching. When fish reach 7 mm or more in body length, marine copepods such as *Tigriopus*, *Acartia*, *Oithona* and *Paracalanus* (or, in their absence, *Moina* and *Daphnia* of freshwater origin) are fed to larvae, together with rotifers, which on their own are too small for larvae of 7 mm. The brine shrimp, *Artemia* (Plate 4.8), is frequently used as food for the larvae of many marine fish, especially when there is a shortage of marine copepods, the advantage being that it is commercially available (chapter 4; section 4.3.2). Larvae larger than 10–11 mm are fed on minced fish, shellfish or shrimps, or on

Plate 7.4 Indoor tanks (40 m^3) for mass culture of rotifers.

artificial diets. The feeding schedule and organism used vary according to the species of fish under cultivation, the type of hatchery and its geographical location. Recently, artificial microdiets have started to be used, partially in place of live foods. When juveniles attain 30−50 mm in length, larval production is considered to be finished.

7.3.5 Growth and production

The growth of some cultured species during the larval and juvenile stages is shown in Figure 7.3. The growth rate of fish in rearing tanks or floating net cages is modified by a number of factors including water temperature, fish density, light intensity, feeding rate, type of feed, etc. Of all these factors, water temperature and the supply of live food are the most important. Water temperature has a close relation to larval growth which is generally faster at higher temperatures, indicating that larvae spawned in the summer grow faster than those spawned in the winter. The nutritional quality of the live food (especially the concentrations of ω3 HUFA) also greatly affects the growth and survival of larval fish.

The productivity of larvae or juveniles is, of course, dependent on their survival rate during rearing in tanks and in net cages. Survival rate is also influenced by many of the factors which affect

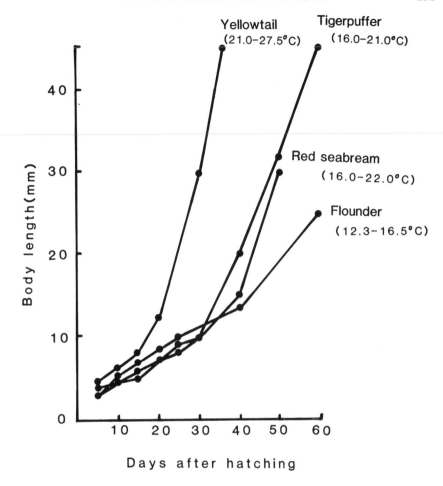

Fig. 7.3 Growth of larvae of main marine species under optimal water temperatures.

larval growth. However, survival is generally higher for fish reared in small tanks than in large ones, and lower for fish kept in tanks than in floating net cages. Juvenile fish of 30−50 mm in total body length are released into coastal waters after intermediary nursing and for cultivation to commercial size in net cages. In the case of red seabream destined for release, fish of 25−30 mm in total length are transferred from hatcheries to floating net cages moored near the releasing point and then grown-on for about 50 days to a size of more than 70 mm before release. The growth of released fish is almost comparable to that of wild fish.

7.4 MARICULTURE OF MAIN SPECIES

Recent technical developments relating to egg-taking, larval rearing and methods for mass culture of live foods have markedly increased the number of fish species in commercial production. At present more than 20 species are mass produced in governmental centres and private hatcheries for release into coastal waters and cultivation to commercial size. Of these, the seabream, yellowtail and Japanese flounder are of major importance.

7.4.1 Red seabream (*Pagrus major*)

Red seabream (Plate 7.5) is arguably the most valuable and popular of all farmed species of fish in Japan. It is cultured in both northern and southern waters. The fish are regarded as a symbol of good fortune and therefore traditionally used for celebratory feasts, such as weddings. The fish grow to nearly 1 m in length and may live for about 20 years, although commercial production is of 20–40 cm fish weighing 1–4 kg and 3–7 years of age. In all, about 50 million red seabream juveniles were produced in 1986 by the 23 prefectural and national fish farming centres. Of these, 27 million juveniles with body lengths of 12–154 mm were used for restocking into coastal waters and 23 million of 10–306 mm size used for on-growing by commercial farms.

The total production of market-size red seabream was more than

Plate 7.5 Red seabream.

60 000 tonnes in 1984, with approximately 50% of the fish being derived from artificial cultivation. Even with this scale of production, commercial demand for the fish still remains very high.

(a) Maintenance of broodstock

Broodstock are usually kept in floating net cages in the sea because of lower costs. They are fed minced fish meat or artificial moist or dry diets. The broodstock are generally transported to the spawning tanks on land from the floating net cages about 4−6 weeks before spawning. A tank of approximately $50-100 \text{ m}^3$ capacity is frequently used for natural spawning. Equal numbers of male and female fish, 3−7 years old, are maintained at densities of 0.7−1.5 individuals per m^3. During the spawning season, high densities of stock and unequal sex ratios decrease successful spawning. Small fresh fish, such as anchovy, sand lance and mackerel, are preferred as feedstuffs. The spawning period extends for 50−70 days during the period March to June. Recently, it has been demonstrated that the spawning period can be extended for more than 6 months by the control of water temperature and by feeding high quality diets such as frozen krill. The onset of the spawning season is earlier in the Pacific region than in the Seto Inland Sea and in the Sea of Japan. In some districts, warm waste water from atomic and steam power stations is used to bring forward the time of spawning.

(b) Egg collection

Fertilised eggs can be obtained from several different sources, as described above. However, as the techniques for spawning in culture have been largely developed for red seabream, viable eggs are now collected from broodstock maintained in tanks.

Although the exact methods used in different locations show some variation, generally the pelagic eggs of red seabream are collected in fine-meshed nets draped across the overflow water from the spawning tanks; the eggs are then transferred to incubation nets and kept in a flow-through system until they hatch. Alternatively, they are introduced directly into a rearing tank at a density of $30-90 000$ eggs per m^3. Eggs spawned during the initial stages of the spawning season are, in general, poorer in quality (as far as hatching rate is concerned) than those obtained during the middle of the spawning season. Between 100 and 400 $\times$ 10^4 eggs can be collected from a single female broodfish of 3$-$7 years of age in one spawning season. When spawning, tanks of 50$-$100 m^3 capacity holding between 40 and 80 fish (equal numbers of male and female) are used, several million eggs can be obtained every day for a month or more.

(c)   Larval rearing

The tanks used for larval rearing generally have a capacity of 10$-$20 tonnes and are aerated. Hatched larvae are maintained at a density of 5$-$50000 per m^3. An optimum stocking density for experimental rearing is recommended to be 4$-$5000 larvae per 500-litre tank. Outdoor tanks are covered with shade nets to reduce the light intensity at the water surface to a maximum of about 5000 lux.

The most suitable diet for hatched larvae is rotifers and these should be maintained at a density of at least 5 individuals per ml in the rearing tanks to avoid larval starvation. The amount of live feed required to supply a rearing tank is estimated to be 1.3 to 1.4 times more than that consumed by the larvae, since the live foods are also carried out of the tank in the outflow water.

(d)   Net-cage culture

Red seabream require similar net-cage conditions to most other maricultured species such as yellowtail. Floating net cages are generally located in small bays or inlets in order to prevent damage from winds and waves. However, as a result of the slow currents, these areas are often poor in water exchange. This frequently results in a deterioration of water quality due to the intensive feeding with trash fish. Water currents of 5$-$15 cm/s in the Spring tide are said to be most suitable for small net cages. However, recently fish culture has been extended into more open sea areas, due to technical advances, in order to prevent the environmental deterioration of farming grounds.

Using current methods it takes about two years (i.e. two winters) to grow red seabream up to a market size of 1 kg. Seawater temperatures should not fall below 10°C or increase beyond 29°C, 17–18°C being the most suitable for rapid growth.

Wild-caught juvenile red seabream of 2–16 cm in length or those reared to 50–60 cm after catching, together with hatchery-reared juveniles, are used as seed for cultivation. Wild-caught fish of one or two years of age weighing 100–120 or 500–600 g respectively are also used to stock net cages. However, in recent years almost all of the seed for culture has come from hatchery-reared stocks.

The size of the net cage ranges from $4 \times 4 \times 3$ to $20 \times 20 \times 5$ m^3, depending on fish size and the cage location. The stocking densities of fish in net cages also vary with fish size (Table 7.4) and water temperature. Low stocking density results in an increase of yield per net cage. A suggested stocking density for 150 g fish is 50–70 fish/m^3 (i.e. 7–10 kg/m^3). In general, a standard density for one-year-old fish is 100 fish/m^3 and for older fish 6–8 kg/m^3.

Generally, most red seabream farms grow fish up to a market size of about 1 kg by feeding one-year-old fish over a 3-year period, the rate depending on the initial size of the fish and time of initial feeding (Fig. 7.4). Growth of fish is not expected when the water temperature drops to less than 12°C.

Locally-available trash fish such as sardine, anchovy, sand-lance and mackerel are frequently used as feed. Minced, chopped or whole trash fish are, in general, mixed with formulated mash diets or fortified with a vitamin mixture and given to fish manually or by automatic feeder. Recently, moist diets prepared by mixing minced fish either with a formulated dry powder in a ratio of 2:3 or 1:1, or with commercially available dry pellets, have become more widely used. The feeding rate depends on fish size, fish density, water temperature, type of feed, size of net cage and feeding methods, etc.

Table 7.4 Stocking density of red seabream in floating net cages

Size of net in m × m × m	Volume in m^3	Fish size in g	Stocking density	
			no. of fish/m^3	kg/m^3
5 × 5 × 5	125	90	40	3.6
6 × 6 × 5	180	130	53	7.4
7 × 7 × 7	340	130	30	3.9
8 × 8 × 7	450	130	22	2.9
10 × 10 × 5	500	100	40	4.0

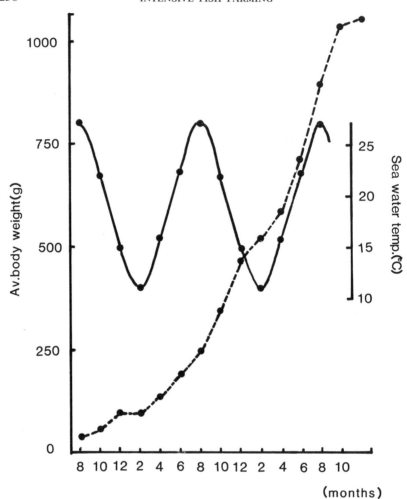

Fig. 7.4 Growth of juvenile red seabream to marketable size by feeding with moist pellets over a two-year period.

(Table 7.5). The survival rate of stock during culture to commercial size, at over 90%, is generally higher than that found with yellowtail.

The pigmentation or colour tone in the skin of cultured red seabream is usually far inferior to that of wild fish, leading to a much lower market price. Several attempts have been made to enhance the pigmentation by feeding with krill, *Mysis*, shrimp and crab waste, and the green alga, *Haematococcus* spp. In 1986 the market price for cultured red seabream was about ¥2000/kg

Table 7.5 Growth and feed conversion efficiency of red seabream fed with trash fish in floating net cages.

Size of net in m × m × m	No. of fish	Feeding period in days	Av. body wt		Daily feeding rate in %	Feed conversion rate
			Initial in g	Final in g		
3.6 × 3.6 × 3.6	1 800	130	33	112	4.3	19.4
8 × 8 × 6	10 000	130	25	132	11.9	9.6
5 × 5 × 5	4 300	150	10	134	15.8	8.0
2 × 2 × 2.5	82	226	68	217	10.7	4.3
5 × 5 × 5	5 000	110	6.7	78	21.6	7.1

($15.68 or £8.50/kg), much cheaper than the ¥4500/kg ($35.15 or £19.00/kg) paid for wild fish.

7.4.2   Yellowtail (*Seriola quinqueradiata*)

Yellowtail (Plate 7.6) is also one of the most popular Japanese fish species and its production is the highest of all the cultured finfish species in Japan. In 1984, total production of yellowtail was about 192 000 tonnes, including both cultured fish (152 000 tonnes) and those from natural waters (40 000 tonnes).

The yellowtail is a migratory fish distributed from Hokkaido to Kyushu and growing up to 2.0 m in length by 8 years of age. Its body size determines the name used, with mojako being less than 10 cm, wakanago 10−20 cm, hamachi 20−30 cm, warasa 30−65 cm and buri more than 65 cm. Commonly, yellowtail culture refers to the cultivation of juveniles up to the hamachi size of fish, although recently the commercial size has gradually increased to include those of buri size. The juveniles used for net-cage culture mostly come from natural supplies and in 1983 their number amounted to 75 million. Recently, artificial spawning and larval rearing have been successfully accomplished and consequently the total number of hatchery-reared larvae has increased from 1.2 million in 1984 to 2.4 million in 1986. These larvae, which range from 17 to 32 mm in

Plate 7.6 Yellowtail.

length, have all been produced by three National Fish Farming Centres for release into coastal waters.

(a) Broodstock and egg collection

Eggs are usually taken from mature broodstock caught by fishing, or from fish reared in net cages after being caught as juveniles. It takes at least 3 years to obtain a well-matured broodstock fish of 65–70 cm and weighing more than 5 kg. Broodstock of 4–5 kg produce 0.5 million eggs, and those of 6–8 kg 1 million eggs. The spawning season starts in February and continues until June, depending upon ambient water temperature. The eggs are collected from mature wild or cultured broodstock by stripping. However, spawning is usually induced by injection of the hormone Synahorin into the dorsal muscle (4 IU/kg of fish). The eggs undergo final maturation within 2 to 3 days of the hormone injection and are then suitable for stripping. The dry method is used for insemination. After fertilisation the batches of eggs are separated into viable floating eggs, which will subsequently develop normally, and those which sink, comprising mainly unfertilised and dead eggs. The floating eggs are transferred to incubation or larval rearing tanks with facilities to control water temperature and aeration. It takes about 70 hours to hatch under the optimum temperature of 18–21°C. The newly-hatched larvae are about 3.5 mm in total length.

(b) Larval rearing

Initial larval feeding starts with rotifers, 3 to 4 days after hatching, which are given to larvae for about 20 days. When fish reach 8–8.5 mm in standard length, *Artemia* nauplii, *Tigriopus* and marine copepods are fed to the larvae, together with rotifers. Larvae larger than 12 mm are fed on *Daphnia* and adult *Artemia* or sometimes even fish eggs. As a result, the larvae grow up to about 23 mm in length during May and survival rates are approximately 10% when stocked into 100 m^3 tanks at a water temperature of 22°C.

Rotifers and *Artemia* nauplii are usually enriched with emulsified lipids rich in ω3 HUFA, as feeding these live foods without enrichment frequently results in high mortalities. At present, mass-produced yellowtail juveniles are all used for release and not for further cultivation. All the seed for net-cage culture is collected from natural waters.

(c) Net-cage culture

The season for yellowtail juvenile collection is, in general, from April to June along the coastal waters, small inlets and bays of southern Japan, except for the Seto Inland Sea. The total catch is higher along the Pacific coast than in the Sea of Japan. The juveniles are found below or around drifting seaweeds and their lengths from 8 to 180 mm, depending on season and location, although juveniles of 30–70 mm in length are the commercial size. A licence issued by the Fisheries Agency is required for catching yellowtail seed. The seed caught by net are roughly separated by size using different mesh nets to prevent subsequent cannibalism. They are then kept in tanks during transportation to the cage farms.

The larvae collected are kept in small floating net cages with stocking densities of 80–200 fish/m^3. They are reared to a fingerling or seedling size of 50–100 mm in length by feeding minced fresh fish, before being grown-on to a market size of 20–30 cm in length in various types of floating net cages. The stocking densities used vary with fish size, being 115–340 fish/m³ for fish of 200–450 g in weight, 10–15 fish/m³ for those of 450–600 g, about 10 fish/m³ for those of 600–1000 g, and less than 5 fish/m³ for those above 1 kg.

Locally available trash fish, such as sand-lance, anchovy and mackerel (minced, chopped or whole depending on fish size), are mainly fed to yellowtail, although this feed is gradually being replaced by moist diets, prepared by mixing fresh fish with a dry formulated powder. The feeding activity is good at water temperatures of 18–27°C, being highest at 24–26°C, and poor at less than 13°C or more than 28°C. The specific gravity of the seawater, together with the amount of dissolved oxygen, also affects the feeding activity of fish. Fish do not accept feed when the specific gravity falls below 1.015 and they begin to die when it drops below 1.006 at 15°C. Marine parasites can be easily removed by temporarily holding yellowtail in freshwater, because they can survive in a freshwater environment for 10 minutes without adverse effects. Fish also show abnormal behaviour at oxygen levels less than 4 mg/litre.

The feeding rate is, in general, about 4–8% of body weight, depending on fish size and water temperature; the frequency of feeding is three to four times per day for fish weighing less than 100 g (Table 7.6). The feed conversion ratio (g feed/g weight gain) is about 5–9 when juvenile yellowtail are cultured from May–June to October–December, and 11–12 when fish weighing 1–2 kg are reared to 5–6 kg from March to January. This is based on the use of wet fish diets and would, of course, be much lower if dry pelleted diets were used.

Table 7.6 Feeding rate of yellowtail in relation to fish size at different water temperatures using raw fresh fish

Average body weight in g	Water temperature in °C	Daily feeding rate in % body weight
10–50	15–22	40–60
50–100	18–24	30–50
100–200	20–25	20–30
200–400	23–36	15–20
400–600	25–28	11–15
600–800	24–28	9–11
800–1000	18–25	8–10
1000–1200	13–20	6–8

Growth of yellowtail is generally much faster than other species. Juvenile yellowtail of 1−60 g in body weight in May−June grow up to 200−700 g (average 550 g) by the end of August, 600−1000 g (average 1.1 kg) by the end of October, and 700−2000 g (average 1.4 kg) by the end of December. They can reach 1.5−4 kg by October and 2.5−6 kg by December (Fig. 7.5). The relationship between fish size and feeding rate of different water temperatures is shown in Table 7.6.

Recently, there has been a trend for the net cages used in yellowtail culture to increase in size and the stocking densities in the cages to be reduced; these changes have resulted in a reduction in the occurrence of disease and an improvement in the health and quality of the product (Table 7.1). Fish culture grounds have also been gradually expanded to more open sea areas by using net cages of area 1600−2400 m^2. The growth rate of yellowtail has also been shown to be better in large cages (Fig. 7.2). In 1986, the market price of cultured yellowtail was about ¥1100/kg ($8.69 or £4.70/kg).

7.4.3 Japanese flounder (*Paralichthys olivaceus*)

Recently the culture of flatfish has shown a rapid increase as a result of improvements in the techniques for the mass propagation of juvenile fish. Amongst the various species of flatfish under cultivation, the Japanese flounder or bastard halibut (Plate 7.7), the 'hirame' or left-eyed flatfish and the 'karei' or right-eyed flatfish (*Limanda* spp.), are the most important. The hirame has one of the highest commercial values, being almost equivalent to that of red seabream. The total production of cultured flounder in 1984 was about 1000 tonnes, much less than the natural production of 264 000 tonnes, which also includes halibut and sole. On the other hand, the total number of juvenile flounder (10−178 mm in length) mass produced by government fish farming centres and private hatcheries in 1985 was about 13.3 million, 60% being used for further cultivation up to commercial size. The price of seedlings ranges from ¥50 ($0.40 or £0.22) each to ¥250 ($2.03 or £1.10), depending on their size.

(a) Broodstock and egg collection

Broodstock used for egg collection are usually caught in natural

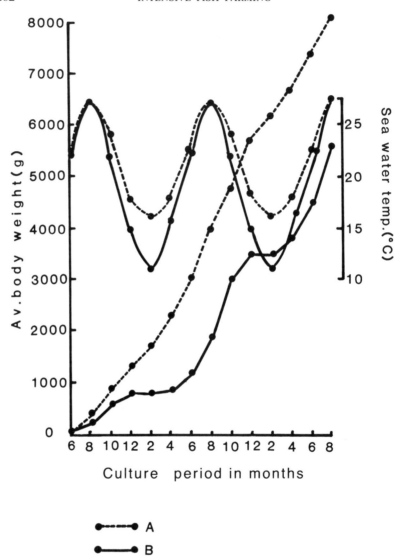

Fig. 7.5 Growth curves of juvenile yellowtail (mojako) to full adult size (buri) during about two years, feeding with moist pellets (A) and trash fish (B).

waters during the spawning season from late March to mid-April. Alternatively, fish caught during fishing may be kept in broodfish tanks for about one year before spawn-taking. The broodfish used are usually 25–30 cm in length. The spawning season varies by locality, ranging from March to April from the Wakayama Prefecture to the Kyushu District, to late April through to mid-May

Plate 7.7 Japanese flounder.

around the northern mainland of Japan. Male and female brood-stock weighing more than 1 kg are stocked in indoor circular tanks, $10-1000$ m^3 in capacity, at a sex ratio of about 2 males:1 female and fed on sand-lance, sardine, mackerel, etc. At water temperatures below 10°C their feeding activity is markedly reduced. The water in the tanks is usually changed between six and eight times a day. The light intensity is reduced by covering the tanks with a dark polyethylene net.

Spawning starts when the water temperature rises above 11°C, with the optimum occurring at $14-15$°C, and ends when the temperature rises above 16°C. From 100 broodstock fish, with a sex ratio of about 1:1 (male:female), about 10 million eggs can be obtained. Eggs are naturally spawned by the fish in the morning and during dusk, and usually no hormone injection is required. Eggs which have been fertilised are pelagic and float, hence they can easily be collected by passing the seawater from the spawning tank through a net bag placed in an additional tank. The collected eggs include a variety of stages of development, from two-cell embryos to morula stage, since they are floating in the broodstock tank for more than half a day before collection. The number of eggs is estimated by weighing a known number beforehand; usually there are about $1800-1900$ eggs/g. Spawning activity usually continues for about 3 months, although it is at its highest for about 10 days in early April.

The eggs which float on the water surface are of high quality and are separated from those of poor quality which accumulate at the bottom of the tank. The poor quality eggs sediment within

10 minutes of being added to the tank, whereas the floating eggs are transferred to indoor hatching tanks, in which small net bags, 30–90 cm in diameter and 40–90 cm deep, are mounted on a wooden frame placed over the tanks. The tank water is aerated (10–20 l/min) 24 hours a day and the water temperature adjusted to 16–20°C. Eggs usually hatch within 48 hours after insemination. The hatching rate at 75–80% is usually higher in the eggs spawned during the first month of spawning and decreases to 30–40% later in the season.

(b)   Larval rearing

The hatched larvae, which are 1.6–2.4 mm in length, are moved to various sizes of larval rearing tanks all equipped with aeration facilities. The water temperature is adjusted to 17–20°C. The larvae are fed on rotifers for the initial feed, followed by *Artemia* nauplii when they reach 5 mm in length. The number of rotifers given to larvae ranges from 10^7 to 12×10^7 individuals per day, being adjusted to as high a level as their feeding activity will allow. *Artemia* nauplii, together with rotifers, are fed to larvae from around the 10–12th days, at an initial level of 10^6 nauplii per 10^5 larvae, gradually increasing to a level of 40×10^6. Both the rotifers and *Artemia* are usually enriched with ω3 HUFA before they are fed to the fish larvae. When the larvae reach about 14–16 mm in length, they settle at the bottom of the tank, at which stage they are fed minced fish, together with *Artemia* nauplii. This is continued until they reach about 30 mm in length. The survival rate of the fish during the period of growth from 15 to 35 mm is approximately 60–70%. Hatchery-reared juveniles are generally the source of seedlings for further cultivation by farms to marketable size and for release into coastal waters.

(c)   Cultivation

Various types of tank are used for cultivation of hirame, including indoor or outdoor, concrete or canvas, and round or square tanks. Floating net cages of various size (5–10 m square) are also used. All tanks (whether indoor or outdoor) or net cages are covered with a dark-coloured net or blackened board to reduce light intensity.

The light intensity at the water surface is one of the important factors affecting the success of culture and represents part of the farmer's know-how.

Water temperature and stocking density vary with the different forms of tank and floating net-cage culture. In the case of canvas tanks, 8 m in diameter and 60–70 cm deep, the water temperature is 16–18°C in Spring, 26–27°C in Summer, 16–18°C in Fall and 14–16°C in Winter. The number of fish stocked in these tanks comprises either 1500 fish 4–15 cm long or 900 fish 15–50 cm long. The net cages located in the bays of the Nagasaki Prefecture are 5 m square with a wooden frame, having a water temperature of 17°C in Spring, 29°C in Summer, 20°C in Fall and 12°C in Winter. About 500 fish, 30 cm long, are stocked in each cage. The water is changed 7 to 8 times per day or sometimes 15 to 16 times in the case of circular canvas tanks. The tanks are customarily cleaned at least once every tenth day and the tank walls scrubbed with a sterile cloth. During this operation the fish are also treated with a nitrofuran drug.

In the case of $6 \times 6 \times 1$ m³ outdoor concrete tanks, with a water depth of 80 cm, about 800 fish 45 cm long (600–800 g) can be harvested in January after stocking with juveniles 4 cm long the previous April. In the canvas tanks (8×1 m², 70–80 cm water depth), juveniles of 4–15 cm in body length, stocked in mid-April, grow up to 35 cm (600 g) by December.

The main food for juvenile hirame is sand eel (minced, chopped or whole according to fish size) given twice a day, in the morning and late afternoon; more recently formulated feeds have become available. Cultured hirame weighing 600–800 g are marketed from late October. They are transported to market as live fish, packed in polythene bags containing cold seawater which is gassed with oxygen. In 1986 the average market price of cultured flounder was about ¥5300/kg ($40.7 or £22.00/kg), this being the second highest value product amongst all of the Japanese cultured finfish, almost equivalent to cultured puffer (*Takifugu rubripes*).

7.5 EXPANSION INTO OTHER SPECIES

The main species of finfish in Japanese mariculture are currently yellowtail, red seabream and Japanese flounder. However, yellowtail culture seems to have reached a plateau in its production at about 152 000 tonnes in 1979. A new candidate for mariculture with a high commercial value is the striped jack (*Longirostrum/Caranx delicatissimus*), which has a very attractive taste. Cultivation of this species has a recent history and in 1984 the total production of cultured striped jack was only about 500 tonnes. This was mainly due to low availability of seedlings for cultivation; only about 500 000 juveniles are estimated to be available from the wild each year. Recently, natural or semi-natural spawning has been replaced by the spawning of captive broodstock at various government and private hatcheries. The total number of hatchery-reared seedlings for cultivation was about 308 000 fish in 1985 and has gradually increased each year. At present the price of juveniles of 10 cm in length is as high as about ¥1000 ($7.84 or £4.23) each. The juveniles are stocked in net cages and fed on trash fish such as sand-lance, anchovy, mackerel, etc. They grow up to a market size of 1.0−1.2 kg (38−42 cm in length) in two years and 2.0−2.5 kg (50−54 cm) after three years. The price of market-size striped jack is ¥3000−4000 ($23.51−31.35 or £12.71− 16.95)/kg for live fish and ¥2300−3300 ($18.03−25.87 or £9.75− 13.98)/kg for dead fish. The production of seedlings and commercial size fish is expected to increase rapidly in the near future. However, further expansion in the cultivation of this species will be dependent on the production of high quality fish which are comparable in terms of taste and texture to wild fish. Striped jack is similar to yellowtail in that lipid accumulates in the body in captivity, which reduces the product quality. This problem will only be remedied by replacing the trash fish diets with formulated feeds which have a well-balanced ratio of protein to energy.

The second candidate species for mariculture is the tiger puffer (*Takifugu rubripes*). This also has a very high market price of ¥4000−7000 ($31.36−54.87 or £16.95−29.66)/kg for cultured fish and above ¥10 000 ($78.39 or £42.37)/kg for wild fish, depending on the season. The total number of hatchery-produced juvenile puffer in 1985 was about 3.3 million. Of these about 58% of the seedlings were used for cultivation to commercial size of around 1 kg, yielding a production of about 500 tonnes of puffer. The amount of puffer caught from natural waters was only about 2000 tonnes, which includes a related species, thus the demand is still very high. However, high mortalities have been encountered in mariculture as a result of a virus disease which usually appears in summer when the water temperature rises to 24−25°C. Avoidance of this disease is an essential prerequisite to the further development of puffer culture.

Japanese grouper (*Epinephelus akaara*) and seabass (*Lateolabrax japonicus*) are also future candidates for mariculture, although their mass propagation has not yet succeeded in Japan.

8 Fish culture in the United States*

8.1 INTRODUCTION

Fish culture in the United States can be traced back to the mid to late 19th century, when pioneers such as Seth Green and Livingston Stone began to supply brood fish, fingerlings and lessons in fish husbandry to aspiring aquaculturists. The new Americans wanted not only native fish, but also brown trout (*Salmo trutta*) and common carp (*Cyprinus carpio*) introduced from Europe. Primitive hatcheries were established to produce the larvae of marine and estuarine species, including striped bass (*Morone saxatilis*), as early as 1880. Larval marine fish were produced and released for several years to augment natural reproduction, but these releases produced no appreciable change in the sport or commercial catch. By 1900 fish culture was well established in both State and Federal hatcheries. Before World War II most cultured fish were coolwater or coldwater species such as trout. Private culture of warmwater species gained momentum in the 1940s as the demand for bait minnows grew. Until the early 1960s commercial fish culture in the United States was principally restricted to rainbow trout (*Salmo gairdneri*), baitfish, and a few warmwater species. Most warmwater fish reared in the United States were stocked in public waters by State and Federal agencies or were used as bait to catch warmwater fish.

The southeastern United States and southern California now support important warmwater aquaculture facilities (Fig. 8.1). The tropical fish industry is restricted primarily to Florida and southern California. Warmwater species constitute about 70% of the fish reared in the United States today (Table 8.1); the most important

---

*This chapter was prepared as part of the official duties of a US Government employee and therefore cannot be copyrighted.

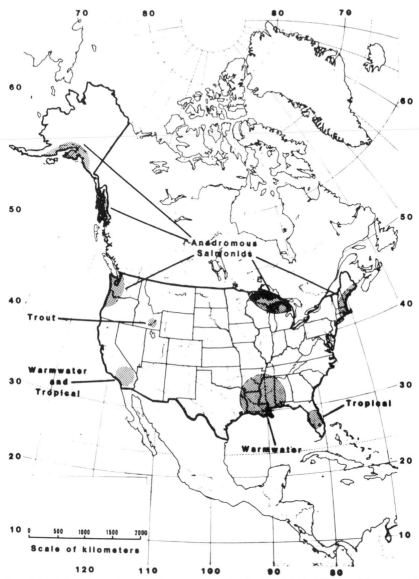

Fig. 8.1 Major sites of aquaculture production in the United States. Marine and estuarine species are grown along all coasts and in Hawaii (not shown), and baitfish, trout, and warmwater species are cultured to some degree in nearly every state.

commercially cultured species is the channel catfish (*Ictalurus punctatus*).

Fish and shellfish production by commercial growers in the United States was estimated to be over 243 500 metric tons in 1985

Table 8.1 Production (metric tons) of fish and shellfish by the United States Fish and Wildlife Service in fiscal year 1981, by State hatcheries in 1980, and by private aquaculturists in 1982

Category	USWF*	States[†]	Private**	Total[‡]
Warmwater	122	545	133 346	134 013
Trout	1638	6314	21 818	29 770
Anadromous salmonids	1029	1243	11 587	13 859
Marine species	0	N/A[§]	12 696	12 696
Other anadromous species	9	27	N/A	36
Total	2798	8129	179 447	190 374

*Source: US Department of Interior, 1981.
[†]Source: Joint Subcommittee on Aquaculture, 1983.
**Based on the 1980 data base of the Sport Fishing Institute and on analysis of that data base by S. J. Cameron, US Fish and Wildlife Service, personal communication.
[‡]Represents best available estimates of fish production in the United States; Federal and State production has not increased significantly since 1980−1.
[§]No data available, but amounts probably insignificant.

(Fig. 8.2). Private aquaculturists produced 94% of the total, and State hatcheries about 4%; US Fish and Wildlife Service hatcheries produced less than 2%, but reared 52 species of finfish in 1981 and expected to rear a similar number in 1985. The total value of privately produced fish was estimated to be $423 (£228) million in 1985.

Sport fishing is an even greater industry; 42.1 million fishermen spent over $17.3 (£9.35) billion on fishing and fishing-related equipment, transportation, food and lodging, and licences and fees in 1980. In addition to the fish produced by aquaculturists and those caught from the wild, the value of edible and non-edible fishery products imported into the United States was $4.5 (£2.5) billion in 1982, $5.1 (£2.76) billion in 1983, $5.9 (£3.19) billion in 1984 and $6.7 (£3.62) billion in 1985. This trade deficit indicates a tremendous potential in the United States for additional aquacultural production to satisfy both the food market and the demands of anglers.

Major crops now include channel catfish, crustaceans (crayfish) and baitfish. Production systems for freshwater prawn (*Macrobrachium rosenbergii*), largemouth bass (*Micropterus salmoides*), penaeid shrimp (*Penaeus* sp.), and striped bass are now being developed. Aquaculture development plans have been completed by the Federal Government's Joint Subcommittee on Aquaculture for 12 groups of species now cultured in the United States; these include five

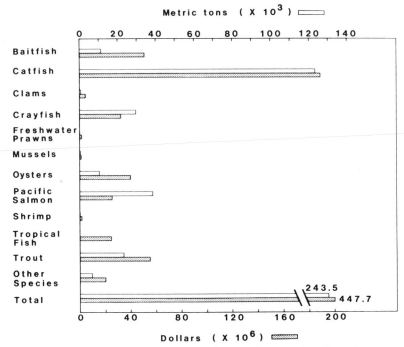

Fig. 8.2 The quantity and value to producers of aquacultural products reared commercially in the United States in 1985. (Data for tropical fish provided by Timothy K. Hennessy, Gibston, Florida and all other data provided by Ben Drucker, National Marine Fisheries Service, Washington DC. Values for oysters and Pacific salmon include some 1984 data when unavailable for 1985).

groups of finfish, three groups of molluscs, and two groups of crustaceans (Table 8.2). These development plans include reviews of the current status of production, the economic investment, development potential, constraints to development, current research and research needed to support the expansion of aquaculture. Included in 24 other aquatic groups (nearly 60 species) that have been identified as potentially important to aquaculture in the United States are reptiles, amphibians, fish, crustaceans, and molluscs other than those treated here, worms and plants.

8.2 CULTURE OF CHANNEL CATFISH

Some of the earliest records of channel catfish production have been traced to Kansas where the State began propagation as early as 1910. In 1949 the Arkansas Game and Fish Commission collected channel catfish from the Red River that were used as broodstock

Table 8.2 Major aquacultural products in the United States*

Category and common name	Scientific name
Finfish	
Baitfish	
golden shiner	*Notemigonus crysoleucas*
fathead minnow	*Pimephales promelas*
white sucker	*Catostomus commersoni*
goldfish	*Carassius auratus*
Catfish	
channel catfish	*Ictalurus punctatus*
blue catfish	*Ictalurus furcatus*
white catfish	*Ictalurus catus*
yellow bullhead	*Ictalurus natalis*
brown bullhead	*Ictalurus nebulosus*
flathead catfish	*Pylodictis olivaris*
Pacific salmon	
chinook salmon	*Oncorhynchus tshawytscha*
coho salmon	*Oncorhynchus kisutch*
chum salmon	*Oncorhynchus keta*
pink salmon	*Oncorhynchus gorbuscha*
sockeye salmon	*Oncorhynchus nerka*
Temperate bass	
largemouth bass	*Micropterus salmoides*
striped bass	*Morone saxatilis*
Trout	
rainbow trout	*Salmo garirdneri*
cutthroat trout	*Salmo clarki*
brown trout	*Salmo trutta*
brook trout	*Salvelinus fontinalis*
lake trout	*Salvelinus namaycush*
golden trout	*Salmo aquabonita*
Dolly Varden	*Salvelinus malma*
Sunapee trout	*Salvelinus aureolus*
Crustaceans	
Crayfish	
whiteriver crayfish	*Procambarus acutus acutus*
red crayfish	*Procambarus clarki*
papershell crayfish	*Orconectes immunis*
virilis crayfish	*Orconectes virilis*
rusty crayfish	*Orconectes rusticus*
signal crayfish	*Pacifastacus leniusculus*
Penaeid shrimp	
blue shrimp	*Penaeus stylirostris*
vannamei	*Penaeus vannamei*
white shrimp	*Penaeus setiferus*
banana prawn	*Penaeus merguiensis*
freshwater prawn	*Macrobrachium rosenbergii*

Table 8.2 Major aquacultural products in the United States* (cont.)

Category and common name	Scientific name
Molluscs	
Hard clams	
northern quahog	*Mercenaria mercenaria*
southern quahog	*Mercenaria campechiensis*
Mussels	
blue mussel	*Mytilus edulis*
Oysters	
American oyster	*Crassostrea virginica*
Pacific oyster	*Crassostrea gigas*
European oyster	*Ostrea edulis*
Olympia oyster	*Ostrea lurida*

*Derived from Joint Subcommittee on Aquaculture, 1983.

and subsequently became the ancestral stock of 95% of the channel catfish cultured in the United States today. By 1960 there were only about 160 hectares (400 acres) of catfish ponds in commercial production. The growth thereafter was astounding. Production expanded to about 950 ha in 1963, 16 000 ha in 1969 and over 49 000 ha by December 1985, an increase of 300-fold in only 25 years. More than 34 000 ha of this total were in the State of Mississippi alone. Some individual fish farms have as many as 2400 ha of ponds and use up to 18 metric tons of feed per day. Aircraft have sometimes been used to distribute food across the water surface of the largest ponds, especially when muddy roads made the ponds inaccessible by truck.

8.2.1 Production systems and site selection

Ponds, raceways, cages and tanks are used for the commercial rearing of channel catfish. The type of system used is often determined by the location. Earthen ponds are by far the most widely used culture system for catfish.

(a) Ponds

Earthen ponds are used throughout the United States to rear warmwater fish, but are most heavily concentrated in the southeastern United States. Ponds are typically constructed by impounding natural basins in areas of uneven terrain or by constructing dikes in areas of relatively flat terrain. Commercial production ponds are 1 to 2 m deep and 2 to 8 ha or larger in surface area. Many catfish ponds have been built to facilitate harvesting of fish by seining the pond without draining it. An adequate water supply, catch basin and drain structure are essential components of productive, manageable ponds (Plate 8.1).

Advances in fish nutrition, the availability of high quality commercial diets and the development of trained and experienced managers have made increases in catfish production possible. As recently as 1973, a production of 2240 kg/ha was achieved by only a few producers, whereas today the average production in commercial ponds is about 4400 kg/ha. Some ponds routinely yield 7500 kg/ha and some producers have reared up to 12 500 kg/ha. Small 0.02 and 0.04 ha research ponds have produced crops of 16 000 kg/ha.

As production in intensively cultured ponds has increased, so have problems associated with water quality. Dissolved oxygen is one of the first factors to become limiting in intensively managed ponds. Emergency aerators are standard equipment on most farms, and range from small electrically operated floating units to large

Plate 8.1 Drained catfish pond in the USA; note standpipe and gate valve and smooth bottom of pond needed for harvesting.

paddlewheels powered by farm tractors. Other paddlewheel aerators are mounted on mobile trailers, and diesel, gasoline or propane engines are used to provide the power. One mobile paddlewheel normally provides emergency aeration for four 8 ha ponds.

Electrically operated floating agitators are commonly used to provide emergency aeration in 0.1−1 ha hatchery ponds. Recently, continuous aeration has been used in smaller ponds to prevent stratification and to improve circulation. Continuous aeration has been provided by surface agitators, by releasing compressed air through aerators or spargers below the water surface, and by using airlift pumps. Production in warmwater ponds is expected to continue to increase.

(b) Raceways

Most raceways in the United States are made of concrete. However, a few are constructed of other materials, such as earth, stone, metal and plastic. Sites suitable for raceway production are primarily limited by the availability of water. A few raceways with geothermal water are used to rear catfish in the Snake River Valley of Idaho, where the major aquaculture crop is trout. Raceways can be placed only where sufficient amounts of high quality flowing water are available. There are few natural sites in the United States where additional raceways can be placed. However, the use of raceways is expected to increase in areas below electric power plants, where the heated effluent can be used as a high quality water source. Raceways receiving water from power plant effluents have been used to culture channel catfish at Gallatin, Tennessee, Trenton, New Jersey, and at a few other sites throughout the country.

(c) Cages and net pens

In waters from which cultured fish cannot be easily captured for harvest, cages and net pens have been used to produce both commercial and non-commercial aquacultural crops. Cages have been used to rear fish in the heated effluents of power plants and in lakes, reservoirs and rivers. Growth and survival of fish in cages is affected by the density of fish per cage, the density of cages per unit of volume or of surface area of water, the species of fish cultured and the quality of the feed. Channel catfish were first grown in cages in the United States in 1966. Few commercial aquaculturists are now raising freshwater fish in cages, and the annual production constitutes only a minor portion of the US production. Cages are more frequently used to hold fish for limited periods before the fish are stocked into ponds or processed for the market. Problems associated with cage and net-pen culture include biological fouling of the mesh material, loss of fish to predators and disease, poor water quality, theft, loss of cages during severe weather, deterioration of cage materials and conflict with navigational and recreational uses of public waters.

Even with these limitations, there has been a recent upsurge of interest in cage culture. Cages are being used by some farmers to rear small quantities of fish that are readily available for sale or for home consumption. The units are normally constructed with plastic or steel wire, with a mesh size of 1.3×2.5 cm^2. Plastic-coated wire can be attached to steel frames to produce a rugged cage that lasts for several years. Cages must have flotation material, a top to exclude predators and a strong anchor or attachment system.

Fish in cages depend solely on the aquaculturist for their food and must be fed a diet containing all nutrient requirements. Floating pellets are usually preferred and are retained in cages by a ring of small-mesh netting placed around the upper end of the cage at water level. Sinking feeds can be used if a small-mesh net is placed above the bottom of the cage to retain feed pellets.

Cages 1.2 m deep $\times$ 2.4 m wide $\times$ 2.4 m long (7.25 m^3) support about 1800–2000 kg of fish when stocked with 350–565 fish/m^3. Although the density of fish per unit of volume (248–276 kg/m^3) of cage is high, the density per unit of surface area of water is not different from that for fish reared in ponds. Cages must be placed so that water freely flows through the cage. Caged fish are obviously unable to swim through the pond to seek high quality water, but must rely solely on water currents to transport dissolved oxygen into the cage and waste metabolites out of it. Also, catfish susceptibility to disease is greater in crowded cages than in fish spread through a pond.

(d) Circular tanks

Circular tanks constructed of plastic, concrete or steel are widely used throughout the United States to culture aquatic species. Small

100−2000 litre tanks are used to spawn fish, to maintain fry and fingerlings and to hold fish before sale. Circular tanks are used as flow-through units and also in water reuse systems.

Water tangentially introduced into circular tanks provides a current against which rheotactic species such as channel catfish orient. A circular flow pattern makes the tanks self-cleaning, since particulate matter is moved across the bottom to the drain in the centre of the tank. As the diameter increases and the depth decreases, circular tanks become hydraulically less suitable for fish culture, owing to the formation of dead water zones. Zones develop where water spins about the tank but does not mix with inflowing fresh water; this section of non-mixing water accumulates metabolic waste-products and is often low in oxygen. Compressed air is commonly used in both large and small circular tanks to increase dissolved oxygen and thus overcome some inherent limitations.

Production of catfish in tanks is typically more expensive than culture in ponds. Fish must be fed a nutritionally complete ration similar to that used for fish in cages, and water must be supplied to the tanks continuously. Water from springs and wells is often too cool for optimum growth of catfish, and surface water may contain wild fish that harbour parasites and pathogens.

The density of catfish in tanks is similar to that of fish in raceways and depends on the flow of freshwater or the quality of the reuse water. Production loads of 213 kg/m^3 were supported in a 1.6 m^3 tank in which the water was exchanged once every 7.6 minutes. Though numerous attempts have been made to develop commercial production units with tanks and water reuse systems, the energy requirements and need for 100% back-up equipment make fish culture much more expensive in such systems than in ponds.

8.2.2 Spawning methods

Channel catfish are typically spawned in ponds, pens or aquaria. The pond method requires the least involvement of the culturist; spawns are collected from 38-litre milk cans or other suitable containers placed in ponds stocked with many brood fish, usually at the ratio of two females per male. The can is normally placed on its side in 0.3−1.5 m of water with its mouth toward the deeper part of the pond. For convenience, the spawning can is usually placed no deeper than an arm's length in the water and is marked with a float so that it can be easily found and checked for eggs

(Plate 8.2). The male parent enters the spawning can and fans out
the debris to form a nest. Eggs deposited by the female and
fertilised by the male form a gelatinous mass on the bottom of the
can.

The male drives the female from the can after spawning is
completed, and cares for the eggs. Males aggressively defend the
nest to drive away predators and readily bite fingers or toes of an
aquaculturist checking the can. The egg mass is rolled and shaken
as the male presses against it with the side of his body, or fans it
with the pelvic fins. This movement of the egg mass helps to
remove debris and dead eggs, and to circulate and thus oxygenate
the water. After the eggs hatch, fry may be removed from the cans
and placed in other ponds or culture units, or the eggs may be
removed from the can before they hatch.

In the pen method of spawning, fish are paired and placed in
earthen-bottom pens about 2 m wide and 3 m long, constructed of
wire or plastic mesh in water about 1 m deep. They are sometimes
constructed with the pond bank forming one end and then require
only one mesh end and two mesh sides.

A spawning can is placed in each pen with the opening of the
can pointed toward deep water. The female should be removed
from the pen immediately after spawning if the eggs are to be left

Plate 8.2 Milk can used as spawning chamber for catfish; note the float used to
locate submerged can.

in the can for incubation by the male. Alternatively, both brood fish and the eggs can be removed after spawning and a new pair of brood fish placed in the pen. When the egg mass is removed from the can, it is placed in a hatching trough for artificial incubation (see section 8.2.3). The pen method of spawning allows the culturist to select the fish to be mated, to remove spent brood fish from the spawning pond more easily, and to control spawn production more closely.

The aquarium method of spawning catfish is rarely used by commercial producers, but is sometimes used by researchers when close control of the spawning event is required. Male and female catfish of similar size are placed in aquaria about $0.6 \text{ m} \times 0.6 \text{ m}$ and 0.3 m deep. Aquaria are continuously supplied with fresh water and are usually aerated with compressed air. Gravid female catfish are induced to spawn by an injection of about 1800 international units of human chorionic gonadotropin per kilogram of body weight. As an alternative, dried pituitaries may be injected into the female at the rate of 4.5 mg/kg of body weight.

8.2.3 Fingerling production

Eggs may be left in spawning cans to hatch, or placed in wire baskets suspended in troughs for artificial incubation. Hatching troughs are typically 0.5–1 m wide, 3–6 m long, and 25 cm deep. Eggs are gently agitated in the baskets by paddles attached to a motor-driven shaft turning at 30 revolutions per minute (Plate 8.3). At 25.5°C channel catfish eggs hatch in 8 days (Table 8.3).

Newly hatched fry, commonly called sac fry because of the conspicuous yellow yolk sac, swim to the bottom of the hatching trough and do not feed. They may then be moved to rearing troughs or to nursery ponds. As they deplete their yolk they turn greyish and begin to swim to the water surface, seeking food. Swim-up fry are fed frequently (6–12 times a day) or may be fed continuously from automatic feeders.

Although channel catfish can be reared to fingerling size in troughs and tanks, most culturists stock fry 2 cm long in rearing ponds. The rate of growth of fry to fingerling size in ponds depends largely on the density at which fish are stocked. Fry stocked at a density of 25 000 fish per hectare will attain a length of 18–25 cm in a 120-day growing season in the southern United States. Those

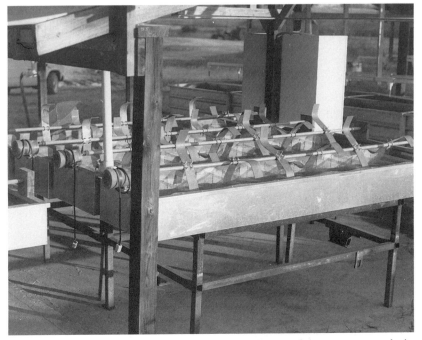

Plate 8.3 Paddlewheel hatching troughs for rolling catfish egg masses during incubation.

stocked at 250 000 per hectare will average only 8−13 cm in the same period. Some culturists stock 2.5 million fry per hectare and grade fish from the pond as needed, several times during the year.

8.2.4 Feeds and feeding

Pelletised feeds formulated for channel catfish are readily available in the United States. Research conducted in the 1960s and 1970s defined many of the nutritional requirements of channel catfish and provided the basis for development of production diets. The protein requirement of channel catfish varies with the age and size of fish. The dietary protein requirements are higher for small than for large fish. Fry require 35−40% protein, fingerlings 25−36% and adults and brood fish 28−32%. The quality of protein is just as important as the quantity. The essential amino acid requirements of channel catfish have been determined, as well as the requirements for vitamins, minerals, carbohydrates and lipids.

Table 8.3 Stages of channel catfish egg development at 25.5°C, as observed with the unaided eye. For each increase or decrease of 3.5°C above or below 25.5°C, subtract or add 1 day to the incubation time*

Distinctive feature	Age in days
No internal pulsation	1
Pulsation visible	2
Bloody streak visible	3
Entire egg bloody in appearance	4
Eyes visible	5
Eyes prominent; embryo turns in shell	6
Embryo complete; no bloody streak present	7
Embryo begins to break from shell	8

*Adapted from Dupree and Huner, 1984.

Commercially produced food pellets are commonly available in two forms, sinking and floating pellets. These diets may be nutritionally complete or incomplete, depending on whether they contain all or only a portion of the nutritional requirements. Complete diets are fed to fry, fingerlings, and larger fish when cultured in tanks, raceways, or other systems where little or no natural food is available. Complete diets may also be required for fish stocked at high densities in ponds. Incomplete diets are used to supplement the natural food in a pond; they usually are deficient in vitamins and minerals, and other nutrients may not be provided in the ratio required by the fish.

The cost of sinking feeds (per unit of weight) is considerably less than that of floating feeds. However, the advantage of observing fish as they feed on floating pellets reduces the risk that they will be overfed and that feed will be wasted. Some farmers feed a mixture of floating and sinking pellets in about a 1:5 ratio.

Fry in ponds are commonly fed starter rations by hand 2 to 3 times daily at a rate of about 10–25 kg/ha per day. Fry in troughs are fed every 2 to 4 hours by hand, or more frequently (or even continuously) with automatic feeders. Feed particles or crumbles are increased in size as the fish grow (Table 8.4) and the feeding rate is decreased.

Feeding rates must also be adjusted for temperature. A feeding rate of 3% of fish body weight is recommended at 27–30°C, the opimum temperature for growth of channel catfish. Other recommended rates are 1% of body weight per day at 32°C and

Table 8.4 Optimum daily feeding rate and particle size for fry and fingerling channel catfish*

Length of fish in cm	Recommended feeding rate[†] in %	Designated particle size in no.	Range in diameter of particles in mm
<1.2	**	0	0.420−0.595
1.2−2.5	6	1	0.595−0.841
2.5−3.7	6	2	0.841−1.19
3.7−6.2	5	3	1.19 −1.68
6.2−10	4	4	1.68 −2.38
10−15	3	5	2.38 −3.36
>15	3	[‡]	9.5

*Modified from Dupree and Huner, 1984.
[†]Feeding rate as percent of body weight per day.
**Due to small size of fish, feeding rate not based on body weight but recommended at 10−25 kg/ha per day.
[‡]Pelletised feed.

higher, 2.5% at 20−27°C, 1.5% at 14−20°C, 0.75−1% on alternate days at 10−14°C and 0.5−1% once every 3 or 4 days at temperatures below 10°C. Feeds offered during winter need contain only about 25% protein, but at least half of the protein should be from animal sources.

The quantity of feed placed in the culture system (pond, raceway, tank, etc.) must never exceed the capacity of the system to remove or oxidise the waste products. It is recommended that daily feeding rates be limited to 30−50 kg of feed per hectare. However, levels of 300 kg/ha have been fed to fish in small ponds (0.02−0.1 ha) in which water was continuously aerated and circulated. A few catfish farmers have fed 100−200 kg/ha per day, but they closely monitored dissolved oxygen and had emergency aeration equipment immediately available.

The most common method of feeding fish in ponds is that of spreading the feed across the water with a truck-mounted blower. A funnel-shaped hopper mounted on a set of scales measures the feed delivered from the hopper. Feed drops from the hopper into the air stream from a high volume blower and is blown across the surface of the pond. Feed is available in bulk or bagged; both types are widely used. Bulk feed is more economical and is more commonly used on large farms (Plate 8.4).

Plate 8.4 Catfish feed bins for bulk transportation and storage.

8.2.5 Food fish production

The average annual yield from catfish ponds has increased from about 1100 kg/ha in the early 1960s to 3300−4400 kg/ha in the 1970s. Today, some farmers produce 5500−7700 kg/ha in intensively managed ponds, raceways, cages and tanks. Most channel catfish are reared in intensively managed ponds of one of two types. Hill ponds are constructed by placing an earthen dam across the valley between two hills. These irregularly shaped ponds, common in Alabama, are filled only by rainwater or surface run-off and production is normally lower than in flat-bottom ponds. Many newer ponds are constructed on former agricultural cropland by placing earthen levees around flat fields. These ponds are typically 8 ha in surface area and arranged in groups so that one well supplies water to four ponds (Plate 8.5). Roads on the levees and around the irregularly shaped hill ponds provide access for feeding and harvest of fish.

Growth of food fish in ponds is related to the density of stocking. Fingerlings averaging 15 g when stocked in March at a density of 5000 fish/ha attained an average weight of 635−816 g in 204 days. However, size varies greatly when fish are stocked at relatively low densities. As stocking rates increase, size variation is reduced and

Plate 8.5 Water level is maintained by a single well drilled at the junction of four ponds; note that flat-topped earthen levees provide vehicle access to all sides of ponds.

the return per unit of surface area increases. At densities of 40 000 fish/ha, 15 g fingerlings stocked in March averaged 318 g in September and 567 g in November, 204 days after stocking. However, such gains are not without risks. Management problems become much more severe as densities increase and ponds must be carefully monitored to maintain adquate water quality and fish health.

Fish are normally fed once or twice per day, small fish more frequently than large ones. The frequency and rate of feeding is reduced during winter. Feed from a truck-mounted hopper is most commonly spread along one side or end of the pond with a blower. However, some farmers in Mississippi use aircraft to feed fish in large ponds or in those inaccessible to trucks during wet weather.

Control of aquatic macrophytes and filamentous algae is a minor problem in catfish ponds. However, nutrient enrichment from uneaten feed and fish metabolic waste supports dense plankton blooms. Although photosynthesis by plankton may elevate dissolved oxygen to more than 20 mg/l on bright sunny days, respiration by plankton, bacteria and fish frequently decreases dissolved oxygen to less than 2 mg/l during the night. During extended periods of cloudy weather or after the dense plankton bloom dies, dissolved oxygen may remain low for several days. Emergency aeration provided by tractor-driven paddlewheels and electric surface agitators is essential on commercial farms. One tractor and paddlewheel are recommended as the minimum emergency aeration equipment available for four 8 ha ponds. On at least one occasion, a farmer who used 9 tractor-driven paddlewheels in one 4 ha pond still lost over 4500 kg of fish.

Some farmers have experimented with air spargers, airlift pumps, axial flow pumps and small surface agitators to provide continuous aeration and destratification in ponds. Although these systems have been effective for research, their application in large 8 ha ponds is still unproven.

In the 1960s it was common to drain a pond and harvest all the fish at one time. Today, ponds may not be drained for 5 to 7 or more years. Fish are seined from the pond several times during the year. Fingerlings may be stocked in the same pond with fish of harvest size. This intensification has produced problems other than low oxygen.

As water quality declines, fish stress increases and fish become more susceptible to pathogens. Other non-pathogenic diseases are directly related to water quality. Brown blood disease is the term used to describe methaemoglobinaemia in catfish. This condition is most prevalent during late summer and early fall, when feeding rates are highest. Ammonium nitrite complexes with haemoglobin to form methaemoglobin. The recommended treatment is salt (NaCl) at 66 g/m^3 of water. The chloride ion competes with the nitrite ion to protect the fish.

Maintenance of water quality is one of the greatest challenges facing aquaculturists. Some farmers use continuously operating dissolved oxygen monitors coupled to emergency aerators to protect fish from low oxygen. Others check dissolved oxygen by hand in the afternoon and throughout the night to protect fish.

In the Delta region of Mississippi, catfish production rate varied little with the size of the farms (Table 8.5). Although the largest farms had 4 times the number of ponds as the smallest farms, average pond size (just over 7 ha) and production per hectare of water (just over 6 metric tons) were virtually constant. However, a comparison of the annual costs and potential profit per hectare of such farms (Table 8.6) clearly shows the commercial advantage of having more ponds. Large farms derive economies of scale in both fixed costs (e.g. taxes and insurance) and variable costs (e.g. labour),

Table 8.5 Size, number of ponds and estimated production of catfish on small, medium and large farms in the Delta Region of Mississippi in 1982. (Modified from Giachelli *et al.*, 1982)

	Farm size		
Item	Small	Medium	Large
Total land area (ha)	65.2	129.2	257.2
Surface area of water (ha)	56.5	103.4	228.7
Production area of water (ha)*	53.7	108.5	217.4
Surface area per pond (ha)	7.1	7.1	7.2
Number of ponds	8	16	32
Total production (kg)[†]	325 334	657 281	1 317 502
Production per pond (kg)	40 667	41 080	41 172
Production per ha of water (kg)	6 058	6 058	6 060

*Reflects a 5% reduction in water available for production due to average annual pond repair.
[†]Includes the initial weight of fish stocked as fingerlings.

so that total production costs per kilogram were $1.46 (£0.79), $1.34 (£0.72) and $1.29 (£0.70) for small, medium and large farms, respectively. Although gross income per hectare was identical, the higher unit costs of the small farms resulted in a net loss, whereas the largest farm achieved nearly double the profit per hectare of the medium-sized farms. Hence it is important to consider the economic as well as the technical factors involved in catfish farming.

8.2.6 Harvest and transport

Catfish removed from ponds as fingerlings or adults may be transported considerable distances, often from State to State, and restocked into other ponds or culture units. Fish transported alive for restocking must be harvested and handled with care to reduce stress and losses. The best method of harvest depends on fish size and the type of culture unit.

Fry and fingerlings are usually harvested from ponds by seining after the water level has been lowered by partial draining or after fish have been attracted by feeding into one corner of a full pond. Because pectoral spines of catfish are prone to entanglement in nets and seines, the latter are usually coated with tar or other similar petroleum compounds to reduce entanglement and facilitate

Table 8.6 Annual cost and potential profit per hectare for channel catfish cultured on small, medium and large farms (as described in Table 8.5) in the Delta region of Mississippi in 1982. (Modified from Giachelli et al., 1982)

| Item | Farm size | | |
	Small	Medium	Large
Annual ownership costs			
Depreciation on ponds, water supply and equipment ($/kg)	0.0957	0.0814	0.0744
Interest on investment ($/kg)	0.1650	0.1503	0.1441
Taxes and insurance ($/kg)	0.0088	0.0057	0.0046
Total annual ownership costs ($/kg)	0.2695	0.2379	0.2231
Annual operating costs ($/kg)			
repairs and maintenance	0.0350	0.0271	0.0229
fuel	0.0667	0.0651	0.0660
chemicals	0.0128	0.0125	0.0125
fingerlings	0.1390	0.1390	0.1388
feed	0.6134	0.6134	0.6134
labour	0.1641	0.0931	0.0649
harvest-hauling	0.0880	0.0880	0.0880
liability insurance	0.0055	0.0037	0.0026
interest on operating capital	0.0662	0.0609	0.0585
Total operating costs ($/kg)	1.19	1.10	1.07
Total costs ($/kg)	1.46	1.34	1.29
Production per hectare (kg)	6058	6058	6060
Gross income per hectare ($1.40/kg)*	8481	8481	8484
Cost per hectare	8846	8123	7822
Profit per hectare	−365	358	662

*Price of catfish peaked at $1.76/kg in March 1985 and resulting profits under these cost conditions would have been $1816 for small farms, $2539 for medium farms and $2844 for large farms.

the removal of fish. Fish are easily stressed by low dissolved oxygen when water temperature is 25°C or higher. During summer, seining is restricted to the early morning, when pond temperatures are low. Fish may also be moved to holding vats containing well water at temperatures 5−8°C cooler than that in the ponds. They are also commonly treated with 1% salt (NaCl), tranquillizers and antibiotics to reduce stress, lower metabolic activity and control bacterial pathogens.

Most catfish harvested for the food-fish market are transported directly to processing plants. Fish of selected sizes may be removed from ponds with a grading seine without draining the pond or removing other fish. Grading seines with a mesh of 3.5 cm remove all catfish weighing 0.34 kg or more; a mesh of 4.1 cm removes those of 0.45 kg or larger. Many diked ponds, common in Mississippi, are drained only once every 5 to 7 years and catfish are removed with grading seines several times during the year. Hill-type ponds, which are common in Alabama, are often too irregular in shape and depth to be seined without draining them. Fish are harvested from these ponds by trapping throughout the year, or in Fall and Winter by draining the ponds. Since most hillside ponds are filled by surface runoff, they can be refilled only during the rainy season (Winter and Spring in the southeastern United States).

Seines used to harvest fish from large ponds (8 ha or more) are usually pulled by mechanical seine-haulers or by tractors. Catfish seines normally have a bundle of small ropes, called the mud line, sewn to the bottom edge to hold the net on the pond bottom and yet keep it above the mud. Nylon ropes attached to the mud line are used to haul the seine. A small boat may be used to push the centre of the seine and lift it from the bottom, to dislodge mud. The mud line quickly settles back to the pond bottom, thus largely preventing the escape of fish under the seine (Plate 8.6).

Fish are corralled in one section of the pond with a large seine and then a smaller seine is used to remove them from the large seine, a truckload at a time. Live cars or large net cages may be used to hold fish until they are shipped. Holding the fish for a day or two allows them to evacuate their intestinal tract and helps to maintain water quality during transport. Fish may also be confined in live cars while quality control tests for off-flavour (discussed later) are conducted.

8.2.7 Processing and marketing

There are two basic markets for catfish, live fish and processed fish. Live fish are normally sold to be stocked in farm ponds or private lakes for sport fishing. Owners of private lakes may charge fishermen an entrance fee to fish, plus a charge based on the weight of fish caught. Fish sold in this market usually command the highest price and may be worth $4.40 (£2.38)/kg of live weight. Such fishing operations are normally more profitable in large metropolitan areas than in rural settings; however, some farmers in rural areas have attracted fishermen from distances of 500 km or more to catch high quality fish.

Plate 8.6 Seining catfish ponds without draining using a boat to place the net while fish are driven out of shallow area; note floating paddlewheel aerators, standpipe and drain in left foreground.

The second-highest return to the farmer is to sell processed fish in a local market. Few fish are sold this way, and it seems to be most appropriate for small family-size operations, by farmers who hire little outside labour. Many southern restaurants specialising in catfish were established to provide this market outlet.

Most of the 123 000 metric tons of catfish reared in 1985 by US farmers were sold to processing plants. The price paid to farmers peaked at $1.76 (£0.95)/kg, whole weight, in March 1985. However, as the supply of catfish increased, the price/kg decreased in 1986 to $1.50 (£0.81) by June and $1.39 (£0.75) by September. Catfish are still considered a southern item, even though recent promotional campaigns have moved them into national markets.

Off-flavour and fatty fish are the major problems in product quality control. Several organic compounds, including the geosmin produced by blue-green algae, concentrate in the flesh of catfish to produce off-flavour. Processors taste test a small sample of fish from each pond the week before harvest, the day before harvest, and then again immediately before processing, to check for off-flavour. Fish with objectionable off-flavour can be moved to fresh water or held in the pond until the flavour improves.

American consumers typically avoid fat, strong-tasting fish, and prefer lean mild-tasting fish. Shelf life is shorter for fish with high concentrations of fat than for lean fish. Many of the organic compounds producing off-flavour concentrate in lipids. Recent research has shown that the lipid content of catfish is inversely correlated with stocking density, and that body fat was lower in fish fed in the morning than in those fed in the afternoon. Additional research is under way to improve quality and shelf life of processed catfish. Although there is a considerable market for whole fish and fish steaks, an increasing proportion of the market is for fillets and other boneless products. Large firms that traditionally processed only beef, pork and poultry have recently expanded their product line to include catfish. These national food marketing firms are developing new value-added products, i.e., rather than selling frozen fillets they are selling seasoned, breaded and precooked fish products that only require heating before thay are served. Although fresh frozen catfish sell for about $6 (£3.24)/kg, value-added, precooked products may sell for $35 (£18.92) or more per kilogram.

8.3   CULTURE OF TROUT AND SALMON

Trout have been cultured in the United States for more than a century. Production facilities (mostly rather small) were established on many springs and water sources before 1900. Although commercial producers are in the New England states, Minnesota, Ohio, Michigan, Tennessee, Wisconsin and a few other states, the larger farms are in California, Idaho, Montana, Oregon and Washington. Commercial farms in the Snake River Valley of Idaho produce 70% of all trout reared in the United States. Trout production (metric tons) was 450 in 1954, 13 400 in 1972 and 23 000 in 1985; the worth to fish farmers in 1985 was $55 (£29.70) million.

Salmon culture began in California in 1872 and now has spread to Oregon, Washington and Alaska on the Pacific Coast, to the Great Lake states of Michigan and Wisconsin, and to states on the northeast Atlantic Coast. Most salmon, reared in hatcheries operated by states or the Federal Government, are released to smolt and migrate to sea. These fish are important to both recreational and commercial fishermen.

Both Oregon and Washington have licensed commercial ocean ranching operations in which fish are released into the ocean as juveniles and harvested when they return as mature adults. Other operators rear salmon in net pens, first introduced to the United States in 1967, and market them as pan-size fish.

On the Pacific coast, chum salmon (*Oncorhynchus keta*), coho salmon (*O. kisutch*) and chinook salmon (*O. tshawytscha*) are the major species being reared for ocean ranching; however, some pink salmon (*O. gorbuscha*) and sockeye salmon (*O. nerka*) are also reared.

8.4   CULTURE OF STRIPED BASS AND HYBRIDS

Anadromous striped bass, *Morone saxatilis*, originally ranged from the St Lawrence River in Canada south along the Atlantic Coast to the St Johns River in northern Florida and into 31 rivers from the Florida Gulf Coast westward to the Mississippi River. Both Atlantic and Gulf Coast stocks of fish have been severely reduced due to environmental degradation, blocked access to spawning grounds by dams and navigational locks, and strong commercial and recreational fishing pressure. However, inland populations of landlocked striped bass have increased dramatically since 1954, when they were first reported to have spawned in Santee-Cooper Reservoir, South Carolina.

Hatchery-reared progeny of Atlantic Coast striped bass and hybrids of striped bass and white bass, *M. chrysops*, have been stocked in more than 400 reservoirs divided among 28 states. Although striped bass mature in inland reservoirs, few inland waters provide suitable spawning grounds to maintain a stable population. Most striped bass and hybrids are produced in hatcheries; the techniques for hormone injection that are used were developed about 20 years ago.

Striped bass weighing 57 kg have been reported, but fish heavier than 15 kg are often considered trophies. Hybrid striped bass may weigh up to 10 or 11 kg, but commonly are only half that size. Brood fish are collected with gill nets, with electro-fishing equipment, with hook and line from inland reservoirs, or from rivers as they migrate from coastal waters to their inland freshwater spawning grounds. Smaller brood fish of 5–10 kg are preferred in the hatchery, since they can be more easily handled than larger fish. However, brood fish from Chesapeake Bay, Maryland, typically weigh 20—25 kg and one fish of 38 kg was spawned in 1985.

Immediately after capture of female brood fish, eggs are collected through an interovarian catheter (3 mm outside diameter) and microscopically examined to predict time of ovulation. As a means of stimulating their ovulation, female brood fish judged eligible for spawning are injected with human chorionic gonadotrophic hormone at the rate of 273—300 international units (IU) per kilogram of body weight, and males with 110—165 IU per kilogram (as needed) to stimulate milt flow. Fish are transported to the hatchery in 1% salt water and separated by sex if gametes are to be taken manually. Gametes may be manually stripped from brood fish at the time of spawning or brood fish may be paired in tanks and allowed to spawn naturally. Fish normally spawn 24—48 hours after injection. Highly buoyant eggs, as from fish in the Chesapeake Bay, are incubated in tanks of 100—11 000 litres, whereas non-buoyant eggs are incubated in McDonald hatching jars. Eggs normally hatch in about 48 hours at 18—20°C.

Most striped bass are reared in ponds, but a limited number are reared in tanks in water recycled through filters to remove metabolic waste products. Fry begin feeding when about 5 days old and are commonly fed *Artemia* nauplii or other live zooplankters for 5—10 days before they are stocked into fertilised ponds. Stocking rates in freshwater ponds (250 000—400 000 fry/ha) are typically less than in brackish water ponds (600 000—750 000/ha). Fry are usually stocked into ponds at night to protect them from strong sunlight. At harvest, ponds are partly drained in the evening so that fish can be removed in early morning when the water temperature is lowest. Fingerlings (2.5—5 cm long; commonly called phase I fish) are harvested from ponds after 35—40 days. They are either stocked into lakes, rivers or estuarine waters, or restocked at lower densities (10 000—25 000 fish/ha) into ponds and reared for 3—5 months to a larger size (15—20 cm; phase II fish) for later stocking. Normally, 50% or fewer of the fry stocked in ponds survive to produce phase I fish. Survival is highest in ponds with 0.5% salinity than in ponds with salt concentrations below this level.

Survival of both fry and fingerlings is improved when fish are handled in 0.5—1.0% salt water. Juvenile striped bass can be transported for 4—8 hours at a density of 20—64 g/l during summer if the temperature is maintained at about 20—26°C. However, stress can be reduced and survival improved if fish are transported when the water temperature is less than 15°C. Fish should not be disturbed if the temperature exceeds 32°C.

8.5 CULTURE OF BAITFISH

Although the 11 000 metric tons of baitfish reared in 1985 made up the fifth most abundant group of fish produced in the United States by volume (Table 8.2), the group ranked third in value, behind catfish and trout. The 1985 values were $51 (£27.56) million for baitfish, compared with $189 (£102.16) million for catfish and $55 (£29.73) million for trout. The average per capita demand for baitfish is about 0.2−0.4 kg per year. Baitfish are produced throughout the country; most are produced in Arkansas, and are shipped throughout the United States.

Of the 20 species of baitfish raised commercially, the golden shiner (*Notemigonus crysoleucas*), fathead minnow (*Pimephales promelas*), and goldfish (*Carassius auratus*) are the most important. Most are cultured in shallow earthen ponds. The ideal size of pond for culturing golden shiners and fathead minnows is about 4 ha, but goldfish are more commonly reared in smaller ponds. Production ranges from 880−1100 kg/ha in fertilised ponds, but may exceed 3300 kg/ha when fish in fertilised ponds are additionally fed commercially prepared diets. Golden shiners may attain a length of 25 cm, but become sexually mature at 1 year of age and at a length of 2.5 cm. Females grow faster and larger than males and because of the ovarian protozoan, *Plistophora ovariae* (which severely reduces egg production in over-yearling fish), broodstock is selected each year from the yearling population. Ungraded golden shiners are stocked at 22−44 kg/ha as brood fish and spawn when the temperature reaches 21°C. Up to 10 000 adhesive eggs per female may be indiscriminately scattered over submerged vegetation or spawning mats made from Spanish moss. Spawning mats of area about 50 × 120 cm are placed level, 2 or 3 cm below the water surface, and golden shiners spawn on them throughout the day. Mats with attached eggs may be moved to nursery ponds; eggs hatch in 4 to 8 days at 24−27°C. Stocking rates for fry vary from 125 000−500 000 fish/ha.

Goldfish production techniques are similar to those for golden shiners. Brood fish are stocked into ponds at rates of 250−750 fish/ha; the number stocked depends on size of fish. Females deposit 2000–4000 eggs per spawn over vegetation or spawning mats when the water temperature reaches 15°C. Spawning activity may be stimulated by a rapid rise in the water level or by the application of potassium permanganate. Most spawning occurs from dawn to mid-morning. For goldfish, 125–375 spawning mats (50 × 125 cm) are normally required per hectare. Mats with eggs may be moved to nursery ponds where eggs hatch in 2 to 8 days, depending on temperature. Stocking rates for fry range from 62 500 to 2.5 million fish/ha.

Fathead minnows spawn on the underside of aquatic plants or other objects in the water and, unlike golden shiners and goldfish, provide parental care for the eggs. Brood fish are stocked at 1250–5000 fish/ha if fry are to be reared in ponds with the adults, and at 50 000–62 500 fish/ha if eggs are to be transferred to nursery ponds. Small pieces of board, 10 × 30 cm, or polyethylene sheets are placed in ponds to provide spawning substrates. Females deposit 200–500 eggs per spawn, beginning when the water reaches 18°C. Males pick up the eggs by mouth, attach them to the underside of any object in shallow water, and guard them until the fry emerge. Females mature at 1 year of age and may spawn as often as 12 times per season.

Baitfish are usually in greatest demand during summer, when many Americans on vacation 'go fishing'; however, there is a demand for fish of certain sizes throughout the year. Harvesting, grading and marketing a quality product is one of the greatest challenges facing the baitfish producer. Fathead minnows and goldfish are much easier to handle, grade and transport than are golden shiners.

Fish are commonly seined, trapped or drained from ponds and moved into holding vats for grading before they are marketed. Holding vats are often 1.2 m × 12 m × 0.6 m deep and supplied with fresh water of 15–21°C. Crowding screens and bar graders are used to separate fish into marketable size classes.

Transport trucks range from those with only one 1136-litre tank used for local markets to those with 6 to 8 such tanks for long hauls. Each tank safely holds about 135 kg of minnows in cool weather, but only 90 kg during warm weather. The ideal temperature of 15.5°C may be maintained during transport with ice or by refrigeration units on long-haul trucks.

Smaller quantities of fish may be shipped in plastic bags by land or air for periods not exceeding 24 hours. Each 46 × 81 cm bag (made of plastic 3 mm thick) will safely hold 2 to 3 kg of minnows in about 6 kg of water. Bags are filled with oxygen, sealed, and placed in a Styrofoam-lined cardboard box. Ice may be placed in the corner of the box to keep the temperature near 15°C. Salt, anaesthetics and ammonia-absorbing compounds are used by some shippers to reduce metabolic stress and to maintain quality of fish during transport.

8.6   CULTURE OF EXOTIC SPECIES

Common carp (*Cyprinus carpio*), the Chinese carp (grass carp, (*Ctenopharyngodon idella*) or hybrids of grass carp and bighead carp (*Aristichthys nobilis*)), silver carp (*Hypophthalmichthys molitrix*) and the tilapias are the major exotic species cultured in the United States as either food-fish or for biological control of vegetation. Researchers at Auburn University, Alabama, who conducted a survey in 1981, located 30 producers rearing Chinese carps in 8 states, primarily in southeastern United States. Only 6 of the 30 reared carp as food-fish, however; by far the largest market for Chinese carps originates from their use in vegetation control or nutrient removal in municipal sewage and animal waste lagoons.

Possession of grass carp is illegal in most states, but sterile hybrids and triploid fish have been more acceptable to authorities for vegetation control. Control of vegetation by hybrids is not as effective as that by grass carp. Now that techniques for producing sterile triploid grass carp have been developed, the market for hybrids has declined considerably.

The Chinese carp can be induced to spawn in the southern states from mid-April to mid-June when water temperature warms to 25°C. Silver carp of both sexes, male grass carp and male bighead carp usually mature at 3 years of age. Most female grass carp and female bighead carp mature at 4 years. Mature females can be identified by their large distended abdomens, which become flaccid just before the fish spawn. Males can be readily identified by the presence of milt easily expressed from the vent during the breeding season. Male grass carp develop protuberances (which feel rough to the touch) on the head, operculum and pectoral fins.

Grass carp are particularly difficult to handle due to their vigorous jumping. Fish taken from the seine can be handled (usually without injury to either the culturist or the fish) if confined in a muslin or porous cloth bag. It is also useful to anaesthetise the fish during transport or whenever they must be handled.

Since the Chinese carp do not spawn in static water ponds, they must be seined from the ponds and induced to spawn by hormone injection. Females are initially injected with 220 IU of human chorionic gonadotropin (HCG) per kilogram, followed by a second injection of 1870 IU after 12 hours. If fish fail to spawn after the second injection of HCG, a third injection of 2.2 mg dried pituitary per kilogram usually induces fish to spawn within 24 hours. Ten hours after the injection of pituitary, the abdomen of females should be lightly pressed to check for ovulation. Eggs flow freely from fish that are ready to spawn.

Eggs of the Chinese carp are stripped into a dry pan to which milt from one or two males is added and mixed with a dry feather or brush. Water is slowly added to the mixture of eggs and milt to activate the sperm and promote fertilisation. Eggs harden after about 10 minutes and can be placed in hatching jars or other incubation chambers. The number of eggs per hatching jar can be estimated volumetrically (1 ml = about 250 eggs). At 25°C, eggs hatch in about 20 hours. Fry can be held in the incubation tanks or moved to other tanks and held for about 5 days before they are transferred to ponds.

Fry of the Chinese carp are first stocked in monoculture at 250 000–1 250 000/ha in well-fertilised ponds. Survival is improved if fry are stocked in the early morning, when temperatures are low. Zooplankton is the first food of the fry, but they accept commercial meals and other finely divided supplemental feeds. Growth rate and maximum size of fish are a function of stocking density and feeding rate. When fish cease to grow they must be restocked into larger ponds at lower densities. At densities of 250 000–300 000 fish/ha, Chinese carp reach lengths of 5 to 6 cm after 20–30 days.

Depending on the market, the several Chinese carp may be restocked into monoculture or polyculture ponds. When the fish are stocked in polyculture, their management is based on the optimum production of grass carp. Bighead carp feed on detritus and silver carp on phytoplankton. Grass carp are commonly fed floating pellets formulated for catfish (36% protein). Daily feeding rates are usually 3 to 4% of the weight of the standing crop of fish.

Outside of the Asian communities in the United States there is only a small market for Chinese carp for food. Most producers market the fish for control of vegetation or maintenance of water quality. Large fish may be reared in polyculture with catfish or other species native to North America.

Tilapia are reared both as food-fish and for vegetation control in the southeastern states, California and Arizona, and in Idaho (in geothermal raceways). They have been well recieved as food-fish and in some markets sold in the round for over $6 (£3.24)/kg. The market for tilapia as vegetation control agents is stable, since in most locations fish die during the winter and must be restocked each year. In some locations the food-fish market has received strong competition from the sale of wild fish taken from lakes in Florida. The wild fish have sold for $0.40 (£0.22)/kg in Florida prior to processing and for $1.32−1.43 (£0.71−0.77)/kg after processing in Florida and shipment to California. About 450 metric tons of wild-caught tilapia were marketed in 1983.

Tilapia are among the easiest fish to culture because they tolerate low dissolved oxygen, high ammonia and generally poor water quality. One of the major problems of culture is their rapid re-production and the crowding and stunting that result. Although tilapia are not predatory, they are aggressive and compete with native species for spawning sites and space. Several states have prohibited the introduction of tilapia.

Culture of tilapia is primarily restricted to southern states or to areas where geothermal water or heated effluents from power plants are available. Tilapia spawn at temperatures above 20°C and prefer water of 26−32°C. They stop feeding at temperatures below 15°C and die at temperatures below 10−12°C. The blue tilapia (*Tilapia aurea*), appears to be the most cold-tolerant of the species cultured in the United States.

Tilapia aurea, *T. hornorum*, *T. nilotica* and *T. zillii* are the most commonly reared tilapias in the United States. Hybrids of female *T. nilotica* and male *T. hornorum* are nearly all male and are popular, because they grow fast. After 100−150 days of culture, hybrids may weigh 225−450 g.

Tilapia hybridise so readily that some authorities have questioned the presence of pure strains in the United States. Some strains have been selected for yellow, gold or red colour through several generations, and appear to be more acceptable in the market as food-fish than are normally pigmented strains.

Stocking rates of 25 000−50 000 tilapia/ha have been used for ponds. Cages and raceways, in which reproduction was limited by crowding and separation of fish from spawning sites, have supported densities of 160–240 kg/m^3. Tilapia are also reared in polyculture with catfish, to provide animal protein for brood fish. Adult and juvenile tilapia control filamentous algae and the young are eaten by the catfish in Summer and Fall.

Monosex culture is used by some producers to prevent reproduction. Males grow faster than females and can be separated by size from the smaller females if fry are graded after a period of growth. Typically, the males are selected and the females are discarded. Sex reversal, induced by feeding male steroids to fry before the time of sex determination, has been somewhat successful, but is not a standard practice.

8.7 CULTURE OF TROPICAL FISH

Between 1200 and 1500 exotic species have been imported into the United States for the hobby fish or aquarium trade. At least 41 of these species have established breeding populations and another 63 species live in public or private waters with native fish. Some tropical fish are cultured in nearly every state, but major commercial operations are restricted primarily to Florida and southern California. The annual retail value of tropical fish imported into the United States from Africa, Ceylon, China, Japan, Malaysia and Central and South America exceeds $500 (£270) million. The annual value to producers of tropical fish grown in the United States was estimated to be $25 (£13.5) million in 1985 and the retail value exceeded $125 (£67.57) million. More than 300 tropical fish farms, the greatest concentration in the United States, are located within a 1-hour drive from the airport in Tampa, Florida.

Depending on the species, tropical fish may be spawned in aquaria or in ponds. Brood fish are maintained indoors in heated tanks and aquaria during winter and often are stocked in spring in small ponds, 6 × 24 m or 24 × 30 m. Some ponds are covered with plastic greenhouse-type structures to protect them from low winter temperatures. Some farms have fewer than a dozen ponds whereas others have 200 or more. Many ponds are dug into the water table and are contained by a 1–1.5 m levee. The ponds cannot be drained and water is pumped from them into a dry well or sump that collects particulate matter and fish.

Both livebearers and egg layers are propagated in tanks and aquaria where environmental conditions can be controlled. Many species are fed special diets containing carotenoids and pigments that enhance coloration. Fry and adults may require live zooplankton, brine shrimp nauplii, phytoplankton or specific species of algae. Although the environmental conditions under which tropical fish survive may be rather broad for some species, conditions conducive to reproduction are usually very restricted. Some rather elaborate systems have been developed to control environmental variables such as temperature, dissolved oxygen, pH and alkalinity. Water reuse systems with biological filters are frequently used.

Some of the more commonly reared tropical fishes include the livebearers of the family Poeciliidae, such as the sailfin molly (*Poecilia latipinna*), guppy (*P. reticulata*) and swordtail (*Xiphophorous helleri*). Egg layers in the carp and minnow family (Cyprinidae) include the zebra danio (*Brachydanio rerio*) and the clown barb (*Puntus everetti*). Other species important in the tropical fish trade are certain killifishes (Cyprinodontidae), characins (Characidae), cichlids (Cichlidae), gouramies (Anabantidae) and Corydoras catfishes (Callichthyidae).

8.8 FUTURE OF AQUACULTURE IN THE UNITED STATES

Although aquaculture has been practised for well over 100 years in the United States, the major increase in production developed during the last 10 to 20 years. Economic conditions on the American farm scene, resulting from overproduction of cereal grains and other agricultural commodities, have stimulated interest in aquacultural products. During the period 1975−85, the per capita consumption of fish and shellfish increased 20% and that of poultry 44%. In contrast, the consumption of beef decreased 10% during the same period (Table 8.7). The per capita consumption of fish is expected to continue to expand.

Fish produced in the public waters of the United States are typically more valuable as recreational or sport fish than as commercial products. The demand for striped bass, red drum (*Sciaenops ocellata*), and other species has already outstripped natural production, and commercial fisheries in several states have been closed. These established markets, now unfilled, are increasingly attracting the attention of aquaculturists. Although catfish and trout are the principal food fishes produced by aquaculturists, others are sure to be added. Trout culture developed slowly over about a century,

Table 8.7 Per capita consumption (kg/year) of fish and shellfish, poultry and beef in the United States, 1975−85*

Year	Fish and shellfish	Poultry	Beef
1975	5.5	18.7	33.9
1976	5.9	20.0	36.4
1977	5.8	20.5	35.4
1978	6.1	21.5	33.6
1979	5.9	23.3	30.1
1980	5.8	23.4	29.5
1981	5.9	24.0	29.7
1982	5.6	24.6	29.8
1983	5.9	25.1	30.3
1984	6.2	25.8	30.3
1985	6.6	27.0	30.5

*Based on figures from the United States Department of Agriculture; data for poultry and beef reduced by 15% to allow for weight of bone.

while markets were being established, husbandry techniques were being refined, and the support industry for drugs, hardware, etc., was being developed. In contrast, the catfish industry and its infrastructure developed within 2 to 3 decades.

The infrastructure exists to expand rapidly production of both traditional and new species, now that feed mills, processing plants, water quality management techniques, chemicals and drugs, transportation equipment and marketing techniques have been developed. Expansion will probably be most pronounced in the southern and southeastern United States where farmers are expected to convert agricultural land to aquatic production. The long growing season and the availability of land and water make further expansion in the south most likely.

Little expansion is expected in the production of coldwater species in fresh water. Most sites with adequate quantities of quality water have either been developed for raceways or are in parks or wilderness areas where development is prohibited. However, major gains in yield are expected as fish are improved by advanced bioengineering technology that includes selective breeding and gene manipulation. Improved environmental monitoring devices, computers and software will allow culturists to monitor, regulate and adjust aquatic systems to increase production in existing facilities.

There is tremendous potential for expansion of aquaculture in brackish and marine waters along the coasts. However, legal constraints and competition for space for recreational, navigational, industrial and municipal uses will continue to be strong impediments. In 1978, when the National Research Council examined constraints and opportunities for aquaculture in the United States, it identified legal constraints as one of the major obstacles to future development. They remain so today and are not expected to improve greatly in the future. Nevertheless, aquaculture has grown dramatically during the past few years and seemingly has entered an exponential growth phase. The future of aquaculture in the United States appears bright and rapid expansion is expected to continue.

9 The development of polyculture in Israel: a model of intensification

9.1 INTRODUCTION

The development of fish farming is determined by a number of factors, in particular the local environment, production technology and economics, and the marketing and distribution aspects. For instance, the environment includes physical, institutional and social factors. The physical environment determines whether a selected species can be grown successfully in a certain area so prevailing temperature, rainfall, water quality and quantity factors are clearly important. Linked in with these is the availability of suitable space: water and land must be obtainable at a suitable cost, taking into account the competing demands of rival user groups. Institutional factors include governmental policy, planning, training, extension services and financial assistance. Political and economical stability on a national level can markedly affect investment decisions. The social factors include tradition, customs and religious beliefs, which may affect fish consumption. The availability of supplies and transportation, i.e. the infrastructure, is also of great importance.

Culture and product technology includes the state-of-the-art for growing selected species, preparing or preserving the resulting products and delivering them to the customers. It includes also the local availability of adequate information concerning culture and product technology, dissemination, training and extension services. Production itself may be divided into four categories, as follows:

1. Planning and management, i.e. development of the concept, species and site selection, capital formation, design, construction and business management.
2. Inputs such as availability of fish seed, feed, water, energy, etc., as well as the logistics of providing the necessary inputs.
3. Operations, including the day-to-day activities of harvesting and preparation for delivery.

4. Costs which are considered separately to emphasise the importance of production costs.

Marketing may also be divided into four categories, as follows:

1. Planning and management, which includes the selection of product forms and marketing strategy. It also covers the business management of the marketing strategy, including scheduling the harvesting of fish farms to provide continuity of supply and minimising seasonal overproduction.
2. Demand – income and price considerations are the major incentives for selecting certain species for production. As the marginal cost of fishing wild stock escalates, fish farming generally becomes more competitive.
3. Operations, including the day-to-day activities needed to process and package fish-farmed products and transport them to selected points in the marketing chain.
4. Revenue, together with production costs, determine the profitability of the fish farm. Logical marketing strategies and market development can increase revenues and high perceived revenues will encourage the expansion of fish farming.

All these factors, except for the physical environment, represent conditions which can be modified to a greater or lesser degree by human intervention. Even the physical environment would technically prohibit fish farming only in extremely unfavourable areas. However, even moderately unfavourable physical environments could make aquaculture uneconomical since expensive controlled habitats might be needed to provide the environmental conditions suitable for growing the selected species.

Given the above context, it is of interest to trace the development and present situation of Israeli aquaculture as a model of a commercial industry. Israel is located in an arid zone with severe water limitations. It has a temperate climate. The annual rainfall is between 200 and 600 mm, concentrated in 4 months of the year, the remaining months being completely dry. There are no permanent streams except the Jordan river on the eastern border and one lake, Lake Kinneret or the Sea of Galilee, with a surface area of 15 000 ha. The total national water consumption is approximately 50 m^3/s. Eight per cent of all the water sources are brackish and only partly suitable for irrigation. At present, the country's water utilisation has reached 95%, including brackish water sources, utilised for fish ponds, as will be described later.

It is perhaps surprising that given such unfavourable conditions, aquaculture has succeeded so well and has become an influential and profitable industry. There were several factors which assisted the development of the industry in its preliminary stages, as follows:

1. The lack of pasture land and suitable conditions for red meat production. One has to remember that in the 1930s, the broiler and turkey industry were not yet developed.
2. The habit of the Jewish population, however tenuously connected with religion, to include fish in their diet at least once a week, with a distinct preference for freshwater fish especially common carp.
3. The possibility of utilising land and water unfit for agricultural use for the production of animal protein.
4. The existence of suitable climatic conditions and skilled and enthusiastic workers who, from the beginning, kept the industry at a highly intensive and efficient level.
5. A well organised and controlled market which absorbed the product.

The results of this can be seen in the rapid development of the industry over the last 45 years (Table 9.1). The development of the industry proceeded through several stages with a gradual increase in intensification, as described below. Since land and water are the limiting factors of agricultural development in Israel, the governing strategy throughout has been to achieve maximum yields from minimum land and water, i.e. continuously increasing intensification.

9.2   STAGE I: CARP MONOCULTURE

At first, the eastern European method of fish farming was introduced utilising the common carp, which was not an indigenous species. Fish farmers realised immediately that improvements must be made for better utilisation of the local climatic conditions and to obtain maximal production from the limited land area and water. These improvements involved mainly the following:

1. Lowering the market size of the carp from 1 kg, which was more common in the eastern European countries, to a size of 400–700 g. This enabled denser stocking rates, shorter growing periods, better utilisation of the seasons and lower costs of production.

Table 9.1 The development of fish culture in Israel (1940−85)

Year	No. of farms	Area in ha	Production in tonnes	National average in tonnes/ha
1940	2	30	50	1.00
1945	47	990	1256	1.26
1950	80	2100	3700	1.76
1955	99	3630	7315	2.01
1960	93	4180	8432	1.97
1965	92	5065	10135	2.00
1970	83	4741	11954	2.49
1975	80	4457	12984	2.91
1980	70	3407	11492	3.37
1985	60	3034	12560	4.14

2. The size of each pond was limited to not more than 10 hectares instead of tens of hundreds of hectares as in the eastern European countries (Plate 9.1). Construction of smaller ponds is certainly more expensive, but allows more efficient intensification, mechanisation, easier manipulation of feeding, fertilising, harvesting and disease control.
3. The supply of fish fresh to the markets (live, in the case of common carp) continued throughout the year, rather than only in a short concentrated season. The supply of fresh fish was organised through a wide net of shops and supermarkets. In this way, pondfish became a popular item in the daily diet of the family.

It was also very soon discovered that the local temperate climate, a long season of optimal temperature and abundance of natural food in the ponds enabled the stocking rates to be increased to 1000−1200 fish/ha, instead of 300−800 fish/ha, which was common in Europe at that time. In planning stocking rates in ponds, one has to remember, in addition to better utilisation of the natural resources, that the stocking density has a two-way influence on individual growth and on the total yield per unit area. It is up to the farm manager to adjust stocking rates to the general planning policy. Tables 9.2 and 9.3, which are based on many years of experience in Israel, illustrate the influence of stocking rates on individual growth and on yields per hectare.

Six years after the pondfish industry started, yields reached a

Plate 9.1 Complex of fish ponds in Israel.

Table 9.2 The influence of different stocking rates on individual growth of carp

Stocking number per ha	Daily individual increment in g	Final weight after 50 days of growth in g
30 000	0.3	17
10 000	1.0	52
5 000	2.0	102
1 000	10.0	502
300	20.0	1002

NB: Basic pond productivity = 10 kg/ha per day; initial weight of fish = 2 g.

national average of 1.0–1.2 tonnes/ha, as compared to 0.3–0.6 tonnes at that time in eastern European countries under similar conditions. However, higher stocking rates per unit area will not necessarily lead to more efficient utilisation of the pond as a whole. The effect will be concentrated on the particular ecological niche in the pond occupied by and specific to the feeding habits of the fish species stocked. Achieving greater overall utilisation of the pond necessitated moving away from monoculture of carp.

9.3 STAGE II: POLYCULTURE

The second stage of development involved the introduction of polyculture. The principle of polyculture is based on the assumption

Table 9.3 Influence of different stocking rates of carp on yield per hectare

Stocking rate per ha × 1000	Initial weight in g	Weight at harvest in g	Daily fish increment in g	Daily fish increment/ha in kg
8.5	460	750	7.25	61.5
15.0	291	670	5.20	78.0
30.0	186	599	3.18	95.0
40.0	119	487	3.17	127.0

that fish growth is an expression of their reaction to, among other things, the sources of natural food in the environment. The variations in daily growth increment per hectare can serve as a measure of changes occurring in the pond and the degree to which fish growth is affected. This can be exploited in various ways to increase yields of farmed pondfish.

9.3.1 Common carp polyculture

Only a few years after introducing European carp monoculture systems, the local farmers realised that under Israeli conditions, populations of carp reared alone in production ponds do not utilise the natural food niches of the pond efficiently. It was also found that different size groups of common carp utilise different sources of natural food. This was the basis of the development of a 'polyculture' system where two size groups of carp were raised in the same pond, serving as a production and nursing pond at the same time. While the larger fish reach marketable size, the fingerlings grow to a suitable size for stocking in production ponds (Plate 9.2). It is considered that this method increases pond productivity by 25–30% and Table 9.4 shows examples of four such ponds.

9.3.2 Carp/mullet polyculture

The second phase of polyculture was the introduction of additional fish species based on their different utilisation of natural feed sources and on the synergism which can exist between different fish species. The total capacity of a pond to produce fish is the sum of production of different ecological niches existing in the pond.

Plate 9.2 Israeli mirror carp; above: 50 g fingerling for stocking grow-out ponds; below: 1.2 kg marketable carp obtained 200 days after stocking as fingerling size.

Table 9.4 Polyculture of two sizes of carp (data from 4 ponds in Israel)

Pond area (ha)	Stocking rate per ha		Initial weight (g)		Harvest size (g)		Yield (kg/ha)		Days of growth
	Large	Nursling	Large	Nursling	Large	Nursling	Large	Nursling	
3.5	1740	2120	190	4	900	320	1098	279	174
10.0	2050	2280	180	0.5	900	210	1242	1024	226
3.0	1870	2500	140	20	550	100	650	175	84
3.5	1610	2570	155	20	675	140	730	272	76

Thus a mixed stock of selected fish species, with complementary feeding habits and different ecological requirements, will fill the different ecological niches in the pond and utilise in the most efficient way the resources available in the pond for fish production. Where there are no differences in growth increment between mono-culture of one species only and polyculture of several fish species within the same pond, this indicates that there is no competition among the fish species. However, in certain instances the daily growth increment of one or more fish species was found to be higher in mixed culture due to a positive mutual influence with each fish species present affecting the environment in such a way as to improve growth conditions for the other species. Such an affect has been called synergism and the degree of influence of one group of fish upon another can be calculated by the formula $C = (a - b)/a$, where C is the competition index, a is the daily yield of the given species in monoculture, and b is the daily yield of this species in polyculture.

This theory that ecologically different species will not compete remains true only within certain limits of stocking density or food abundance. With increased stocking density inter-specific and intra-specific competition will eventually increase, and fish production will slow down. Consequently, the various inter-relationships between different fish groups must be understood and appropriate stocking densities for each species must be worked out.

On this basis the introduction of the grey mullet (*Mugil cephalus*) was of great importance. Mullet fry are abundant in great numbers along the eastern coast of the Mediterranean. Of the five species of *Mugilidae* occurring it was found that grey mullet in particular adjust easily from seawater to freshwater and grow very fast in ponds. In 12–16 months this species increases from 0.2 g when caught along the coast, to a weight of 500–800 g without using supplementary food, whereas in the sea it takes 3 to 4 years to reach the same weight.

The addition of grey mullet to production ponds gave an additional yield of 150–200 kg/ha. The total quantity of mullet in pondfish production has varied from 500·to 800 tonnes annually, representing 4–7% of total production and approximately 40% of total mullet supply on the market. Mullet production in ponds depends on the number of fry caught when they appear in the estuaries of the Mediterranean coast and this depends on various environmental and climatic factors influencing their movements. The price of mullet on the local market is more than twice that of common carp or tilapia. This high market value, coupled with the fact that they are cultivated without supplementary feed, makes the contribution of mullet to the profitability of the industry much higher than their 4–7% contribution by harvest weight. Considerable progress was made in pond cultivation of mullet after Israeli scientists established the correct seasonal appearance along the coast of shoals of the two main species suitable for farming *Mugil cephalus* and *Liza ramada*, and once an easy method of identification had been developed for fish of 18–22 mm in length.

The fry are transported in tanks for distances of more than

300 km without any damage and they adjust easily to brackish water ponds with salinities of 1−1.5 parts per thousand. In the first phase the fry are nursed in polyculture with carp, at stocking densities of 20000−30000/ha, reaching 15−40 g after 150−200 days of growth. Survival is approximately 70% and the daily individual growth increment is about 0.25 g (Table 9.5). In the second phase the nursed fingerlings are grown on again in polyculture to marketable size (500−800 g) without any supplementary feed. In recent years, average yields in ponds have increased by 80% compared with the mid-1960s due to various technical improvements which will be considered later. Nevertheless, on farms specialising in the cultivation of grey mullet in a polyculture system, the proportion of mullet has remained at about 20% of the total yield (Table 9.6).

9.3.3   Use of tilapia in polyculture

Israel is situated at the northern tip of the Syrian−African rift and is almost the northern limit of the natural range for tilapia. The minimum temperature for tilapia species cultivated in Israel is 12°C. During the early years of the development of Israeli fish farming, tilapia were considered a nuisance in the fishponds. Two of the most common tilapia in Israel, *T. zillii* and *Oreochromis aureus*, are highly prolific and attain sexual maturity at the end of

Table 9.5 Nursing of mullet fry in polyculture in fishponds

| | Pond number | | | |
	2	4	6	8
Pond area (ha)	3.6	3.0	3.0	3.0
Days of growth	207	167	145	190
Fingerlings harvested (1000/ha)	17.5	11.7	18.7	17.0
Survival (%)	57	70	70	69
Weight at harvest (g)	35	40	33	15
Increment (kg/ha per day) mullet	2.9	2.3	2.6	1.3
carp	8.2	6.8	2.6	2.5
total	11.1	9.1	5.2	3.8
Proportion of mullet in total yield (%)	26	25	53	34

NB: Initial size = 18−25 mm; 0.25 g.

Table 9.6 Raising mullet to marketable size in polyculture

	Pond number				
	1	2	3	4	5
Pond area (ha)	3.5	3.3	8.0	10.0	3.3
Days of growth	175	215	240	187	230
Number of fish harvested/ha	2130	2120	1310	2090	1830
Survival (%)	85	100	100	100	100
Initial weight (g)	90	100	111	163	166
Weight at harvest (g)	500	650	870	500	640
Yield (kg/ha) silver carp	806	497	804	495	651
mullet	857	1196	1006	718	870
carp	1664	3075	4464	2424	2237
tilapia	172	1009	803	349	416
total	3552	5777	7077	3986	4174
Proportion in					
total yield (%) mullet	24	21	14	18	21
carp	48	53	63	61	54
tilapia	5	17	11	9	10
silver carp	23	9	12	12	15
Food conversion ratio	0.9	1.4	1.3	1.3	1.7

their first year of life. Under Israeli conditions this means they may breed at the rate of almost 6 spawnings per season. As a result, one-year-old tilapias (sometimes 50–100 g in size) contaminate the ponds with millions of fry from wild spawnings, which impede the normal growth of the commercially stocked fish populations.

The situation changed drastically in the mid 1960s when, following research by Israeli scientists, the hybridisation of two tilapia species resulting in almost 100% male progeny was achieved and adopted by the pondfish farmers. In addition, mass production of males by sex reversal methods has also taken place recently. The tilapia raised in Israeli fishponds are hybrids from crossing *Oreochromis aureus* × *O. niloticus* (Plate 9.3). The possibility of cultivating all-male stocks of tilapia in polyculture without endangering the ponds by wild spawnings boosted tilapia cultivation and this species became the second most important fish, after carp, in the Israeli polyculture system. The production of pond tilapia has increased during the last 20 years almost five-fold and its contribution to total pondfish production has increased from 5% in 1963–7 to 23% in 1983–4 (Table 9.7).

Plate 9.3 A tilapia hybrid (*Oreochromis aureus* × *O. niloticus*) male.

Table 9.7 Marketing of tilapia from fishponds during 1963−84

Years	Tonnes	Proportion of total pondfish marketed in %
1963−67	2 402	5.0
1968−72	5 450	9.5
1973−77	8 547	13.0
1978−82	11 838	19.8
1983−84	5 169	23.1

9.3.4   Current polyculture systems

In recent years polyculture in Israel has utilised from 3 to 5 different species. In general terms the common carp represents 65%, tilapia 23%, grey mullet 7% and silver carp 5% of the total, each species utilising a different ecological niche of natural food in the pond. Supplementary food is used to supply the needs of common carp and tilapia only. Consequently, as can be seen from Table 9.1, the national average yield at the beginning of the 1970s reached 2.5−3.0 tonnes/ha (almost twice as much as 20 years previously) and food conversion rates had decreased to about 2.0. It has been found that male tilapia (in polyculture or monoculture), when fed with protein-rich pellets, give excellent individual daily growth rates of 2.0−3.5 g at high stocking rates and have the capacity of producing high yields per unit of area. Another advantage of stocking tilapia is that in most cases, where they are reared together with carp, the growth of carp has consequently increased.

This may be associated with the improvement of the plankton balance in the ponds due to grazing by the tilapia, which controls the negative effects of heavy algal blooms, such as an increase in organic load in the sediment and the development of anaerobic conditions, oxygen depletion and an increase in toxic levels of ammonia and nitrite.

The local market requires tilapia in the size range 300–700 g and a continuous daily supply throughout the year. This demands spawning as early as possible and fry should be nursed as quickly as possible and at high stocking rates in order to reach a suitable size for early stocking in grow-out ponds. Two methods exist for nursing tilapia fry in earth ponds: the first uses monoculture and heavy stocking rates with a saving in pond space, but fingerlings reach a smaller size at the end of the season; and the second uses lower stocking rates, and in polyculture with common carp. Table 9.8 gives several examples of the two methods in terms of production per hectare. The common combination adopted for tilapia to market size is in polyculture with common carp (60:40 by ratio) with the addition of 10% silver carp and/or grey mullet. Such a combination gives a maximum utilisation of most of the natural food niches of the pond. In addition, as already mentioned, there is an improvement in the water quality of the pond, involving factors such as oxygen, ammonia and nitrite, which under certain instances can be very dangerous to the fish, especially at heavy stocking rates. As an example, Table 9.9 shows the results of such

Table 9.8 Nursing of tilapia fry (data calculated for 1 ha)

	Monoculture		Polyculture		
	Pond 1	Pond 2	Pond 3	Pond 4	Pond 5
Area of pond (ha)	0.15	0.12	0.7	3.0	3.5
Days of growth	48	72	58	107	72
Number at harvest ($\times 10^3$)	700	460	185	90	31
Initial weight (g)	0.8	0.8	1	1	1
Weight at harvest (g)	8	24	15	54	33
Daily average individual increment (g)	0.15	0.32	0.25	0.50	0.40
Daily increment (kg/ha)	105	148	45	43	14
Number of carp in polyculture* ($\times 10^3$)	—	—	3	2	2.5

*The addition of common carp in polyculture prevents the development of filamentous algae which complicate the harvest.

Table 9.9 Growth of tilapia in polyculture (data calculated for 1 ha)

	Pond number			
	1	2	3	4
Total number of all fish species harvested	8800	5500	6500	7400
Number of tilapia harvested and (%) of total	1810(20)	3030(54)	4000(61)	4900(66)
Initial weight of tilapia (g)	100	190	175	150
Weight at harvest after 100 days (g)	480	470	505	390
Daily individual increment of tilapia (g)	3.8	2.8	3.3	2.4
Daily individual increment of carp (g)	4.2	5.0	7.3	5.3
Total daily increment (kg/ha)	36.6	21.0	31.5	24.3
Percent of tilapia in daily increment (%)	19	40	42	46

a polyculture system, whereas Table 9.10 shows the overall influence of polyculture on fish yield in Israel when compared to monoculture.

9.4   STAGE III: MORE INTENSIVE SYSTEMS

In the past decade the fish farming industry has encountered an increase in production costs of water and energy. Using the existing infrastructure of ponds, equipment, etc., the farmers have chosen to increase yields by the introduction of so-called 'intensive' or 'semi-intensive' cultivation systems in order to obtain higher yields from the same area, water and labour inputs (see chapter 10; section 10.3.1). These systems are based on the following principles:

1. Aeration of the pond by various artificial methods. This is an alternative to using increased water flow as a means of enriching the pond with oxygen, which is impossible in Israel but vital for higher stocking rates and hence higher yields.

2. Feeding with protein-rich pellets and with controlled automatic feeders enabling the fish to take food on demand but quantitatively controlled by the farmer according to feeding charts. The need for protein-rich feed is a direct result of the decreased quantities of natural food per fish with increasing

Table 9.10 Influence of polyculture on fish yield in comparison with monoculture

	Monoculture in tonnes/ha	Polyculture (carp, tilapia, silver carp or grey mullet) in tonnes/ha
Fertilized, without feeding	1.2	3.0
Fertilized, with supplement feed	3.5	9.4

stocking rates. Observations over many years have shown that as long as the total weight of the fish stock (fish biomass) is in the range of 1–1.2 tonnes/ha, there is no need for increased supplementary protein rates. Carbohydrate-rich feeds, such as sorghum, corn or wheat, are preferable to balance the fish diet as the natural food base is sufficient to supply all the necessary protein. Only when the fish biomass exceeds 2 tonnes/ha is enrichment of the supplementary feed by protein-rich ingredients required.
3. A three- to six-fold increase in stocking rates of various fish species, whereas the supplementary feed is calculated for carp and tilapia only.

Depending on the various local conditions, the intensification of different farms has been carried out using either semi-intensive or intensive systems. The difference between the two systems is mainly in the aeration facilities, which also determine the stocking rates of the different fish species. In intensive ponds the aerated area covers at least 50% of the pond enabling higher stocking rates. For this purpose, it is recommended that ponds not larger than 1.5–2.0 ha be used. In contrast, the semi-intensive system is based on aeration of 10–15% of the pond area only, giving an increase in stocking density of fish by only 2 to 3 times compared with conventional ponds. As existing fish farms already contained ponds of 3–10 ha in size they have been modified to the semi-intensive system, i.e. partial aeration and installation of controlled demand feeders (Plate 9.4).

When planning a new site for a fish farm, however, only the intensive system is recommended if the aim is to maximise fish yield. Thus the increased demand for tilapia on the local market and the export potential of processed products, such as tilapia fillets, following the construction of processing plants, have urged

Plate 9.4 A controlled demand feeder and paddlewheel aerators.

the fish farmers to investigate more intensive methods for tilapia culture. The first trials with high stocking rates of male tilapia in monoculture and polyculture started in the second half of the 1970s. Daily increments of up to 160 kg/ha were obtained in monoculture. By extrapolation this means yields of more than 30 tonnes/ha in 200 days of growth. Fairly similar results were also achieved in polyculture together with common carp (Table 9.11).

Two obstacles which were sometimes encountered in farm ponds had to be overcome, namely high food conversion ratios which indicated bad utilisation of the additional food, and the build-up of

Table 9.11 Cultivation of tilapia at high densities in monoculture and polyculture

	Monoculture of tilapia	Polyculture		
		Tilapia	Carp	Total
Number of fish at harvest	80	10.0	9.5	19.5
Initial weight (g)	150	290	387	—
Weight at harvest (g)	354	500	665	—
Daily individual increment (g)	3.0	5.5	7.3	—
Daily increment per ha (kg)	163	55	69	124
Extrapolated yield for 200 days (kg/ha)	32.6	—	—	24.9
Food conversion ratio	3.1	—	—	2.5

metabolites and chronic oxygen deficiency due to heavy fish biomass and food use. To overcome this, a modification was made in the existing small ponds by adding an additional outlet in the centre of the pond, which enabled frequent flushing from the pond of excess metabolites. The sludge can be removed very rapidly in this way. For instance, in 20 seconds sludge containing 80 mg/l suspended matter has been flushed out of such ponds. By doing this several times during the day, ponds are cleared, fish growth is accelerated and food utilisation improved. Resulting yields can reach up to 50 tonnes/ha with reasonable food conversion rates.

A further step has been the cultivation of fish in concrete tanks. Such a system requires high inputs and a high turnover of water, but it is capable of producing large quantities of marketable size fish from very small units. Results obtained under pilot-plant conditions in concrete tanks of 50 m² gave yields of 9–21 kg/m² in 164 days of growth (Table 9.12). Experience over the last 10 years from commercial fish farms shows that yields of 30–35 tonnes/ha can be obtained from intensive ponds and 10–12 tonnes/ha from semi-intensive ponds. The increase in yield has reduced the quantity of water needed to produce 1 tonne of fish from 12 000 m³ to 5000 m³, which considerably reduced production costs as water represents more than 20% of total expenditure. In addition, thanks to the significant increase in yields, there has been a 35% reduction of total pond area in the last 14 years producing almost the same

Table 9.12 The results of intensive tilapia culture in 50 m² concrete tanks

	Stocking density of fish per m²	
	50	100
Initial weight (g)	140	140
Weight at harvest after 164 days (g)	600	523
Daily individual increment (g)	2.76	2.35
Daily increment per m² (g)	138	235
Initial biomass of fish (kg/m²)	7	14
Fish biomass at harvest (kg/m²)	29.6	52.5
Increment in 164 days per m² (kg)	226	38.5
Food conversion ratio	2.5	2.4

NB: Aeration was achieved with a 0.5 HP paddle-wheel aerator; feeding with 30% protein-containing pellets at a rate of 2% of fish body weight via automatic feeders 6 times a day; the water volume of the tank was changed twice a day.

quantity of fish and releasing more than 1600 ha of land for other agricultural purposes.

Table 9.13 compares the yields from conventional, semi-intensive and intensive ponds. However, as discussed further in chapter 10, increased yields per unit of area are generally justified only if they also increase the profitability of production. Therefore, the different yields resulting from different levels of intensification have to undergo economic evaluation (chapter 10; section 10.3.2). A comparison of costs and returns from the three different systems is therefore presented in Table 9.14 (based on 1983 data) and the results are clear-cut. Yields of 3 tonnes/ha in the conventional system were not economical at all. The semi-intensive system resulted in a net profit of almost $300 (£162) per tonne, after deducting the cost of construction capital and depreciation. However, the intensive system was not economically justified unless the yields exceeded 20 tonnes/ha due to the very high cost input for construction. As mentioned before, this system was not introduced until recently, whereas existing ponds could, in the meantime, be modified with relatively small investment to the profitable semi-intensive system.

Since the beginning of the 1980s all the fish farms in Israel have been modified to the semi-intensive system. The precise details of such farms vary somewhat, depending mainly on factors such as the sources of water supply (whether continuous throughout the

Table 9.13 Comparison of yields in intensive, semi-intensive and conventional ponds

		Intensive	Semi-intensive	Conventional
Area of pond (ha)		0.5	4.3	4.3
Days of growth		80	120	100
Carp	number per ha	15 040	6 140	2770
	daily increment (kg/ha)	72.5	29.5	10.0
Tilapia	number per ha	21 310	12 800	3000
	daily increment (kg/ha)	40.2	5.7	1.5
Grey mullet	number per ha	—	1 000	370
	daily increment (kg/ha)	—	1.6	1.7
Silver carp	number per ha	—	650	690
	daily increment (kg)/ha)	—	7.7	1.1
Total increment per day (kg/ha)		112.7	44.5	14.3
Total number of fish per ha		36 350	10 590	6830
Food conversion rate		2.6	2.5	2.3
Extrapolated yield for 200 days of growth (tonnes/ha)		22.5	8.9	2.9

Table 9.14 Average costs and returns from intensive, semi-intensive and conventional ponds (in US$)

	Intensive	Semi-intensive	Conventional
Yield per ha (tonnes)	20	9	3
Data for the production of 1 tonne			
area (ha)	0.05	0.11	0.33
water (m^3)	2250	5000	12 000
cost of construction ($)	5000	3300	5000
labour (man-days/tonne)	5	5	4
food conversion rate	2	1.7	2
Annual fixed costs ($/tonne)			
cost of construction capital	750	495	750
(at 15% including depreciation)			
annual variable costs ($/tonne):			
feed @ $250/tonne	500	425	500
fingerlings @ $0.045 per fish	163	48	30
labour @ $35/day	175	175	140
water @ $0.04/$m^3$	90	200	480
pond management costs	40	60	50
Total costs	1718	1403	1950
Sales revenue per tonne of fish			
(average of mixed stock)	1700	1700	1700
Net income or loss per tonne ($)	−18	+297	−250

season or during the rainy season only); the type of polyculture (how many fish species); and whether the farm supplies its own stocking material or purchases it. The following is an outline of a production plan for a typical semi-intensive farm based on a production of an average of 5 tonnes/ha:

1. *Basic data* Total annual production of marketable fish = 250 tonnes.
   Mean yield/ha = 5 tonnes. Total area = 50 ha.
   Fish species cultivated: common carp, tilapia, grey mullet and silver carp.

2. *Planning of area* Nursing ponds = 15 ha (30%); grow-out ponds = 32 ha (64%); storage and service ponds = 3 ha (6%).

3. *Size of ponds* Six ponds of 0.5 ha each = 3 ha; 10 ponds of 1.5 ha each = 15 ha; 10 ponds of 3.2 ha each = 32 ha.

4. *Nursing the stocking material* (including 30% loss)
   Stocking of nursery ponds:
   > carp: 220 000 of 2−5 g; end of May
   > tilapia: 250 000 of 2−3 g; end of June
   > grey mullet: 65 000 of 1−2 g; April
   > silver carp: 10 000 of 300 g (purchase from a hatchery for direct stocking in grow-out ponds)

   The total number of stock = 545 000 fingerlings.

5. *Stocking the nursery ponds*
   22 000 carp/ha reaching a size of 200−300 g at harvest in polyculture together with 6500 grey mullet reaching 150−200 g at harvest in 10 ha of nursery ponds.
   50 000 tilapia reaching 70−100 g at harvest in 5 ha of nursery ponds.

6. *Expected yields of marketable fish*

Fish species	Total in tonnes	Mean weight in g	Stocking number per hectare	Number of fish harvested
Carp	125	750	5 200	167 000
Tilapia	75	400	5 900	187 000
Grey mullet	30	600	1 600	50 000
Silver carp	20	2 500	250	8 000
Total	250		12 950	412 500

Stocking and marketing are undertaken throughout the year. The same also applies to all the other operations like fertilising, stocking and harvesting.

9.5 STAGE IV: INTEGRATED FISH FARMING

The fish farming industry has now had to face a new crisis due to increased production costs, mainly water and energy. In addition, new methods of crop irrigation have enabled the use of brackish water with salinities of up to 35 parts per thousand for growing cotton and other crops. Consequently, these profitable alternative crops have competed economically with fish production. Many fish farms overcame this problem by revamping the ponds for use as irrigation reservoirs and integrating them into the farms general water scheme. In such a system the water is used during the first part of the Summer for growing fish and from July onwards for irrigation of the surrounding fields. In practice, the cost of production of the fish in these reservoirs does not include the expenses involved in the supply of water which are incurred by the irrigated crops. As an example, Table 9.15 depicts the water balance of such a reservoir in which almost 80% of the accumulated water was utilised for irrigation and feeding the surrounding fish ponds, and only 20% served the fish produced from this reservoir.

More than 20% of the fish pond area in Israel was adapted in 1982 to the dual purpose of fish farming and irrigation and two types of reservoirs are currently in use. The long-season reservoirs are located in areas where brackish water is available throughout the year and at salinity levels suitable for partial-use irrigation as well as for fish production. At the end of the irrigation season in October, the reservoirs are refilled and fish production resumes. In these reservoirs, production lasts for about 250 days.

The short-season reservoirs depend on the flood during the rainy season. The water is used for irrigation until September at which time the fish are finally harvested and the reservoirs then remain dry until the next rainy season. The fish production in these reservoirs lasts for 150–190 days only.

Both types of reservoirs range in size from 8 to 30 ha with a maximum depth of 3–8 m. In all reservoirs, the classic Israeli system of polyculture of 3 or 4 fish species is practised. Table 9.16 gives various data on stocking policy, yields and food conversion ratios for 6 reservoirs of both types.

Experience has shown that the growth capacity and yields of these reservoirs, under the temperate climatic conditions in Israel, are positively influenced by the large volume of water. In terms of surface area, production of the target species is much higher than in shallow conventional fish ponds. Moreover, the accumulation of

Table 9.15 Water budget of an aquaculture-irrigation reservoir

Size of the reservoir (ha)		17
Volume of water accumulated in the period October–May ($m^3 \times 10^3$)		832
Utilisation of the water:		
for irrigation of 107 ha crops ($m^3 \times 10^3$)	628	
for feeding surrounding fish ponds ($m^3 \times 10^3$)	33	
remaining in the reservoir at final fish harvest ($m^3 \times 10^3$)	70	
loss through evaporation and seepage ($m^3 \times 10^3$)	101	832

Table 9.16 Yields of different fish species and food conversion ratios in long- and short-season reservoirs

	Long-season			Short-season		
	Pond 1	Pond 2	Pond 3	Pond 4	Pond 5	Pond 6
Carp						
number per ha	4 000	3600	6 900	4300	6900	3980
yield (tonnes/ha)	5.26	8.08	6.04	2.78	4.72	2.72
percentage in total yield	41	64	54	42	76	51
Tilapia						
number per ha	13 250	2500	10 000	1000	—	1000
yield (tonnes/ha)	4.82	2.03	2.83	2.18	—	1.58
percentage in total yield	37	16	25	33	—	30
Grey mullet						
number per ha	2 000	1800	2 000	2000	2000	1990
yield (tonnes/ha)	1.80	0.75	0.57	0.64	0.84	0.70
percentage in total yield	14	6	5	9	14	13
Silver carp						
number per ha	500	550	500	300	200	300
yield (tonnes/ha)	1.03	1.79	1.80	1.08	0.60	0.31
percentage in total yield	8	14	16	16	10	6
Total						
number per ha	19 750	8450	19 400	7600	9100	7270
yield (tonnes/ha)	12.91	12.65	11.24	6.68	6.16	5.31
food conversion ratio	1.23	1.74	1.33	1.50	2.40	2.60

metabolites, found to be a growth-limiting factor in conventional intensive pond systems, does not take place in these deeper reservoirs due to the large volume of water. Higher stocking rates and greater fertilisation with poultry manure, which enhances the primary productivity in the reservoirs, both contribute to greater final yields.

During reservoir construction, adequate access to the water for heavy vehicles and equipment is necessary at all times and complete discharge of all the water into the system is vital. Similarly, the automatic feeders must be installed using a pontoon system so that they remain operational despite changes in water level. However, the main problems associated with such a system involve coordinating the needs, not only of the fish farmer, but also of the crop grower. There is also a need for a considerable number of service ponds for nursing the fish to a suitable size for stocking and subsequently holding market-size fish following harvest from the reservoir (Plate 9.5). Due to a relatively shorter growing season, the size

Plate 9.5 Harvesting an irrigation reservoir using a vacuum pump and sorting different species and sizes of fish.

and stocking density are critical in order to obtain fish of suitable size when irrigation necessitates the draining of the reservoir. Management strategies must be directed to accurate stocking and timely 'thinning' of the fish stock, as the water used for irrigation reduces the area in the pond suitable for fish growth, in order to maintain adequate fish biomass to water volume ratios.

9.6 FARMING NEW SPECIES

9.6.1 Introduction

Three recent developments have encouraged the authorities responsible for the planning of the industry to examine the possibility of introducing new species suitable for export. These developments were the saturation of the local market with pondfish, the dramatic increase in yields/ha which made available spare production capacity in the existing fish pond area, and also the national interest in developing exportable products. The resulting attempts to diversify into farming new species have included ornamental fish as well as different food-fish and crustacea. Thus 12 farms are now engaged in production and export of various exotic crossbreeds for ornamental purposes. Such farms vary in size from several tanks up to 3−4 hectares and some utilise the existing pond area. As regards aquaculture for human consumption, emphasis has recently focused on farming eels (*Anguilla anguilla*) and freshwater prawns (*Macrobrachium rosenbergii*). Although this text is primarily concerned with finfish cultivation, freshwater prawns are now being farmed in polyculture with finfish in Israel (and elsewhere) and brief details will therefore be given.

9.6.2   Freshwater prawn farming

The giant freshwater prawn was introduced in the late 1970s. One central hatchery and several smaller ones produce postlarvae in intensive recirculating and heated facilities. Israel's climate means that only 6−7 months have a suitable ambient temperature for this tropical prawn. Therefore, postlarvae are sold to commercial nurseries heated by geothermal, solar or conventional means, which then hold juveniles throughout the winter.

The advantages of releasing nursed juveniles into production ponds, as opposed to newly metamorphosed postlarvae, are as follows:

1. Higher production and greater final mean size.
2. Juveniles are more suitable for maximising production in a climatically limited growing season.
3. Losses during the first stages of grow-out are reduced due to the fact that juveniles are more resistant to the wide range of fluctuating conditions in an open grow-out pond.
4. Stocking with nursed juveniles enables polyculture with fish.

In ponds stocked with nursed postlarvae, more than 50% of the total yield was medium sized (30 g) or larger, compared with 21% in ponds stocked with non-nursed juveniles. The yields of marketable (>30 g) prawns generally decreased as stocking density increased for any given density within the range examined. In ponds stocked with nursed juveniles, however, the yields were significantly higher than those obtained in ponds stocked with postlarvae. It was therefore evident that efficient commercial nursery operations were vital for the further development of the freshwater prawn industry.

In Israel the grow-out ponds are filled with water during the winter. Temperature becomes suitable for stocking prawns in May. In November the temperature decreases to below 20°C and ponds are totally harvested. Nursed prawns are therefore raised to market size, subject to the constraints of annual water exchange and during a period of 6−7 months. Results of commercial polyculture in regular fishponds are given in Table 9.17. The yields were 90−375 kg/ha within a grow-out period of 150−175 days. Although the proportion by weight of prawns in the total yield of the pond was only 2−6%, in terms of value it was quite a lot higher due to the fact that the net income from prawns is 3 to 5 times higher than that from pondfish. Stocking densities for prawns in these systems were 0.2−1.5/m². the weight of the prawns at harvest was 32−70 g and survival rates varied from 40 to 95%. About 90% of these prawns were above the large size product (>45 g). No feed was added specifically for the prawns, although the pondfish were fed by automatic feeders situated in one corner of the pond. This demonstrates the ability of the prawns to derive nutrition from the pond's natural productivity.

9.6.3   Eel farming

The long season of more than 250 days of temperatures in the range 18−25°C is ideal for eel culture and the European market for the product looks promising, particularly in West Germany and Holland (see chapter 3; section 3.6). Encouraging results have been obtained from nursing elvers, which are imported from Europe where they are caught as they enter rivers during the period from February to June. Because of the suitable temperatures in March, elvers imported in early February reach a size of 2−5 g as early as the beginning of June, when they are transferred from the hatcheries to grow-out ponds. The results obtained in hatcheries, showing a survival rate of 25−30% of 2.5 g fingerlings, are recognised as a considerable achievement from the economic standpoint. At the time of writing, however, the techniques of further cultivation in grow-out ponds in monoculture are still on a pilot-plant basis. Nevertheless, in 1984 15 tonnes of marketable eels of 300−500 g reached the market.

Table 9.17 Polyculture of various fish and prawns in commercial ponds

Size of pond in (ha)	Species	Stocking per ha	Days of growth	Av. weight stocked in g	Av. weight harvested in g	Loss in %	Biomass stocked in kg/ha	Biomass harvested in kg/ha	Food conversion ratio
7.0	Common carp	4 000	230	15	850	1	60	3 366	
	Tilapia	5 500	205	150	450	2	825	2 426	
	Mullet	2 000	250	130	950	2	260	1 710	
	Prawn	15 000	165	0.25	55	50	37	412	
	Total						1182	7 914	1.4
6.0	Common carp	5 600	220	25	850	1	140	4 726	
	Tilapia	8 000	70	265	420	3	2048	3 268	
	Mullet	2 500	220	150	1000	5	375	2 370	
	Silver carp	390	220	27	2000	0	11	780	
	Prawn	7 500	160	0.2	32	25	1	180	
	Total						2575	11 324	1.7
5.0	Common carp	3 700	237	22	900	1	81	3 303	
	Mullet	3 800	210	100	550	2	380	2 046	
	Silver carp	350	237	43	2000	0	15	700	
	Grass carp	400	240	10	800	0	4	320	
	Prawn	2 000	175	0.9	70	32	2	96	
	Total						482	6 465	1.8
8.0	Common carp	3 000	250	9	800	0	27	2 400	
	Tilapia	13 000	80	4	75	2	54	994	
	Mullet	3 000	255	40	830	4	120	2 382	
	Prawn	4 500	153	0.5	60	56	2	120	
	Total						203	5 896	0.9

9.7   THE CONTRIBUTION OF FISH FARMING TO ISRAELI FISH CONSUMPTION

The annual per capita consumption of fish in Israel has been fairly stable at around 10 kg (wet weight), with more than 40% of this being imported. The importance of pondfish in the total production is greater, representing 50% of the total local fish production from all sources and 30% of the total consumption. The tendency on the part of the Israeli population is to purchase fish as 'live, fresh or frozen'. More than 60% of the total demand for the last 5 years belongs to this category. The influence of pondfish in the total consumption is illustrated by the fact that it provides almost 50% of the demand for fresh fish (Table 9.18). Of those species popular on the local market, pondfish are dominant in supply. The common carp which comprise 19% of the total consumption are supplied totally from fish ponds. Tilapia comprise 7% of the consumption and 86% come from ponds while 55% of mullets, which comprise 3% of the consumption, are also supplied from ponds.

The result of the continuous intensification of the fish farming industry during the various phases mentioned above has been an increase in average yields per hectare. As market demand did not increase accordingly, a wide gap developed between production and marketing. Consequently, as the potential of the existing pond areas could not be utilised, the area of ponds and the number of farms decreased, especially small-scale farms and those which, because of technical reasons (i.e. cost of energy, water, labour, etc.) were on the threshold of profitability. From 80 farms covering an area of 4500 ha in 1975, only 62 farms covering an area of 3100 ha were operating in 1984, as illustrated by Table 9.19.

Table 9.18 The annual per capita consumption of fish in Israel in total and according to use (mean of the years 1980–5)

Source	Consumption		Live, fresh and frozen in %	Canned and processed in %
	kg	%		
Pond fish	2.9	29.6	46.8	3.7
Sea and lake – local	1.5	15.3	10.4	39.6
Atlantic deep sea	1.1	11.2	22.2	7.8
Imported	4.3	43.9	20.6	48.9
Total	9.8	100	100	100

Table 9.19 Fish pond area and resulting yields of marketed pond fish in the years 1975–84

Year	Net area in ha	Number of farms	Fish marketed in tonnes	Portion of carp in total		Average yield in kg/ha per year
				tonnes	%	
1975	4457	80	12 984	9072	70.0	2910
1976	4384	80	13 291	8700	65.5	3030
1977	4153	79	13 454	8139	60.5	3240
1978	?	77	13 117	8583	65.5	?
1979	3529	77	12 619	7604	60.2	3491
1980	3407	70	11 492	7226	62.9	3373
1981	3425	72	11 419	7579	66.4	3334
1982	3331	68	11 343	7155	62.7	3400
1983	3100	61	10 936	7701	70.4	3500
1984	3082	62	11 342	6857	60.4	3680

Economic evaluation has shown that in modern fish farming, where the infrastructure input costs like land, water, energy and labour are high, profitability is a direct result of increased yields per hectare or per water unit (chapter 10; section 10.3). Hand in hand with the reduction of pond area, the average national yield has increased continuously to cope with the total domestic demand as well as the growing export demand for processed pondfish products. The national average yield during the last 10 years increased by 770 kg/ha, from 2910 to 3680 kg/ha. However, the national average yield per ha does not reflect the real change in productivity of the ponds (as can be seen in Table 9.20) because fifty per cent of the total pond area yielded from 3.5 to more than 4.0 tonnes per hectare.

9.8   INSTITUTIONAL FRAMEWORK

The main governmental and private organisations which influence the production and marketing of fish farms are as follows:

- The National Board of Aquaculture.
- The Fisheries Department of the Ministry of Agriculture.
- The Extension Service of the Ministry of Agriculture.
- The Fish Breeders' Association.
- The Agricultural Research Organisation (Volcani Institute).
- Independent universities and private enterprises.

Table 9.20 Average market yields of pond fish from farms by productivity groups (1980 and 1984)

| Productivity group in kg/ha | 1980 | | | 1984 | | |
| | Number of farms | Area | | Number of farms | Area | |
		ha	%		ha	%
Up to 1500	3	48	1.4	6	218	7.3
1510–2000	7	258	7.6	6	218	7.3
2010–2500	9	456	13.4	4	232	7.7
2510–3000	11	507	14.9	8	410	13.7
3010–3500	9	477	14.0	6	240	8.0
3510–4000	14	803	23.5	9	470	15.8
Over 4000	17	858	25.2	23	1204	40.2
Total	70	3407	100	62	2992	100

The National Board of Aquaculture is a committee appointed by the Minister of Agriculture. The members are delegates from three main sectors: research, commercial farming and administration. The research sector is represented by prominent scientists affiliated to this field. The farmers are delegates of the Fish Breeders' Association and the administration is from various governmental offices. The main aim of this committee is to define the national policies and priorities for the development of aquaculture and to evaluate the research programmes, for which national funds are required. The definition of these aims is made known to other organisations which orientate their efforts and funds according to the priorities defined by the Board.

The Fisheries Department of the Ministry of Agriculture is divided into three main sections, Marine Fisheries, Fisheries of Lake Kinneret and the Aquaculture Division, which is the largest of the three. Each one of these sections deals with subjects related to research, development programmes, administrative and financial aspects, marketing, statistics, etc. The Lake Kinneret authorities are also responsible for the stocking of the lake. The aquaculture section operates two research centres: one in Ginosar and the other in Nir-David. The Ginosar station is responsible for the introduction of intensive fish farming which is now in operation on most of the Israeli fish farms. Until recently the main research programmes dealt with such problems as pond operation methods, manure and manuring, and nutritional requirements of carp and tilapia. The main current interests are the operational methods and nutritional requirements of freshwater prawn and eel culture.

The Ginosar station operates under a Board of Directors who represent most of the main aquaculture organisations. The station's personnel are competent fish farmers volunteered by various kibbutzim to work there for a period of 3 or 4 years. The station collaborates closely with various researchers from other universities or institutions who are allowed, under certain conditions, to use its facilities.

The Nir-David Laboratory for Fish Diseases gives free veterinary services to all the fish farms and also maintains a research programme in association with the Hebrew University of Jerusalem. Its research programme includes the study of pathological factors, such as parasitic, bacterial and fungal diseases, that influence the fish directly or indirectly, together with environmental conditions in ponds which inhibit growth or cause mortalities, etc. The editorial office of *Bamidgeh*, the quarterly journal on aquaculture (in English) of the Israeli fish breeders, is also housed in the Nir-David Laboratory offices.

The Extension Service of the Ministry of Agriculture provides free technical assistance to all farmers. Each area has its extension office with a special regional adviser. The task of these workers is to help the farmers prepare their annual and periodic production plans, to introduce innovations and to propagate the information of any technical progress made in the research stations or on other farms. The workers from all the regions are periodically summoned to discuss their professional problems and to establish the best guidelines for the various cultures. The chief extension officer is responsible for the realisation of the extension programmes throughout the country. He is expected to organise various professional courses for fish farmers and is also responsible for the implementation of all the development programmes. The chief extension officer also serves as the chairman of the National Board of Aquaculture

The Fish Breeders' Association is the organisation of all commercial fish farmers. Its main aims are:

1. The marketing of all fish farm products through the central 'Tnuva' cooperative, as well as arranging all the book-keeping related to this operation, and establishing pricing policy.
2. The respresentation of fish farmers vis-à-vis governmental or other authorities and institutions (research, insurance, etc.).

3. The import of necessary products and equipment.
4. The development and implementation of insurance policies for mass fish mortalities.

The organisation of marketing is undoubtedly the most important task of all. Each farm has a marketing quota voluntarily accepted by all the farmers with the aim of avoiding any possible over-production of fish farm products. Every week, all the interested farmers are asked to place a demand, specifying the quantities of fish they would like to sell. The marketing organiser is responsible for allocating the marketing requests coming from the cooperative market outlet (Tnuva) between the various producers according to their quota. This marketing system is strictly observed by all farmers because it is in their interest to do so. Thus, very few marketing opportunities are lost and the farmers can usually obtain the best possible prices for their products. Tnuva cooperates very closely with the Fish Breeders' Association on issues such as the establishment of prices for all fish farm products, marketing strategies and quality control.

The Agricultural Research Organisation (Volcani Institute) serves as the national centre for agricultural research. This centre operates two stations for aquaculture programmes, one in Dor and the other in Beit-Dagan. Dor was the first research centre for fish culture in Israel and its workers have taken part in most of the aquaculture developments during the last 30 years. The main areas of research are intensive polyculture, genetic selection for tilapia broodstocks used for hybridisation purposes, genetic selection for *Macrobrachium rosenbergii* and nutritional requirements of tilapia.

Researchers from various universities and other private enterprises are directly involved in aquaculture research programmes. This sort of cooperation between the various organisations has proved to be very fruitful. Some of the main subjects studied by these sectors include freshwater prawn culture (in association with the Hebrew University of Jerusalem), identification of tilapia populations and gynogenesis of fish (Bar-Ilan University), endocrinological research (Tel Aviv and Jerusalem Universities), research in pond ecology and growth-inhibiting factors (Hebrew University), pond soil ecology (Technion, The Technological Institute in Israel) and others.

The Department of Research and Development of the Ministry of Science keeps a desk for aquaculture sciences and is sponsoring research projects from a special ministerial fund as well as through various bilateral research funds.

Independent fish hatcheries have also contributed to the development and implementation of various programmes, such as the mass production of tilapia hybrids, mass production and export of various carp species, ornamental fish production and commercial freshwater prawn hatcheries.

10 Commercial development and future prospects

10.1 PREAMBLE AND CURRENT STATUS

About 10% of the world fish harvest is now produced by aquaculture, the cultivation of aquatic organisms under controlled conditions. Over 4 million tonnes of finfish (excluding farmed shellfish and seaweeds) are farmed annually of which 15−20% is the result of intensive farming methods. Intensive fish farming relies on high stocking rates and formulated diets to achieve high yields of marketable fish within purpose-built rearing facilities by means of close supervision over the entire production cycle. Freshwater farming of rainbow trout for human consumption was the first intensive industry to become widely established as an off-shoot of trout production for stocking angling waters. Recent emphasis in the farming of salmonids, the salmon and trout family, has switched more towards salmon because of market preference and the greater availability of seawater. However, the variety of rearing systems adopted for salmonids and their stringent environmental requirements offer a useful model of the technical factors involved in fish farming under temperate conditions. Hence the emphasis on salmonid husbandry in earlier chapters, as well as the particular attention paid to the key husbandry areas of reproduction, nutrition and disease control, as they relate to intensively farmed fish in general.

The greatest diversity of finfish species intensively farmed is in the Far East where fish consumption is highest. In particular Japan has pioneered intensive farming of an increasing range of high priced marine fish, such as yellowtail and red seabream. Japanese marine finfish farming has therefore been described here in detail, together with the development of culture-based fisheries in which hatchery-reared juvenile fish are released into the sea for subsequent recapture as adult fish, obviating the need for costly ongrowing farms. This is not to underestimate the contributions of

Table 10.1 Pattern of Japanese protein consumption (grams per capita per day) since 1965

	1965	1970	1975	1980	1983	1984	Percentage change	
							1965–75	1975–84
Animal protein	198.4	239.9	303.3	313.5	334.2	327.1	+ 53	+ 8
comprising fish	76.3	87.4	94.0	92.5	93.4	91.5	+ 23	– 3
meat	29.5	42.5	64.2	67.9	70.7	71.3	+118	+11
eggs	35.2	41.2	41.5	37.7	40.7	40.3	+ 18	– 3
dairy produce	57.4	78.8	103.6	115.2	129.4	124.0	+ 80	+20
Vegetable protein	479.8	442.1	408.5	383.0	383.1	374.3	– 15	– 8
Total	678.2	682.0	711.8	696.5	717.3	701.4	+ 5	– 1

Source 'Protein in the Japanese diet' by K. Yamaji published by the Japan International Agricultural Council, Tokyo (cited in Eurofish report No. 248, Agra Europe (London) Ltd, 1987).

Plate 10.1). However, these changes in fish consumption have been greatly overshadowed by the dramatic increase in poultry consumption. Thus per capita consumption of chicken has doubled in Europe and the USA over the last 25 years due to a combination of lower prices resulting from intensive production and the promotion of various chicken-based convenience products. In the USA only turkey and catfish seem likely to surpass this growth rate, but from much smaller base levels.

10.2.2 Supply factors

Fish farming now offers to remove the uncertainty of supply associated with traditional fishing, enabling food retailers to place contracts and plan forward sales, as is customary for conventional livestock products. Indeed the bulk of sales for certain scarce species (e.g. Atlantic salmon, yellowtail) now comprise farmed fish and as product quality has improved, so initial price discrimination against the farmed product has begun to disappear. However, in

Plate 10.1 Generic promotion of farmed rainbow trout emphasising health advantages.

Plate 10.2(a) Processing farmed Atlantic salmon: gutting.

order to capitalise on the inherent advantages of a farmed source of material, it is clearly important to organise harvesting, packing, processing and distribution of the product to achieve maximum sales volume and value. Plates 10.2 show the interior of a salmon farm's processing unit where fish are hand-gutted, weighed and packed on ice in polystyrene boxes before despatch.

Plate 10.2(b) Processing farmed Atlantic salmon: weighing and packing.

There is usually a range of supply options open to any fish farming concern and Figure 10.1 illustrates the different outlets and distribution channels open to a trout farm. Direct farm gate or farm van retail sales are generally small and in the UK the major proportion of the trout industry output is either supplied to caterers via the wholesale fish markets or processed for supply into retail supermarkets. The latter is retailed either as a frozen pack of gutted trout or, increasingly, as the fresh product in a controlled atmosphere pack containing gutted trout or trout fillets; other product options include smoked trout and trout paté (Plate 10.3). A similar trend has occurred in catfish marketing in the USA which has changed from a whole fish commodity towards greater emphasis on catfish fillets and steaks. More recently, major US

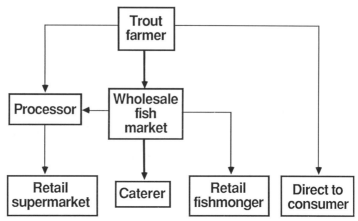

Fig. 10.1 Simplified diagram of farmed trout distribution.

food retailers have been further processing catfish to give new value-added products (e.g. breaded; burgers; pre-cooked; see chapter 8; section 8.2.7). This has been partly in response to the greater security of supply from what is now a well-established large-scale industry.

In order to generate more interest on the part of food processors and retailers, fish farmers have therefore had to alter management routines in favour of year-round availability of fish. Quality factors of particular importance are accurate size, grading, appearance, lack of any off-flavours, professional packing and freshness. The early years of intensive fish farming were often characterised by disorganised marketing with seasonal gluts of fish. Small producers in particular were prey to disastrous price falls during periods of high availability. As a result of these problems, cooperative farmed fish sales organisations have become established in many countries with major industries. For example, the Norwegian Fish Farmers' Association helps to organise farmed salmon sales and stabilise prices while the Norwegian Trout Sales Organisation buys and sells trout and guarantees minimum prices to its members. An equally important function of these trade bodies is to establish and maintain appropriate product quality standards to which members must adhere. For instance, members of the Scottish Salmon Growers' Association display a trademark on their goods, signifying adherence to the Association's quality assurance scheme (Plate 10.4). The

Plate 10.3 A range of added-value rainbow trout products (e.g. smoked trout; trout paté).

resulting discipline among producers then makes it easier to under-take sales promotions for the industry as a whole. Within this overall framework the more enterprising individual producers will continue to seek out and exploit particular market niches and outlets offering greater reward.

SCOTTISH SALMON GROWERS ASSOCIATION

QUALITY APPROVED
No. 008102

Plate 10.4 A quality mark used for farmed Atlantic salmon by an association of growers.

10.3 FISH PRODUCTION

10.3.1 Production indices

Production indices for intensively farmed fish are often quoted to illustrate that yields of animal protein are comparable to conventional livestock production. Intensively reared broiler chickens are currently stocked at approximately 23 kg/m^2 and a typical broiler house averages 5.5 crops per year, corresponding to an annual production of 1265 tonnes per hectare. This is more than a ten-fold increase over the highest yields per unit area achieved by means of standing water intensive pond farming of fish, although some trout, carp and catfish farmers have on occasion obtained similar yields in running water tanks and raceways. Highly automated pig and poultry farms also produce well in excess of the yields per unit labour achieved by fish farms which approach 100 tonnes per man year on certain mechanised catfish and salmonid farms. If water is scarce, fish farming is unlikely to represent the best use of resources, although Israeli experience has shown that the water requirement per tonne of pondfish (at least 5000 m^3) can be shared out between other crops by means of irrigation reservoirs doubling as fish farms. Modern compounded diets enable food conversion ratios for both chicken and fish to be in the region of 2:1 (feed:liveweight gain) or better, although fish feed formulations generally contain a minimum of 15% animal protein for complete carp or catfish diets and are therefore more costly than poultry diets with only 7−8% animal protein. Thus most farmed fish tend to be better protein converters but poorer energy converters than broiler chickens which receive lower quality diets; also their slower growth than chickens results in relatively lower land and labour productivity.

By contrast with fish, chicken production indices suffer when corrected for what is lost during processing and for the proportion of the final product discarded as inedible. For instance, eviscerated yield is about 72% of the unplucked whole bird and a typical broiler carcase comprises 34% meat (excluding 8% skin), whereas eviscerated yield in salmon is about 88% and the carcase has over 60% edible meat. Thus in filleting salmon, the discarded frame represents 37% and a further 5% weight loss occurs in the production of smoked salmon as moisture evaporates during smoking to give the final product plus skin. Unit size has an influence and smaller salmonids, such as 200 g rainbow trout, have an eviscerated yield (guts and gills) of only 80%. Also fish with longer digestive tracts (e.g. carp, tilapia) have somewhat lower eviscerated yields, but the proportion of edible meat in the whole carcase is still superior to poultry and other livestock and it is virtually all high quality protein.

10.3.2   Costs and returns

Intensive fish farming is only appropriate if the additional costs of intensification are more than outweighed by the additional returns. In order to enable comparative assessments of efficiency, it is necessary to assign monetary values to the costs and returns of fish farming. The balance of income over expenditure can then be expressed as a financial return on the original investment. A number of technical and commercial variables influence the rate of return from investment in fish farming. To illustrate this, Figure 10.2 depicts the sequence of decisions involved in investment appraisal of fish farming. The first question is whether a suitable market opportunity exists, expressed as potential fish sales volume, price and harvest pattern. Assuming this is favourable, the next key variable is water, which directly influences production capacity and potential growth rate. The flow chart shows how technical site factors influence the appropriate choice of fish farming system and hence construction costs. Input costs for purchase of young stock, fish feed, labour, insurance, etc., determine running costs. The sales revenue achieved from the projected harvest yield is then compared with total costs to project cash flow and hence the overall financial yield on investment. Of particular importance is the recognition that these technological economic factors are site-dependent and any appraisal should therefore be undertaken with detailed site information. Also, comprehensive step-wise analysis should determine not only whether fish farming is feasible at a particular site but, if so, how this can be exploited to optimum commercial advantage.

Chapter 8 described how the highest profit margins achieved by US catfish farmers result from supplying live fish via fish-out ponds or 'put-and-take' fisheries where the public pay an entrance fee plus a price for any fish caught. The second highest return is achieved by selling processed fish to a local market such as specialised catfish restaurants. However, the major proportion of the US catfish farming industry supplies processing plants at somewhat lower prices. Such outlets may offer lower profit margins but larger market volumes and hence the largest potential profit provided individual farms operate on a large enough scale. Studies of the trout farming industry in the UK and elsewhere have clearly shown the effect of farm size on profitability. In general, small family-run operations appear to do well, particularly if they can sell most of their output direct to the public and local hotels at favourable prices, sometimes with an ancillary put-and-take fishery (cf catfish). Above production levels of about 30 tonnes per annum, trout farmers costs increase because they have to hire outside labour, but more important the local retail markets become saturated and

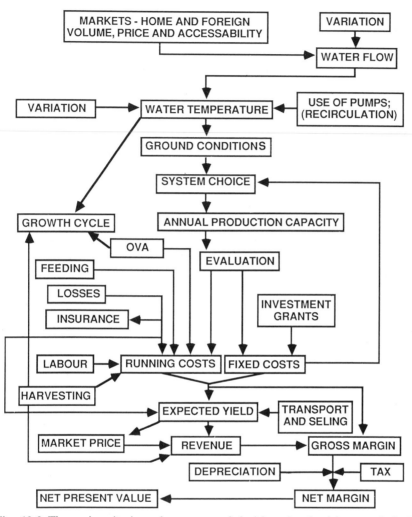

Fig. 10.2 The main criteria and sequence of decisions involved in appraisal of intensive fish farming investment.

they have to turn to the local and national wholesale fish markets which offer much lower prices. Large trout farms producing several hundred tonnes per annum are able to achieve economies of scale, such as keen bulk purchasing terms for fish feed, stock, etc., sometimes also with proportionately lower construction costs, so that unit production costs are low. Provided production and harvesting are planned, the greater quantities of fish available mean they are usually less vulnerable to market fluctuations and adverse wholesaler activity. Also, the largest producers are becoming involved in processing farmed trout for direct supply to large multiple retailers and supermarkets on a contract basis, thus by-passing the traditional wholesale fish markets.

If market outlet and scale of operation are important in determining profitability, so is the intensity of production. Chapter 9 illustrates how costs and returns vary under different conditions for standing water polyculture systems in Israel, with semi-intensive ponds yielding a higher return than intensive ponds and conventional ponds being the least attractive of the three stocking systems. Notwithstanding the benefits which have accrued from progressive intensification of the Israeli pond fish industry, maximum intensity clearly does not represent the economic optimum and over-capitalisation can readily threaten fish farming viability in Israel and elsewhere. Arguably the most important profit-determining factor is the management skill of the farmer as this determines choice of market, farm site and size, farming system and production intensity as well as the all-important operating efficiency throughout the production cycle. Thus it is the skill of the farmer to decide the optimum farm size and yield for a particular fish farming technique at a particular site. Experience has shown Israeli farmers that the optimum yield is about 20 tonnes per hectare and further increases in yield, entailing more elaborate pond construction, are not viable. Table 10.2 compares the costs and returns for US catfish production under different conditions of scale and stocking rate. As can be seen, returns to management, risk and family labour are much lower for the low stocking rate systems compared to the more intensive ones, with negative returns being recorded for the smaller size units with only 1000 catfish per acre. Not suprisingly, those farmers who produce larger fish with fewer losses in a growing season and achieve greater food conversion efficiency than their competitors are clearly at an advantage. They will produce even more efficiently if they buy feed and stock more cheaply, use less labour and harvest the fish more cheaply.

10.3.3 Economic case study

Salmon farming is an example of a high cost/high value operation with separate freshwater hatchery and marine on-growing production facilities. In order to provide more realistic details of the economics

Table 10.2 Comparative summary of estimated costs and returns for catfish production in South Carolina, 1986 (in US dollars)

			Gross receipts	Variable costs	Total costs	Returns to mgr., risk + family labour	Breakeven price cash/all
Low stocking rate							
1000 fingerlings per acre							
1	acre	Year 1	963.75	821.72	1 121.46	(157.71)	0.64/0.87
		Year 2	750.00	590.33	871.56	(121.56)	0.59/0.87
2.5 acre		Year 1	2 409.38	1 948.80	2 409.48	(0.11)	0.61/0.76
		Year 2	1 875.00	1 362.10	1 778.01	96.99	0.54/0.71
5	acre	Year 1	4 818.75	3 783.60	4 893.89	(75.14)	0.59/0.76
		Year 2	3 750.00	2 609.10	3 625.43	124.57	0.52/0.73
10	acre	Year 1	9 637.50	7 570.10	9 304.90	406.29	0.59/0.72
		Year 2	7 500.00	5 291.30	6 770.10	729.90	0.53/0.68
20	acre	Year 1	19 275.00	15 071.70	17 929.98	1 345.02	0.59/0.70
		Year 2	15 000.00	10 514.10	12 987.77	2 012.23	0.53/0.65
High stocking rate							
3500 fingerlings per acre							
1	acre	Year 1	3 375.00	2 585.90	3 096.83	275.17	0.57/0.69
		Year 2	2 625.00	1 774.28	2 220.28	404.72	0.51/0.63
2.5 acre		Year 1	8 437.50	6 351.21	7 356.35	1 081.15	0.56/0.65
		Year 2	6 562.50	4 304.61	5 146.02	1 416.48	0.49/0.59
5	acre	Year 1	16 875.00	12 552.23	14 585.82	2 289.18	0.56/0.65
		Year 2	13 125.00	8 564.33	10 278.89	2 846.11	0.49/0.59
10	acre	Year 1	33 750.00	25 190.68	28 833.73	4 916.27	0.56/0.64
		Year 2	26 250.00	17 214.88	20 219.86	6 030.14	0.49/0.58
20	acre	Year 1	67 500.00	50 271.36	57 063.11	10 436.89	0.56/0.63
		Year 2	52 500.00	34 284.66	39 797.47	12 702.53	0.49/0.57

Assumptions: existing pond; land is owned; price $0.75 per lb.
Source R.S. Pomeroy *et al.* (1987).

of intensive finfish farming, the options available to an Atlantic salmon farmer in Northern Europe and the projected costs and returns under different circumstances are analysed in Appendix I. The key points which emerge may be applied more generally to other types of intensive finfish farming, but in particular to marine cage farms, as follows:

(a) Capital costs

In the case of freshwater salmon smolt production, construction costs vary widely depending on site conditions. For example, if insufficient gravity flow is available, it may be necessary to construct a dam and/or to install pumping equipment with a stand-by generator. Also, excavation costs will be much higher if rock blasting is needed. In general there will always be some economy of scale in capital costs, if only because certain items such as on-site housing for staff and feed storage are required whatever the size of farm. The use of shore-based on-growing salmon farms using pumped seawater may be less vulnerable to storm damage and enable easier fish access compared with floating marine cage systems. However, the latter have become generally adopted due to flexibility in use and lower initial capital costs estimated at less than £1000 ($1850)/tonne production capacity. The useful economic life of a marine cage system is limited to a maximum of about 5 years depending on local conditions and further capital injection is therefore needed subsequently.

(b)   Operating costs

Operating costs are dominated by feed costs and to a lesser extent by purchase of replacement young stock. Salmon require a high cost diet based on fishmeal, but the average overall food conversion ratio is efficient at about 2:1, with some farms claiming as low as 1.7:1. Although volume discounts are available for large-scale buyers, average feed cost per tonne liveweight gain is currently about £900 ($1665), which represents 36% of total operating costs for a farm producing 250 tonnes of salmon over a two-year cycle. Attempts have been made to reduce feed costs by using minced-up trash fish diets, but this entails extra equipment including a cold store to ensure continuous feed availability. Compounded diets can be easily fed automatically and are nutritionally consistent, yielding a final product of standard quality, including the required level of flesh pigmentation (due to in-feed addition of carotenoids as in chicken layer diets). It is of interest that even in less capital intensive systems where much lower specification fish diets are used, fish feed costs still represent the major item of operating costs. Thus in Israel where freshwater pondfish feed costs less than half the price of salmon feed, this represents 42–52% of total operating costs depending on circumstances (Table 9.14).

Unlike the variations in salmon feed price which reflect the international fishmeal commodity market, variations in smolt price to date have been related to the excess of demand over smolt supply. The current shortage enables producers to charge £1.30–£1.40 ($2.40–2.59) per fish on contract or more on the 'spot' market. As smolt production increases it seems likely these prices will fall as large-scale producers incur total production costs of only about 60p ($1.11) per fish. In-house smolt production to cover the requirements of a 250 tonne salmon farm has distinct advantages is terms of securing stock supplies derived from on-farm selection of brood-stock with known health status and sea-water performance potential. However, smolt production is a specialist activity and the additional costs of hatchery investment are likely to balance the saving in reduced smolt costs compared with third party purchase. This is based on production of 120 000 smolts at about 89p ($1.65) each to produce 250 tonnes of salmon with a 15% overall seawater mortality. However, hatchery investment becomes a more attractive option to the larger producer achieving economies of scale at, say, the 500 000 smolt output level, either to service larger in-house sea farming requirements or to supply surplus smolts into the market for other farmers.

Other operating costs include labour (12%), stock insurance (8%), packaging and distribution (12%), fuel, electricity and administration costs, etc. (6%). For a 250 tonne farm it is assumed that fish are harvested, packed on ice and delivered to wholesale fish markets (at about £300 ($555)/tonne) without further processing or sales effort. Hence total operating costs, including stock and feed, are estimated at approximately £2500 ($4625)/tonne at this production level. Addition of depreciation and interest on borrowed capital at 14% per annum is estimated to increase total production costs over the two-year cycle to £3400 ($6290)/tonne.

(c)   Sales revenue

Although most salmon mature and are therefore harvested after two years in the sea, a proportion mature after only one year as 'grilse'. Smaller fish generally command a lower price per kilo than large fish but many farmers harvest as much as 50% of stock after 12 months in order to generate earlier cash inflow and to empty pens for replacement smolts. Judging the optimum harvest weights is therefore a complex matter of equating the potential increase in value of a growing fish stock with the potential increase in production costs (including the finance costs of borrowing the additional capital) against the background of fluctuating market prices.

At a 1986 average delivered wholesale salmon price of £3.98 ($7.36)/kg and assuming the above cost profile, a 250 tonne salmon farm harvesting 50% of stock after one year in the sea could expect to obtain 73% pretax profits over the two-year production cycle on an initial capital cost of £200 000 ($370 000) (i.e. an annual return on investment of 36.5%).

(d)   Sensitivity, risk and reward

In the above example a mere 15% reduction in average salmon selling prices to £3.40 ($6.29)/kg would cause the farm only to break even financially with income and expenditure in balance. Overall profitability is clearly sensitive to relatively small changes in value of the end-product and is therefore highly sensitive also to stock losses due to storms, predators, disease, etc. Approximately 15% losses are normally budgeted for during the first few months but an additional loss of only 15% of fish occurring just prior to the final harvest would also bring the project to breakeven point. An increase in feed costs of about 50%, due to either higher feed prices or poor conversion efficiency, or both, would have the same effect after taking account of the resulting increase in interest charges. By the same token, some highly efficient producers are now claiming overall food conversion ratios as low as 1.7:1, which would result in an increase in the annual return on investment from 36.5% to 46% assuming other factors remain constant.

Although many farmers take out stock insurance, at the present time most cover is prohibitively costly unless the insured is prepared to meet the cost of losing the first 20% or so of total fish value insured in a single incident (cf vehicle insurance 'excess' clause). A prudent salmon farmer may therefore prefer to manage a number of smaller cage operations at different sites without necessarily paying stock insurance in order to spread the risk of losing stock from site-specific problems. It is because of such considerations that most individual salmon farming units do not exceed a production level of 200–300 tonnes per annum.

Clearly salmon farming is a relatively high-risk activity compared with conventional livestock production, but the industry has grown as farmers have judged the likely return on investment sufficient to compensate for this risk element. Very high salmon prices were available during the 1970s which more than offset the frequent losses encountered, but as increased production has begun to restrain prices so the industry has had to prune costs and place greater emphasis on reducing technical and commercial risks in order to achieve adequate reward.

10.4 INDUSTRIALISATION OF FISH FARMING

10.4.1 Specialisation and integration

The history of intensive farming of different finfish species in most countries generally started with independent producers exploiting small but profitable local markets of high unit value, such as farmgate retail sales or live fish sales. As more producers became established, some saw merit in specialising in particular aspects, such as fry or fingerling production which requires particular skills, high quality water, etc. On-growers could then choose to concentrate on producing for the table market although many preferred an integrated hatchery-cum-farming operation in order to secure all or part of their requirements for young replacement stock on an in-house basis. Economies of scale encouraged farms to build more ponds or cages and even to acquire neighbouring farms. This search for lower unit production costs has encouraged some large trout and catfish farming concerns to cooperate in setting up fish feed mills in order to achieve closer control over their largest operating cost centre. The logic of such 'vertical integration' is greatest if fish feed manufacture, the supply of young stock and the bulk retailing of farmed fish are in relatively few hands. As intensive fish farming has developed into distinct industries, integration has therefore occurred back up the production chain as well as forwards into the marketplace.

One means of reducing exposure to adverse price movements is by investment in processing facilities and by establishing a separate marketing and distribution operation. For a species like salmon, processing losses if fish are eviscerated and bled average 12%, hence c. 287 tonnes of salmon at harvest yield 250 tonnes of gutted fish. It may not be possible to recoup the additional farming costs and the substantial capital investment in processing plant, together

with allied operating costs (e.g. power and labour), by achieving
sufficient of a premium over the equivalent price for whole fish.
Moreover, the heavy fixed costs in running a processing factory
necessitate continuity of throughput for cost-effective operation. In
practice this means salmon farming output would probably need to
exceed 1000 tonnes per annum before in-house processing becomes
worthwhile. However, it gives the operation greater market flexibility
with an ability to supply outlets which would not ordinarily purchase
whole fish. The improved shelf life of gutted fish facilitates servicing
more remote markets which entail long-distance haulage, whereas
other product possibilities would include further processing (e.g.
smoked salmon fillets). Provided the scale of operation is sufficient
to make it viable the rationale of processing may be as much to
achieve more marketing options with less vulnerability and risk as
to achieve added value in terms of enhanced net sales price. The
logical extension from an integrated fish farming industry is therefore
to develop a fish marketing company or consortium, which contracts different farms to supply its raw material. By this means the
fish supplier and/or processor can concentrate on improving and
satisfying demand for his product by creating new outlets, while
sourcing from farms onto which is placed all the technical risks of
production.

10.4.2   The parallel with broiler chicken

Since the 1950s the development of the broiler industry has trans-
formed chicken from a high-priced luxury meat to a staple food
with consumption more than doubling in most of the developed
world as a result. In order to achieve this, chicken producers have
had to make radical changes to the traditional ways of farming
and marketing chicken. The initial thrust came from widespread
adoption of intensive rearing methods and was accelerated by the
economic result of technical advances in husbandry.

Purpose-built rearing units and scientifically formulated diets
enabled chicken producers − and also now some fish farmers − to
adopt intensive stocking regimes. However, efficient preventive
medicine is still at an early stage of development for farmed fish
compared with poultry farming. The emphasis is on prevention
rather than cure in modern chicken production where a comprehensive range of vaccines is administered routinely and production

systems are designed to avoid introducing disease organisms. The link between health and nutrition is also better understood than in fish, hence the risk of 'metabolic' diseases caused by continuing emphasis on better performance from cheaper rations has been mitigated by dietary improvement. With certain exceptions (e.g. catfish, salmonids) the poor understanding of nutritional requirements of fish means that, by contrast, fish diets are still too often empirically formulated, based on a fixed recipe of raw materials, and are therefore over-specified and costly. However, the greatest difference between intensive chicken and fish farming lies in the genetic potential of the stock used and hence the performance achieved. Within a period of some 20 years, the production cycle for broiler chicken has fallen from about 10 to about 6 weeks for what is now a larger bird, due almost entirely to genetic selection work. By contrast, fish farmers still rely on virtually wild stock and little systematic genetic improvement has taken place.

In the last 10 years broiler producers have therefore continued to reduce the real cost of production, but they have also improved the end-product value. Chicken has consequently moved away from being seen as a high-priced single product commodity because competing food firms have worked with the broiler industry to develop new products thus creating increased demand. The resulting diversity of further processed chicken products means that the industry is less vulnerable to fluctuations in the general economy. However, it would not have happened without heavy development and promotional expenditure, aided by the increasing concentration of production with more birds being produced by fewer companies. The modern broiler industry is therefore not only in control of its input cost base but increasingly in control of its end-markets as well.

In seeking to assess whether industrial parallels exist between broiler production and intensive fish farming it is important to distinguish between the different finfish species involved. Output of farmed channel catfish and Atlantic salmon has grown at average rates of 33% and 45% per annum respectively over the period 1980−5, while growth in farmed eels, trout and yellowtail has been lower and the Israeli pondfish industry has marginally declined. Clearly trout farming is much longer established than, say, salmon farming and, unlike salmon, farmed trout represents the only trout commercially available. With the exception of eel farming, which still relies on wild-caught elvers for replacement stock, the farming of all six groups has a degree of industrial structure in most major producer regions with increasing backward control over input supply and/or well established supply industries competing for farm sales of feed, stock, equipment, etc. The stagnation of Israeli output illustrates a highly efficient industry which has saturated the domestic market to the extent that 30% of total fish consumption in Israel represents farmed fish. Virtually all this farmed production is sold as live, fresh or frozen fish and no large-scale attempts have been made to differentiate the product. Yet compared with the intensive farming of eels, salmonids or yellowtail, Israeli pondfish offer a lower cost technology due to the shorter production cycle and cheaper feed. It is of interest that tilapia yield a large bland white-fleshed fillet which could be further processed into a variety of convenience added-value fish products. By contrast, the US catfish industry has switched dramatically from commodity sales of live, fresh and frozen catfish towards processed catfish products, which comprised 7000 tonnes in 1980 and 87 000 tonnes by 1985, being now the dominant product sector. Thus catfish has successfully traded up its image from a 'poor man's' food restricted to the southern states of the USA in favour of an array of added-value products with much wider consumer appeal. The structure of the catfish farming industry now clearly has much in common with that of the broiler industry. At the other extreme, the farming of salmon and other high value luxury fish is now bringing market prices down and profit margins will come under increasing pressure against a higher cost base than catfish.

10.4.3 Ranching and culture-based fisheries

An important new phase in the industrialisation of fish farming is now starting with the trend towards release and recapture. Reference has already been made (chapter 1; section 1.5) to the long-standing importance of reservoir stocking with cultured fry in relation to whether the resulting yields of carp and other freshwater fish should be included in (extensive) fish farming statistics. Clearly this involves release into a small enclosed body of water, whereas by contrast cod fry have been released into Northern European waters for many years with little apparent benefit to the cod fishery. Until recently, commercial emphasis was therefore on releasing migratory fish, such as salmonids, for 'ranching' purposes.

If the salmon hatcheryman could guarantee that sufficient numbers of the smolts released would return to the hatchery after seawater life in the wild as adult salmon, then clearly this would enable the industry to dispense with the costly on-growing phase completely. The breakeven salmon return rate obviously depends on smolt costs and salmon prices, but is unlikely to be more than 10% under present market conditions which could well become achievable in the future at selected sites (e.g. smolt production costs of 60p ($1.11) per fish; average sales price of £3 ($5.55)/kg for 2-kg fish returning). The expansion of culture-based marine fish production in Japan has demonstrated the scope for mass propagation of juvenile marine fish for release into tightly controlled fishing zones (e.g. the Seto Inland Sea) where they are left to grow up to a suitable harvest size under natural conditions. It is of interest that the Japanese fishing cooperatives own these coastal hatcheries and have a vested interest in respecting each others' fishing rights. Provided similar agreements can be achieved elsewhere, it would seem very likely that new industries will emerge for particular species of fish suitable for exploitation on this culture-based fishery principle.

10.5   CONSTRAINTS ON DEVELOPMENT

10.5.1   Market constraints

Ill-informed market comment about 'battery' farming of fish, or the suggestion that farmed fish might contain artificial pigments or hormones could ruin the entire industry. It is not enough for individual fish farmers to act responsibly; the industry must be seen to do so collectively. Product quality must therefore be at least as good as the wild equivalent to avoid discrimination against farmed fish, either by price discounting or even outright condemnation. Until fish farmers learn to be disciplined and ensure that poor, malformed, badly pigmented fish or specimens showing signs of scale loss and fin erosion are graded out at harvest, they will continue to put the entire industry at risk. The same applies to those farmers who do not pay close attention to the need for good presentation of the product once it has left the farm. Even if fish are sold ungutted this means accurate size grading, professional packaging and rapid chilled transportation, ensuring freshness to the end-consumer.

The marketplace is interested only in fish products of the right quality being available at the right time and right price. Recent history has shown that where particular fish farming industries have learned this lesson, there is no reason why farmed fish cannot rapidly come to dominate the total market for a particular species. Other things being equal, fish buyers (especially those acting for large food retailers or restaurant chains) would prefer to place long-term contracts with suppliers having the means of production securely within their control, hence the inherent advantage of intensive fish farming industries over conventional fishing industries relying on the uncertainty of wild catches (see section 10.2.2). To take maximum advantage, individual fish farmers need to extend their perspective beyond the farm gate and ensure their primary goal is to identify and satisfy the particular needs of the individual market being served. It is unlikely that this can be achieved if they operate in isolation from other producers in the same sector with whom they share a common interest in maintaining quality standards and price levels. Nor should they underestimate the intelligence of their customers and end-consumers, who may prefer to receive convenience packs of ready-to-cook fillets in preference to purchasing unprocessed whole fish on an ex-farm basis. Probably the greatest current curb on further expansion of intensive fish farming is the unwillingness of many producers to forego a 'production-led' in favour of a 'market-led' orientation to their business activity.

10.5.2 Technical constraints

As has already been noted (section 10.4.2), most fish farming industries lag far behind modern poultry production in technological achievements. Thus little scientific stock improvement has taken place on fish farms, other than simple selection of faster-growing strains for breeding purposes and sex manipulation for all-male or all-female stock. However, further extension of spawning control to produce seed 'on demand', together with recent developments in cryopreservation and genetic manipulation (e.g. triploidy) suggest that rapid progress should be possible. Although genetic improvement is not simple or cheap (particularly in the case of temperate species of fish which take far longer to reach sexual maturity than chickens), there is clearly great scope here for reducing production costs by improving survival, daily weight gain and the efficiency of feed conversion. As with poultry genetics, this emphasis on better performance may also be allied to altering genetic makeup in line with market requirements for fish conformation, flesh quality, better yield at processing, etc. Depending on the species and technique in question, other genetic aims of commercial value could include selection for delayed sexual maturity and disease resistance.

Clearly the lack of improved stock, coupled with a relatively early stage of development in veterinary and nutritional expertise, represents important technical constraints on the establishment of intensive fish farming industries. However, an even more fundamental constraint is posed by the fish farmer's inability to control reproduction and early-rearing. Despite the advances in artificial spawning due to hormonal manipulation, fish such as eels still cannot be spawned in captivity. Equally important is the difficulty of larval rearing, particularly when farming those species which produce very small eggs and have larval stages requiring microscopic diets. To date research effort has focused mainly on the mass culture of live foods comprising phytoplankton and zooplankton (e.g. nauplii of brine shrimp, *Artemia salina*) as a way of overcoming this bottleneck to the farming of certain species, particularly of marine fish. There is obvious scope for large-scale controlled production of algae using process fermentation instead of conventional illuminated tank systems. However, recent progress in developing artificial microencapsulated larval diets of the right particle size and nutritional composition will undoubtedly pave the way for further developments in controlled rearing of such fish. A parallel development has been the recently discovered need for adequate inflation of the swim-bladder in certain larval marine fish, such as gilthead bream (*Sparus aurata*) and sea bass (*Dicentrarchus labrax*) to prevent heavy losses and malformed adult fish. Overcoming these hatchery constraints has a double significance. Firstly, it extends the scope for intensive fish farming to include many species with complex larval stages. This means that it provides a truly secure basis for those industries which currently rely on catching wild supplies of young stock: lack of control over stock supply has been a major disincentive to further investment in well established fish farming industries, such as the farming of milkfish (*Chanos chanos*) and has prevented large-scale investment in others such as the farming of bluefin tuna (*Thunnus thynnus*) or turbot

(*Scophthalmus maximus*). Secondly, it facilitates the further development of culture-based fisheries to include a wider variety of marine fish.

An increasing constraint on the expansion of intensive fish farms, particularly those sited on freshwater streams, is the problem of pollution from fish farm effluent. Already this is causing countries, such as Denmark, to insist on costly settlement tanks for new farms and as a result Danish freshwater trout production has reached a plateau. This trend will undoubtedly gather momentum, particularly in the wealthier economies whose populations are increasingly unwilling to adopt a *laissez-faire* attitude to environmental issues including clean rivers and natural habitats. It is of interest that marine farming of trout has now commenced in Denmark, partly to avoid anti-pollution laws. Depending on factors such as species and site, a switch to seawater farming may not necessarily be feasible, but in general the abundance of seawater and the effects of tidal exchange obviously offer an attractive means of overcoming the pollution constraint. Since there is usually greater competition for freshwater than seawater, any general switch to seawater farming would obviously find favour with other users while releasing land for alternative development.

It follows from the foregoing that a number of different factors favour greater exploitation of marine fish farming, including advances in rearing of marine fish larvae, pollution avoidance, scope for culture-based fisheries, etc. Hitherto the main risk in marine farming has been the exposure to storms, typhoons, etc. The increased costs associated with shore-based marine farms would appear to rule out this option as a general solution, although the development of more efficient pumping systems may make it feasible under certain circumstances. So in practice marine farmers have chosen sheltered floating-cage sites where long term self-pollution problems are probably greater than in more exposed sites. Considerable engineering effort is currently being expended to resolve these difficulties and it seems likely that a combination of much deeper cages (e.g. 10 m) and the use of expanded food with a high digestibility should help to reduce investment costs per unit volume as well as reducing pollution. Innovations in cage design to reduce storm exposure are likely to involve an increasing trend away from conventional small net pens with flotation collars towards larger units, including some which are capable of being submerged beneath the surface and hence require specialised adaptations for buoyancy and access.

10.5.3   Organisational constraints

Given the relative novelty of intensive fish farming techniques, it is easy to understand why they have not been adopted more quickly in spite of favourable technical and economic conditions. Even where national governments have decided to encourage fish farming, there is often the very real administrative problem of choosing which government departments are most appropriate to undertake the task. Strong arguments can be put forward for the administration of fish farming being placed in the hands of either Agriculture or Fisheries Departments. Frequently there is an existing tradition of extensive freshwater cultivation being administered by Inland Fisheries Departments, but does that make them the appropriate administrators for intensive freshwater or brackish water farming, still less of intensive marine farming? Whatever the decision, it is unlikely that existing legislation will be sufficient to safeguard the rights of fish farmers without some modification. The lack of a suitable legal and institutional framework will tend to obstruct efforts to establish fish farms and also prejudice the chances of obtaining financial assistance. These institutional constraints reflect ignorance or unwillingness on the part of government and other bodies to recognise fish farming as a primary producing industry akin to agriculture.

Integral to the concept of fish farming is ownership of a specific stock of fish, yet this may not be easy to substantiate in law, particularly for marine farms. Problems of common ownership may also complicate the assignation of rights to moor cages and to gain access across the foreshore. Marine farming in Japan flourishes in part because the Prefectural governments designate suitable areas which the local fishermen's cooperative association then apportion free of charge to individual farmers. Potential conflicts may arise if a fish farm is likely to interfere with traditional fishing methods or with the passage of vessels in navigable waters. By contrast, the legal position is more straightforward with land-based farms, where planning controls over water abstraction, pollution, etc., are usually well established for other purposes. But even here it is obviously important to secure a sufficiently long lease to warrant investment in developing a fish farm, and restrictions on lease periods for fish ponds have been identified as a major constraint on expansion of this sector in India. The introduction of live fish stock to a farm may also come under government control, particularly if it involves species which are not native to the particular region. Whereas governments may well have a duty to protect wild fish populations from any ecological threat posed by the escape of exotic species or of fish disease organisms, in practice fish farming has not posed any such threat and it would be particularly unfortunate if unnecessary licensing procedures stifled investment. Health certification and import controls on live fish are, however, of proven value in safeguarding a developing industry from certain fish diseases, notably viruses. Experience from various countries shows that the practical implementation of such statutory regulations should be devolved to a technically qualified government inspectorate (e.g. the veterinary service) rather than shared jointly between different private and public sector bodies.

Hence a prerequisite for secure long-term expansion is an appropriate legal and administrative framework taking into account the special features of fish farming. The advantages of being allied to Agriculture Departments of central government tend to increase with the use of more intensive methods. Thus government research and development on intensive fish farming is most needed in the areas of genetics, nutrition and health, which are likely to rely more on animal husbandry expertise than fishery science. The same goes for training and extension services, which are important in disseminating knowledge and practical skills in a rapidly changing industry. The advantages of grant aid and government incentives to promote fish farming are more debatable, at least in the developed world. For instance, Southern Italy has been the major recipient of European Community aid for fish farming with little tangible result to date. By contrast, a country like Norway, which is not in the Community and which does not give fish farms more favourable treatment relative to agriculture and fisheries, has undergone major aquaculture expansion during the last 20 years. In general, fish farms are located in more remote rural localities and although this will tend to increase distribution and other costs, it is often partially offset by their consequent eligibility for regional development allowances. The latter are perhaps most important in ensuring that investment is not discouraged due to lack of a suitable infrastructure such as roads and services.

It would also seem appropriate to extend to fish farming the same general concessions which apply to other forms of intensive livestock production, such as waiving certain industrial taxes to promote agricultural production (cf agricultural derating). A key element in organising and promoting fish farmers' interests is the establishment of trade associations and producer groups. Although the rationale is partly to coordinate effective marketing (section 10.2.2), it also enables the industry to represent its members' views to government, international agencies and counterpart bodies in other countries. Certain fish farming trade associations have played a major role in moulding government policy and institutions towards a more favourable environment for expanding their business. At the same time they represent a powerful collective forum for bulk purchase of inputs. The latter include services such as stock insurance and credit facilities. Many credit agencies are still reluctant to lend for what is often regarded as a novel industry with an uncertain risk profile. However, the involvement of trade associations in spreading the risk and highlighting the potential reward has encouraged a more positive approach to intensive fish farming on the part of external financial institutions.

10.6 THE WAY AHEAD

Intensive fish farming encompasses a diverse array of fish species, each with different biological and environmental requirements and therefore posing different problems for the fish farmer. However, it is clear that, depending on the maturity and experience of the particular fish farming industry, technical risks are now generally much better understood and that the farmer's management skill, together with a market-led approach to intensive finfish farming and marketing, are key determinants of commercial success. As regards economic viability, this will remain unproven for some species, such as Dover sole, until husbandry techniques are improved, although the farming of related flatfish (e.g. turbot) has now moved beyond the experimental stage to commercial production. At the other extreme, high profits are being made from species such as Atlantic salmon and the challenge ahead will be in reducing production costs and improving global distribution. Thus by 1990 it has been estimated that farmed salmon production will dominate world trading in fresh and frozen premium products and will represent 15−20% of total salmon supplies. But increased production will lead to significantly lower profitability and the challenge will be to adjust production to the needs of the market. New markets will open up as the price of farmed salmon decreases.

Some of the links in the complex marketing chain will disappear and there will be a move towards the development of production centres convenient to the larger markets in which national salmon farmer associations will play a role.

The technical areas of ignorance now being addressed will widen the range of species to be farmed and lower production costs of those already being farmed. In particular, techniques of marine larval production are now being pioneered which will become widely adopted, whereas further expansion in eel farming will be curbed unless it becomes possible to reproduce eels in captivity. Knowledge of fish health and nutrition has improved considerably, although it is concentrated mainly on salmonids and catfish. The obvious gaps to be filled include a comprehensive range of vaccines against the commercially significant diseases of farmed finfish, preferably administered via the feed. Also as the farmed supply of high value fish such as Atlantic salmon, red seabream and yellowtail now dominates the total world harvest of these species, it is vital that more attention is paid to least-cost formulation of commercial diets. These are currently overspecified while feeding tables sometimes bear insufficient attention to optimising economic results; the resulting cost improvements will help to maintain profitability in the face of declining market prices for the end-product.

Perhaps the greatest area of potential husbandry advance, however, is in stock improvement. Farmed fish are essentially undomesticated and performance under intensive farming conditions will be greatly improved as genetic selection seeks to eliminate undesirable traits, as has happened with such spectacular success in the broiler chicken industry. Such work has only started seriously in the last decade and progress will be slow but the rewards will be very real. Advances in biotechnology have brought the synthesis of fish growth hormone close to reality and promise to revolutionise aspects of fish breeding, as well as fish nutrition (e.g. algal production) and health (e.g. vaccines).

The increased competition for unpolluted sources of freshwater and the organic polluting potential of large-scale intensive fish farming will increasingly hinder further exploitation of freshwater sites. Attempts to develop reliable marine finfish farming systems are becoming increasingly successful although some hazards may prove difficult to resolve, notably the occurrence of toxic plankton blooms in some regions. It is anticipated, however, that the bulk of further expansion will be in intensively farmed marine finfish, giving rise to industrial production of demersal round fish (e.g.

gilthead bream) and flatfish (e.g. turbot) and pelagic fish such as tuna. Simultaneously, there will be much greater attention focused on the 'ranching' of migratory fish (e.g. salmon) and culture-based fisheries which obviate the need for a costly on-growing stage. It is difficult to extrapolate globally from recent success with culture-based marine fisheries under the special conditions offered by the Inland Sea of Japan. Although cod ranching has so far met with little obvious success, provided the hatchery can be assured that it will properly benefit from the subsequent landings of stocked fish (perhaps by means of a royalty arrangement for tagged fish with close governmental control over fishing grounds), then the technical problems do not appear insurmountable for certain locations and species.

For intensive finfish farming to make a greater impact on the supplies of the largest mass volume fish products, instead of being restricted mainly to high priced species of limited market size, certain factors will need to be borne in mind. A number of finfish species can be intensively farmed using relatively lower cost diets and techniques, for example tilapia will grow fast in floating cages under tropical or subtropical conditions without the need for a high cost diet. The successful marketing of catfish products in the USA has shown what can be done by imaginative product development and promotion based on a farmed raw material which had a traditionally poor image.

Per capita consumption of fish is generally increasing as its health advantages relative to red meats are becoming understood. Food processors and retailers wish to exploit this trend and, other factors being equal (e.g. price), would prefer a secure supply of consistent raw material supplied to their specification rather than having to rely on the vagaries of wild-caught fish. Farmed tilapia offers a bland white fillet with a variety of product options but is only sold currently as whole fish to a very limited market in the developed world. Thus the effective marketing of intensively farmed finfish is crucial in determining whether its overall current growth rate continues to accelerate beyond the 10% average for the aquaculture industry as a whole. At the same time, the reluctance of governments and institutions to acknowledge the growing status of fish farming as a primary producing industry strictly comparable to conventional livestock production means the legal, administrative and investment framework is often inappropriate to its development needs. Nor has the recent expansion of this aquaculture sector owed much to the subsidies and grant aid which protect most of conventional livestock production in the developed world. Despite this need to find its own way, the clear signs are that the global status of truly intensive finfish farming is now being recognised. Although projections must be somewhat speculative, by the year 2000 AD it should have multiplied to over 3 million tonnes per annum from a current annual base of about 700 000 tonnes.

Further reading

CHAPTER 1

Bardach, J.E., Ryther, J.H. and McLarney, W.O. (1972) *Aquaculture: the Farming and Husbandry of Freshwater and Marine Organisms*. New York: Wiley Interscience.

Bilio, M., Rosenthal, H. and Sindermann, C.J. (eds) (1986) 'Realism in aquaculture: achievements, constraints, perspectives'. *Proc. World Conference Aquaculture*, Venice, 21–25 September, 1981. European Aquaculture Society.

Chua, T.–E. (1986) 'World aquaculture production'. NAGA. April Newsletter of ICLARM (International Center for Living Aquatic Resources Management). Metro Manila: ICLARM.

FAO (1983) *Yearbook of Fishery Statistics for 1981*. Rome: Food and Agriculture Organisation.

FAO (1985) 'Aquaculture statistical data'. ADCP unpublished data. Rome: Food and Agriculture Organization.

Halley, R.J. (ed.) (1985) *Agricultural Notebook*, 17th ed. London: Butterworths.

Hickling, C.F. (1962) *Fish Culture*. London: Faber and Faber.

Huet M. (1986) *Textbook of fish culture*, 2nd ed. Farnham: Fishing News Books Ltd.

Huisman, E.A. and Machiels, M.A.M. (1985) 'Fish production for food in the tropics' In *Proceedings of XIII International Congress on Nutrition* (eds T.G. Taylor and N.K. Jenkins). London and Paris: John Libby, pp. 892–6.

Nettleton, J.A. (1985) *Seafood nutrition*. New York: Osprey Books.

Rosenthal, H. and Sarig, S. (eds) (1984) 'Research on aquaculture'. Proceedings of a seminar held in Hamburg in March 1984. *Special Publication No. 8*. European Aquaculture Society.

Shang, Y.C. (1976) 'Economic comparison of milkfish farming in Taiwan and the Philippines, 1972–1975'. *Aquaculture* 9: 229–36.

CHAPTER 2

Bond, C.E. (1979) *Biology of Fishes*. Philadelphia: Saunders College Publishing/ Holt, Rinehart and Winston.

Boyd, C.E. (1979) 'Water quality in warmwater fish ponds'. *Agric. Exp. Sta. Rep.* Auburn University, Auburn, Alabama.

Cushing, D.H. (1981) *Fisheries Biology. A Study in Population Dynamics*, 2nd ed. Madison: University of Wisconsin Press.

Everhart, W.H., Eipper, A.W. and Youngs, W.D. (1975) *Principles of Fishery Science*. London: Cornell University Press.

Forster, J.R.M. (1975) 'Recirculation as an answer to limited water supply.' Unpublished paper presented to the Institute of Fisheries Management (Limiting factors in freshwater fish farming), 8–9 July 1975, London.

Hoar, W.S. (1973) *General and Comparative Physiology*, 3rd ed. New Jersey: Prentice Hall Inc.

Hoar, W.S. and Randall, D.J. (eds) (1984) *Fish Physiology*, vol. X, A. and B. Gills. New York: Academic Press.

Houlihan, D.F., Rankin, J.C. and Shuttleworth, T.J. (1982) *Gills*. London: Cambridge University Press.

Lackey, R.T. and Nelsen, L.A. (eds) (1980) *Fisheries Management*. London: Blackwell Scientific Publication.

Lagler, K.F., Bardach, J.W., Miller, R.R. and Passino, D.R.M. (1977) *Ichthyology* 2nd edn. New York: John Wiley and Sons.

Liao, P.B. (1971) 'Water requirements of salmonids'. *Prog. Fish-Cult.* **33**: 210–15.

Love, R.M. (1980) *The Chemical Biology of Fishes*, Vol. II. New York: Academic Press.

Marshall, N.B. (1965) *The Life of Fishes* London: Weidenfield and Nicholson.

Moyle, P.B. and Cech, J.J. (1982) *Fishes: An Introduction to Ichthyology*. New Jersey: Prentice Hall.

Norman, J.R. (revised by Greenwood, P.H.) (1963) *A History of Fishes*. London: Ernest Benn Ltd.

Pitcher, T.J. and Hart, P.J.B. (1982) *Fisheries Ecology*. Connecticut: AVI Publishing Co.

Royce, W.F. (1984) *Introduction to the Practice of Fishery Science*. New York: Academic Press.

Stickney, R.R. (1979) *Principles of Warmwater Aquaculture*. New York: John Wiley and Son.

Trussell, R.P. (1972) 'The percent un-ionized ammonia in aqueous ammonia solutions at different pH levels and temperatures'. *J. Fish. Res. Bd. Canada* **29**: 1405–1507.

Wedemeyer, G. (1977) cited by Piper, R.G., McElwain, I.B., Orme, L.E., McCraren, J., Fowler, L.G. and Leonard, J.R. (1982) *Fish Hatchery Management*. Washington DC: US Dept of the Interior, Fish and Wildlife Service.

CHAPTER 3

Ackefors, H., Munroe, A., Müller, F. and Querellou, J. (1986) *Development of Aquaculture in Europe*. Fifteenth FAO Regional Conference for Europe, Istanbul, Turkey (28 April–2 May 1986). Rome: Food and Agriculture Organization.

Alabaster, J.S. and Lloyd, R. (1982) *Water Quality Criteria for Freshwater Fish*. London: Butterworth Scientific.

Edwards, D.J. (1978) *Salmon and trout farming in Norway*. Farnham: Fishing News Books.

Hawkins, A.D. (1981) *Aquarium Systems*. London: Academic Press.

Koops, H. and Kuhlmann, H. (1980) 'Eel farming in the thermal effluents of a conventional power station in the harbour of Emden'. In *Symposium on New Developments in the Utilization of Heated Effluents and of Recirculation Systems for Intensive Aquaculture*. Stavanger, Norway, 28–30 May 1980.

Kuhlmann, H. (1979) 'The influence of temperature, food, initial size, and origin on the growth of elvers (*Anguilla anguilla* L.)'. Rapp. P.–V. Reun. *Cons. Int. Explor. Mer.* **174**: 59–63.

Leitritz, E. (1960) 'Trout and salmon culture (hatchery methods)'. *Fish. Bull. No. 107*. State of California, Dept of Fish and Game.

Lemercier, P. (1981) 'Intensive eel farming in a nuclear power plant effluent'. In *Proceedings of the Bio-Engineering Symposium for Fish Culture* (eds L.J. Allen and E.C. Kinney) Washington DC: Fish Culture Section, American Fisheries Society, pp. 235–7.

Liao, P.B. and Mayo, R.D. (1972) 'Salmonid hatchery water re-use systems'. *Aquaculture*, 317–35.,

Lockwood A.P.M. (ed.) (1976) *Effects of Pollutants on Aquatic Organisms*. London: Cambridge University Press.

Mayo, R.D. (1971) 'Hatchery design and operation information'. In *Aquaculture – a New England perspective* (ed. T.A. Gaucher). Portland: Trigom, The New England Marine Resources Information Program.

Milne, P.H. (1972) *Fish and Shellfish Farming in Coastal Waters*. Farnham: Fishing News Books.

Muir, J.F. (1981) 'Management and cost implications in recirculating water systems'. In *Proceedings of the Bio-Engineering Symposium for Fish Culture* (eds L.J. Allen and E.C. Kinney). Washington DC: Fish Culture Section, American Fisheries Society, pp. 116–27.

Muir, J.F. (1988) *Recent Advances in Aquaculture*, Vol. III. London: Croom Helm (in press).

Spotte, S.H. (1979) *Fish and Invertebrate Culture: Water Management in Closed Systems*. New York: Wiley Interscience.

Tesch, F. W. (1973) *Der Aal*. Hamburg and Berlin: Verlag Paul Parey.

Tiews, K. (ed.) (1981) *Aquaculture in Heated Effluents and Recirculation Systems*, Vols I and II. Berlin: Heinemann.

Westers, H. (1970) 'Carrying capacity of salmonid hatcheries'. *Prog. Fish-Cult.* **32**: 43–6.

Wheaton. F.W. (1977) *Aquacultural Engineering*. New York: Wiley Interscience.

CHAPTER 4

Bagenal, T.B. (1978) 'Aspects of fish fecundity'. In: *Ecology of Freshwater Fish Production* (ed. S.D. Gerking). Oxford, London, Edinburgh: Blackwell Scientific, Melbourne, pp. 75–101.

Bayer, H. von (1950) 'A method for measuring fish eggs'. *Prog. Fish-Cult.* **2**: 105–7.

Bromage, N. (1982) 'Fish culture and research into reproduction'. In *Proc. Inst.*

Fish. Mgmt (ed. M. Bulleid). Reading: Commercial Trout Farming Symposium, pp. 165–76.

Bromage, N.R. and Cumaranatunga, P.R.C. (1988) 'Egg production in the rainbow trout'. In *Recent Advances in Aquaculture*, Vol. 3, (eds R.J. Roberts and J.F. Muir). London: Croom Helm, (in press).

Bromage, N.R. and Duston, J. (1986) 'The control of spawning in the rainbow trout using photoperiod techniques'. *Rept Instit. Fres. Res. Drottn. 63* pp. 26–35.

Bromage, N.R., Harbin, R. and Whitehead, C. (1979) 'Problems of salmonid maturation'. *Proc. 11th Two Lakes Conference*, October 1982, Romsey, England. Kent: Janssen Services, pp. 95–101.

Bromage, N.R., Elliott, J.A.K., Springate, J.R.C. and Whitehead, C. (1982) 'Current developments for the all-year-round supply of eggs and improvement of egg quality in salmonid culture'. *Proc. 14th Two Lakes Conference*, October 1982, Romsey, England. Kent: Janssen Services, pp. 82–90.

Bromage, N.R., Elliott, J.A.K., Springate, J.R.C. and Whitehead, C. (1984) 'The effects of constant photoperiods on the timing of spawning in the rainbow trout'. *Aquaculture* **43**: 213–23.

Dodd, J.M. and Sumpter, J.P. (1984) 'Fishes'. In *Marshall's Physiology of Reproduction* (ed. G.E. Lamming), Vol. 1, Reproductive cycles of vertebrates. Churchill Livingstone, pp. 1–126.

Donaldson, E.M. and Hunter, G.A. (1983) 'Induced final maturation, ovulation and spermiation in cultured fish'. In *Fish Physiology*, (eds W.S. Hoar, D.J. Randall and E.M. Donaldson), Vol. IX, Reproduction, Part B: Behaviour and fertility control. New York: Academic Press, pp. 351–403.

Grimaldi, E. and Rosenthal, H. (eds) (1986) 'Trends and problems in aquaculture development'. *Proc. 2nd Int. Conf. Aquafarming*, Aquacoltora '84, Verona.

Hunter, G.A. and Donaldson, E.M. (1983) 'Hormonal sex control and its application to fish culture'. In *Fish Physiology*, (eds W.S. Hoar, D.J. Randall and E.M. Donaldson), Vol. IX, Reproduction, Part B: Behaviour and fertility control. New York: Academic Press, pp. 223–303.

Idler, D.R. and Ng, T. Bun (1983) 'Teleost gonadotropins: isolation, biochemistry and function'. In *Fish Physiology* (eds W.S. Hoar, D.J. Randall and E.M. Donaldson), Vol. IX, Reproduction, Part A: Endocrine tissues and hormones. New York: Academic Press, pp. 187–222.

Jhingran, V.G. and Pullin, R. (1985) 'A hatchery manual for the common, Chinese and Indian major carps'. Manila: ICLARM.

Kagawa, H., Young, G., Adachi, S. and Nagahama, Y. (1985) 'Estrogen synthesis in the teleost ovarian follicle'. In *Salmonid Reproduction* (eds R.M. Iwamoto and S. Sower). Washington Sea Grant Program, University of Washington, pp. 20–5.

Kapur, K. (1981; 'Spawning of Indian major carps – a review'. *Acta hydrochim. hydrobiol.* **9**: 163–82.

Lam, T.J. (1982) 'Applications of endocrinology to fish culture'. *Can. J. Fish. Aquat. Sci.* **39**: 111–37.

Lam, T.J. (1984) 'Artificial propagation of milkfish: present status and problems'. In *Advances in Milkfish Biology and Culture* (eds J.V. Juario, R.P. Ferraris and L.V. Benitez). Manila: Id. Pub House, pp. 21–30.

Leitritz, E. and Lewis, R.C. (1976) 'Trout and salmon culture'. *Calif. Fish and Game Fish Bull.* **164**: 1–197.

MacCrimmon, H.R. (1971) 'World distribution of rainbow trout (*Salmo gairdneri*)'. *J. Fish. Res. Bd. Can.* **28**: 663–704.

Nagahama, Y., Young, G., Veda, H., Kagawa, H. and Adachi, S. (1985) 'Endocrine control of final gamete maturation in salmonids'. In *Salmonid Reproduction* (eds R.N. Iwamoto and S. Sower). Washington Sea Grant Program, University of Washington, pp. 8–19.

Ng, T. Bun and Idler, D.R. (1983) 'Yolk formation and differentation in teleost fishes'. In *Fish Physiology*, (eds W.S. Hoar, D.J. Randall and E.M. Donaldson) Vol. IX, Reproduction, Part A: Endocrine tissues and hormones. New York: Academic Press, pp. 373–404.

Peter, R.E. (1983) 'The brain and neurohormones in teleost reproduction'. In: *Fish Physiology* (eds W.S. Hoar, D.J. Randall and E.M. Donaldson, Vol. IX, Reproduction, Part A: Endocrine tissues and hormones. New York: Academic Press, pp. 97–135.

Springate, J.R.C. and Bromage, N.R. (1984) 'Broodstock management: egg size and number – the 'trade off'.' *Fish Farmer* **7**(4): 12–14.

Springate, J.R.C. and Bromage, N.R. (1985) 'Effects of egg size on early growth and survival in rainbow trout (*Salmon gairdneri*)'. *Aquaculture* **47**: 163–72.

Springate, J.R.C., Bromage, N.R. and Cumaranatunga, R. (1985) 'The effects of different rations on fecundity and egg quality in the rainbow trout (*Salmo gairdneri*)'. In *Nutrition and feeding in fish* (eds C.B. Cowey, A.M. Mackie and J.G. Bell). London: Academic Press, pp. 371–91.

Yamazaki, F. (1982) 'Sex control and manipulation in fish'. *Aquaculture*, 329–54.

CHAPTER 5

Cho, C.Y., Slinger, S.J. and Bayley, H.S. (1982) 'Bioenergetics of salmonid fishes: energy intake, expenditure and productivity'. *Comp. Biochem. Physiol,* **73B**: 25–41.

Cho, C.Y., Cowey, C.B. and Watanabe, T. (1985) *Finfish Nutrition in Asia– Methodological Approaches to Research and Development*. Ottawa: IDRC.

Lall, S.P. and Hines, J.A. (1985) 'Manganese bioavailability and requirements of Atlantic salmon (*Salmo salar*) and brook trout (*Salvelinus fontinalis*)'. IUNS Satellite Symposium, XIII International Congress of Nutrition, Brighton, UK.

NRC (1981) *Nutrient Requirements of Coldwater Fishes*. Washington DC: National Academy Press.

NRC (1983) *Nutrient Requirements of Warmwater Fishes and Shellfishes*. Washington DC: National Academy Press.

Ogino, C. (1980) 'Protein requirement of carp and rainbow trout'. *Bull. Japan. Soc. Sci. Fish.* **46**: 385–8.

Ogino, C. (ed.) (1985) 'Fish Nutrition and Feeds'. Tokyo: Koseisha-Koseikaku.

Satoh, S., Takeuchi, T. Tabata, K. and Watanabe, T. (1985) 'Effect of dietary tricalcium phosphate levels on the availability of zinc to rainbow trout.' Abstr. No. 246, Annual Meeting of Japan. Soc. Sci. Fish., Tokyo.

Takeda, M. (1985) In *Fish Nutrition and Diets* (ed. Y. Yone). Tokyo: Koseisha-Koseikaku, pp. 111–3.

Takeuchi, T., Watanabe, T. and Ogino, C. (1978) 'Optimum ratio of protein lipid in diets of rainbow trout'. *Bull. Japan. Soc. Sci. Fish.* **45**: 1521–5.

Watanabe, T. (1982) 'Lipid nutrition in fish'. *Comp. Biochem. Physiol.* **73B**(1): 3–15.

Watanabe, T., Takeuchi, T., Satoh, S., Niimi, T. Toyama, K. and Okuzumi, M. (1982) 'Effect of histamine and its related compounds on rainbow trout.' Abstr. No. 427, The Annual Meeting of Japan. Soc. Sci. Fish., Tokyo.

Watanabe, T., Nanri, H., Satoh, S., Takeuchi, M. and Nose, T. (1983) 'Nutritional evaluation of brown meals as a protein source in diets for rainbow trout.' *Bull. Japan. Soc. Sci. Fish.* **49**: 1083–7.

Watanabe, T., Kitajima, C. and Fujita, S. (1983) 'Nutritional values of live organisms used in Japan for mass propagation of fish: a review.' *Aquaculture* **34**: 115–43.

Watanabe, T., Thonglod, S., Satoh, S., Takeuchi, T. (1986) 'Requirements of yamame and white fish for essential fatty acids.' *Annual Meeting of Japan. Soc. Sci. Fish., Octobes.* Kochi (abstr.).

CHAPTER 6

Amlacher E. (1970) *Textbook of Fish Diseases* (translated by D.A. Conroy and R.L. Herman). Neptune: TFH Publications.

Aoki, T., Egusa, S., Kimura, T. and Watanabe, T. (1971) 'Detection of R factors in naturally occurring *Aeromonas salmonicida* strains'. *Appl. Microbiol.* **22**: 716–17.

Bullock, G.L., Conroy, D.A. and Snieszko, S.F. (1971) 'Bacterial diseases of fishes'. In *Diseases of Fishes* (eds S.F. Snieszko and H.R. Axelrod), Book II. Neptune: TFH Publications.

Davis, H.S. (1953) *Culture and Diseases of Game Fishes*. Berkeley: University of California Press.

Herman, R.L. (1970) 'Prevention and control of fish diseases in hatcheries'. In *A Symposium on Diseases of Fishes and Shellfishes* (ed. S.F. Snieszko). *Special Publication No. 5*, American Fisheries Society, Washington, DC, pp. 3–15.

B.J. Hill (1986) 'Vaccination of fish'. In *Trends and problems in aquaculture development* (eds E. Grimaldi and H. Rosenthal), *Proc. 2nd Int. Conf. on Aquafarming, Aquacoltora '84*, Verona.

Hoffman, G.L. (1967) *Parasites of North American Freshwater Fishes*. Berkeley and Los Angeles: University of California Press.

Kabata, Z. (1970) 'Crustacea as enemies of fishes'. In *Diseases of Fishes* (eds S.F. Snieszko and H.R. Axelrod), Book I. Neptune: TFH Publications.

Klontz, G.W. and Anderson, D.P. (1970) 'Oral immunization of salmonids: a review'. In *A Symposium on Diseases of Fishes and Shellfishes* (ed. S.F. Snieszko). *Special Publication No. 5*, American Fisheries Society, Washington, DC, pp. 16–20.

Manning, M.J. and Tatner, M.F. (eds) (1985) *Fish Immunology*. London: Academic Press.

Pickering, A.D. (ed.) (1981) *Stress and Fish*. London and New York: Academic Press.

Reichenbach-Klinke, H. and Elkan, E. (1965) *The Principal Diseases of Lower Vertebrates*, Book I: Diseases of fishes. New York: Academic Press.

W.E. Ribelin and G. Migaki (eds) (1975) *The Pathology of Fishes*. Madison: University of Wisconsin Press.

Roberts, R.J. (ed.) (1978) *Fish Pathology*. London: Bailliere Tindall.

Roberts, R.J. (ed.) (1982) *Microbial Diseases of Fish*. London: Academic Press.

Roberts, R.J. and Shepherd, C.J. (1986) *Handbook of Trout and Salmon Diseases*, 2nd edition. Farnham: Fishing News Books.

Sarig, S. (1971) 'The prevention and treatment of diseases of warmwater fishes under subtropical conditions, with special emphasis on intensive fish farming'. In *Diseases of Fishes* (eds S.F. Snieszko and H.R. Axelrod), Book III. Neptune: TFH Publications.

Selye, H. (1950) 'Stress and the general adaptation syndrome'. *Br. Med. J.* i: 1382−92.

Sindermann, C.J. (1970) *Principal Diseases of Marine Fish and Shellfish*. New York and London: Academic Press.

Vivares, C.P., Bonami, J.R. and Jaspers, E. (1986). 'Pathology in marine aquaculture'. *Special Publication No. 9*, European Aquaculture Society.

Wedemeyer, G. (1970) 'The role of stress in the disease resistance of fishes'. In *A Symposium on Diseases of Fishes and Shellfishes* (ed. S.F. Snieszko) *Special Publication No 5.*, Washington DC: American Fisheries Society, pp. 30−5.

Wedemeyer, G.A., Meyers, F.P. and Smith, L. (1976) 'Environmental Stress and fish diseases'. In *Diseases of Fishes* (eds S.F. Snieszko and H.R. Axelrod), Book V. Neptune: TFH Publications.

Wood, J.W. (1974) *Diseases of Pacific Salmon, their Prevention and Treatment*, 2nd edition. Olympia, Washington: Dept. of Fisheries, Hatchery Division, State of Washington.

CHAPTER 7

Fujita, S. (1973) 'Importance of zooplankton mass culture in producing marine fish seed for fish farming'. *Bull. Plankton Soc. Japan* **20**: 49−53.

Fujita, S. (1975) 'Rearing of a large amount of larval fish'. In *Feeding and Development of Larval Fish*. (ed. Japanese Society of Scientific Fisheries, JSSF). Tokyo: Koseisha-Koseikaku, pp. 100−13.

Fujita, S. (1979) 'Culture of red seabream, *Pagrus major*, and its food'. In *Cultivation of Fish Fry and its Live Food* (eds E. Styczynska-Jurewicz, T. Backiel, E. Jaspers and G. Persoone). European Mariculture Society, *Special Publication No. 4*. Bredene, Belgium: EMS, pp. 183−97.

Fujita, T., Satake, M., Watanabe, T., Kitajima, C., Miki, W., Yamaguchi, K. and Konosu, S. (1983) 'Pigmentation of cultured red seabream with astaxanthin diester purified from krill oil'. *Bill. Japan. Soc. Sci. Fish.* **49**: 1855−61.

Kitajima, C. (1979) 'Swim bladder deformity and lordosis in hatchery-reared black seabream (*Acanthopagrus schlegeli*)'. *Bull. Nagasaki Pref. Inst. Fish.* **5**: 27−32.

Kuronuma, K. and Fukusho, K. (1984) *Rearing of Marine Fish Larvae in Japan.* Ottawa: IRDC.

Watanabe, T., Kitajima, C. and Fujita, S. (1983) 'Nutritional values of live organisms used in Japan for mass propagation of fish: A review'. *Aquaculture* **34**: 115–43.

Watanabe, T., Ohta, M., Kitajima, C. and Fujita, S. (1982) 'Improvement of dietary value of brine shrimp, *Artemia salina*, for fish larvae by feeding them on ω3 highly unsaturated fatty acids and fat-soluble vitamins'. *Bull. Japan. Soc. Sci. Fish.* **48**: 1775–82.

Watanabe, T., Tamiya, T., Oka, A., Hirata, M., Kitajima, C. and Fujita, S. (1983) 'Improvement of dietary value of live foods for fish larvae by feeding them ω3 highly unsaturated fatty acids and fat-soluble vitamins'. *Bull. Japan. Soc. Sci. Fish.* **46**: 499–503.

Yamaguchi, M. (1978) *Tai-Yoshoku no Kiso-to-Jissai.* Tokyo: Koseisha-Koseikaku.

CHAPTER 8

Bowen, J.T. (1970) 'A history of fish culture as related to the development of fishery programs'. In *A Century of Fisheries in North America* (ed. N.G. Benson). *Special Publication No. 7.* Washington DC: American Fisheries Society, pp. 71–93.

Boyd, C.E. (1979) 'Water quality in warmwater fish ponds'. Auburn University, Agricultural Experiment Station, Auburn, Alabama.

Broussard, M.C., Jr., and Simco, B.A. (1976) 'High-density culture of channel catfish in a recirculating system'. *Prog. Fish-Cult.* **38**(3): 138–41.

Burrows, R.E., and Chenoweth, H.H. (1955) 'Evaluation of three types of fish rearing ponds'. Research Report No. 39, United States Department of the Interior, Fish and Wildlife Service, Washington DC.

Courtenay, W.R., Jr., and Stauffer, J.R. Jr. (1984) *Distribution, Biology, and Management of Exotic Species.* Baltimore: Johns Hopkins University Press.

Dupree, H.K., and Huner, J.V. (eds) (1984) 'Third report to the fish farmers: The status of warmwater fish farming and progress in fish farming research'. US Fish and Wildlife Service, Washington DC.

Ensminger, M.E., and Olentine, C.G. Jr. (1978) *Feeds and Nutrition-Complete.* Clavis: The Ensminger Publishing Co.

Giachelli, J.W. Coats, R.E. and Waldrop, J.E. (1982) 'Farm-raised catfish: January 1982 cost-of-production estimates'. Report No. 134, Department of Agricultural Economics, Mississippi Agricultural and Forestry Experiment Station, Mississippi State, Mississippi.

Giudice, J.J., Gray, D.L. and Martin, J.M. (1981) 'Manual for bait fish culture in the South'. Joint publication of the US Fish and Wildlife Service, Jackson, Mississippi and Stuttgart, Arkansas and the University of Arkansas Cooperative Extensive Service, little Rock, Arkansas.

Godfriaux, B.L., and Stolpe, N.E. (1981) 'Evolution of the design and operation of a freshwater waste heat aquaculture facility at an electric generating plant'. In *Proceedings of the Bio-Engineering Symposium for Fish Culture* (eds Allen,

L.J. and Kinney, E.C.) Bethesda: Fish Culture Section, American Fisheries Society, pp. 259–65.

Hill, T.K., Chesness, J.L. and Brown, E.E. (1973) 'Growing channel catfish, *Ictalurus punctatus* (Rafinesque), in raceways'. *Proceedings of the Annual Conference of the Southeastern Association of Game and Fish Commissioners* **37**: 488–99.

Joint Subcommittee on Aquaculture (1983) *National Aquaculture Development Plan*, Vol. I'. Washington DC: US Department of Agriculture, US Department of Commerce and US Department of Interior.

Meyer, F.P., Sneed, K.E. and Eschmeyer, P.T. (eds) (1973) 'Second report to the fish farmers: The status of warmwater fish farming and progress in fish farming research'. Resource Publication 113. Washington DC: US Department of the Interior, Fish and Wildlife Service.

National Academy of Sciences (1978) *Aquaculture in the United States: Constraints and Opportunities*. Washington DC: National Academy of Sciences.

National Oceanic and Atmospheric Administration (1983) 'U.S. fish imports are up'. *Fisheries* (Bethesda) **8**(4): 36.

National Research Council (1983) *Nutrient Requirements of Warmwater Fishes and Shellfishes* (revised edn). Washington DC: National Academy of Sciences.

Noeske-Hallin, T.A., Spieler, R.E., Parker, N.C., and Suttle, M.A. (1985) 'Feeding time differentially affects fattening and growth of channel catfish'. *J. Nutrition* **115**(9): 1228–32.

Parker, N.C. (1976) 'A comparison of intensive culture systems.' In *Proceedings of the 1976 Fish Farming Conference and Annual Convention, Catfish Farmers of Texas* (ed. J.T. Davis). Texas: Texas A&M University, College Station, pp. 110–20.

Parker, N.C. (1979) 'Channel catfish production in continuously aerated ponds'. In *Proceedings of the 1979 Fish Farming Conference and Annual Convention, Catfish Farmers of Texas*. Texas: Texas A&M University, College Station, pp. 39–52.

Parker, N.C. (1981) 'Performance of four aeration devices in channel catfish ponds'. In *Proceedings of the 1981 Fish Farming Conference and Annual Convention, Catfish Farmers of Texas* (ed. S.K. Johnson). Texas: Texas A&M University, College Station, pp. 36–8.

Parker, N.C. (1983) 'Air-lift pumps and other aeration techniques'. In *Water Quality in Channel Catfish Ponds* (ed. C.S. Tucker). *Southern Cooperative Series Bulletin 290*, Mississippi Agriculture and Forestry Experiment Station, Mississippi State University, Mississippi State, Mississippi, pp. 24–7.

Parker, N.C. (1985) 'Status and overview of fish culture systems and techniques in the United States' (in English). In *Aquaculture in the USSR and USA: Proceedings of a Soviet-American Symposium on Aquaculture* (title translated from Russian) (ed. E.P. Yakovleva). Moscow: Rotaprint VIRNO.

Parker, N.C. and Simco, B.A. (1973) 'Evaluation of recirculating systems for the culture of channel catfish'. *Proceedings of the Southeastern Association of Game and Fish Commissioners* **27**: 474–87.

Parker, N.C., Simco B.A., and Strawn, K. (1975) 'Recirculating raceways and mariculture: university research for the producer'. In *Proceedings of the 1975 Fish Farming Conference and Annual Convention, Catfish Farmers of Texas*.

Texas Agricultural Extension Service, Texas A&M University, College Station, Texas, pp. 67–9.

Schmittou, H.R. (1970) 'The culture of channel catfish, *Ictalurus punctatus* (Rafinesque), in cages suspended in ponds'. *Proceedings of the Annual Conference Southeastern Association Game and Fish Commissioners* **23**: 226–44.

Smitherman, R.O., and Dunham, R.A. (1985) 'Genetics and breeding'. In *Channel catfish culture* (ed. C. Tucker). New York: Elsevier, pp. 283–321.

Stickney, R.R. (1986) *Culture of nonsalmonid freshwater fishes.* Boca Raton: CRC Press Inc.

Tennessee Valley Authority (1978) 'State of the art: waste heat utilization for agriculture and aquaculture'. National Fertilizer Development Center, Bulletin Y-132. Division of Forestry, Fisheries and Wildlife Development, Technical Report B-12. Tennessee Valley Authority and Electric Power Research Institute. Muscle Shoals, Alabama.

Tomasso, J.R., Simco, B.A. and Davis, K.B. (1979a) 'Chloride inhibitions of nitrite-induced methemoglobinemia in channel catfish (*Ictalurus punctatus*)'. *J. Fish. Res. Bd. Can* **36**: 1141–4.

Tomasso, J.R., Simco, B.A. and Davis, K.B. (1979b) 'Inhibition of ammonia and nitrite toxicity to channel catfish'. *Proceedings of the Annual Conference of the Southeastern Association of Fish and Wildlife Agencies* **33**: 600–5.

US Department of Commerce (1985) 'Fisheries of the United States, 1984'. *Current Fisheries Statistics No. 8360. National Marine Fisheries Service, Washington, DC.*

US Department of Commerce (1986) 'Fisheries of the United States, 1985'. Current Fisheries Statistics No. 8580. National Marine Fisheries Service, Washington DC.

US Department of the Interior (1981) 'Propagation and distribution of fishes from national fish hatcheries for fiscal year 1981'. *Fish Distribution Report No. 16.* US Department of the Interior, Fish and Wildlife Service, Division of Hatcheries and Fishery Resource Management, Washington DC.

US Department of the Interior, US Fish and Wildlife Service and US Department of Commerce, Bureau of Census (1982) *1980 National Survey of Fishing, Hunting and Wildlife-Associated Recreation.* Washington DC: US Government Printing Office.

CHAPTER 9

Arieli, Y., Sarig, S. and Bejerano, Y. (1981) 'Observations on pond growth of *Macrobrachium rosenbergii* at the Ginosar Fish Culture Station in 1978 and 1979'. *Bamidgeh* **33**(2): 57–68.

Arieli, Y. and Rappaport, U. (1982) 'Experimental cultivation of the freshwater prawn *Macrobrachium rosenbergii*'. *Bamidgeh* **34**(4): 140–3.

Avnimelech, Y. and Lacher, M. (1979) 'A tentative nutrient balance for intensive fish ponds. *Bamidgeh* **31**(1): 3–8.

Bograd, L. 1951 'Occurrence of *Mugil* in the rivers of Israel'. *Bull. Res. Counc., Israel*, **B9**(4): 169–90.

Cohen, D., Raanan, Z., Rappaport, U. and Arieli, Y. (1983) 'The production of the freshwater prawn *Macrobrachium rosenbergii* (DeMan) in Israel: Improved conditions for intensive monoculture'. *Bamidgeh* **35**(2): 31–7.

Halevi, A. (1979) 'Observations on polyculture of fish under standard farm pond conditions at the Fish and Aquaculture Research Station, Dor, during the years 1972–1977'. *Bamidgeh* **31**(4): 96–104.

Mires, D. (1969) 'Mixed culture of tilapia with carp and grey mullet in Ein-Hamifratz fish ponds'. *Bamidgeh* **21**(1): 25–32.

Mires, D. (1970) 'Preliminary observations on the effect of salinity and temperature of water changes on *Mugil capito* fry'. *Bamidgeh* **22**(1): 19–24.

Mires, D. (1977) 'Theoretical and practical aspects of the production of all-male tilapia hybrids'. *Bamidgeh* **29**(3): 94–101.

Mires, D. (1983) 'The development of the freshwater prawn (*Macrobrachium rosenbergii*) culture in Israel'. *Bamidgeh* **35**(3): 63–72.

Mires, D. (1985) 'Genetic problems concerning the production of tilapia in Israel'. *Bamidgeh* **37**(2): 51–3.

Palewski, N. and Sarig, S. (1955) 'Mixed nursing as a means of raising pond productivity'. *Bamidgeh* **6**(2): 71–3.

Perlmuter, A., Bograd, L. and Pruginin, Y. (1956) 'Use of estuarine and sea fish of the family *Mugilidae*, grey mullets, for pond culture in Israel'. *GFCM Proceedings and Technical Papers*. Rome: Food and Agriculture Organization.

Raanan, Z. and Cohen D. (1982) 'Production of the freshwater prawn, *Macrobrachium rosenbergii* in Israel. Winter activities 1980/81'. *Bamidgeh* **34**(2): 47–58.

Rappaport, U. and Sarig, S. (1975) 'The results of tests in intensive growth of fish at the Ginosar (Israel) Station ponds in 1974'. *Bamidgeh* **27**(4): 75–82.

Rappaport, U. and Sarig, S. (1979) 'The effect of population density of carp in monoculture under conditions of intensive growth'. *Bamidgeh* **31**(2): 26–34.

Reich, K. (1975) 'Multispecies fish culture (Polyculture) in Israel'. *Bamidgeh* **27**(4): 85–100.

Robinson, M.A. (1984) 'Trends and prospects in world fisheries'. *FAO Fisheries Circular No. 722* Rome: Food and Agriculture Organization.

Rimon, A. and Shilo, M. (1982) 'Factors which affect the intensification of fish breeding in Israel. I. Physical, chemical and biological characteristics on the intensive fish ponds in Israel'. *Bamidgeh* **34**(3): 87–100.

Sarig, S. (1954) 'Methods of exterminating wild fish in ponds'. *Bamidgeh* **6**(1): 7–9.

Sarig, S. (1981a) 'The Mugilidae in polyculture in fresh and brackish water fish ponds'. In *Aquaculture of Grey Mullets* (ed. O.H. Oren). *Int. Biol. Prog.* **26**: 391–409.

Sarig, S. (1981b) 'The status of warm-water aquaculture in Israel. A model of a commercial industry'. In *II Conbep. Recife (Brazil)*, pp. 3–10.

Sarig, S. (1984a) 'A review of tilapia culture in Israel'. In *Proceedings of International Symposium on Tilapia in Aquaculture*. Tel Aviv University, pp. 116–22.

Sarig, S. (1984b) 'Trends in Israeli aquaculture'. In *Research on aquaculture* (eds H. Rosenthal and S. Sarig). *Europ. Maricult. Soc.* **8**: 1–6.

Sarig, S. (1984c) 'The integration of fish culture into general farm irrigation systems in Israel'. *Bamidgeh* **36**(1): 16–20.

Sarig, S. (1985) 'Fisheries and fish culture in Israel in 1984'. *Bamidgeh* **37**(3): 63–76.

Sarig, S. and Marek, M. (1974) 'Results of intensive and semi-intensive fish breeding techniques in Israel in 1971–1973'. *Bamidgeh* **26**(2): 28–50.

Sarig, S. and Arieli, Y. (1980) 'Growth capacity of tilapia in intensive culture'. *Bamidgeh* **32**(3): 57–65.

Shilo, M. and Rimon, A. (1982) 'Factors which affect the intensification of fish breeding in Israel. II. Ammonia transformation in intensive fish ponds'. *Bamidgeh* **34**(3): 101–14.

Tal, S. and Ziv, I. (1978) 'Culture of exotic species in Israel'. *Bamidgeh* **30**(1): 3–11.

Yashouv, A. (1963) 'Increasing fish production in ponds'. *Trans. Am. Fish. Soc.* **92**(3): 292–7.

Yashouv, A. (1966) 'Mixed fish culture. An ecological approach to increase pond productivity'. In Proceedings of FAO World Symposium on Warmwater Pond-fish Culture. *FAO Fisheries Report No. 44* **4**: 258–74. Rome: Food and Agriculture Organisation.

Yashouv, A. (1966) 'Breeding and growth of grey mullet (*Mugil cephalus* L.). *Bamidgeh* **27**(4): 85–100.

Yashouv, A. (1969) 'Mixed fish culture in ponds and the role of tilapia in it'. *Bamidgeh* **21**(3): 75–92.

Zohar, G., Rappaport, U., Avnimelech, Y. and Sarig, S. (1984) 'Results of the experiments carried out in the Ginosar Experimental Station in 1983. Cultivation of tilapia in high densities and with periodic flushing of the pond water'. *Bamidgeh* **36**(3): 63–9.

Zohar, G., Rappaport, U. and Sarig, S. (1985) 'Intensive culture of tilapia in concrete tanks'. *Bamidgeh* **37**(4): 103–11.

CHAPTER 10

Aitken A., Mackie, I.M., Merritt, J.H. and Windsor, M.L. (1982) *Fish Handling and Processing*, 2nd edn. Edinburgh: HMSO.

Anon. (1986) 'An extra 1.1 mt of poultrymeat and 0.6 mt of eggs a year'. *Poultry International* **25**(10): 153–6.

Allen, P.G., Botsford, L.W., Schuur, A.M. and Johnston, W.E. (1984) *Bio-economics of Aquaculture*. Developments in Aquaculture and Fisheries Science No. 13. Amsterdam: Elsevier.

F.A.O. (1987) 'Future economic outlook for aquaculture and related assistance needs'. *ADCP/REP/1987/25*. Rome: Food and Agriculture Organisation.

Halver, J.E., & Tiews, K. (eds) (1979) *Finfish Nutrition and Fishfeed Technology*. Berlin: H. Heenemann GmbH and Co.

Lewis, M.R. (1979) 'Fish Farming in Great Britain: an economic survey with special reference to rainbow trout'. *Misc. Study 67*. Reading: Dept of Agricultural Economics and Management, University of Reading.

Nash, C.E. (1987) 'Aquaculture attracts increasing share of development aid'. *Fish Farming International* **14**(6): 22–4.

Pomeroy, R.S., Luke, D.B. and T. Schwedler (Jan/Feb 1987) 'The economics of catfish production in South Carolina'. *Aquaculture Magazine*, Asheville, North Carolina, pp. 29−35.

Rosenthal, H. and Oren, O.H. (eds) (1981) 'Research on intensive aquaculture'. *Special Publication No. 6*, European Mariculture Society.

Shang, Y.C. (1981) *Aquaculture Economics: Basic Concepts and Methods of Analysis*. London: Croom Helm.

Shang, Y.C. (1986) 'Coastal aquaculture development in selected Asian countries: status, potential and constraints'. *FAO Fish. Circular No. 799*. Rome: Food and Agriculture Organisation.

Shepherd, C.J. (1974) 'The economics of aquaculture − a review' (ed. H. Barnes). *Oceanogr. Mar. Biol. Ann. Rev.* **13**: 413−20.

Shepherd, C.J. (1978) 'Aquaculture: some current problems and the way ahead'. *Proc. Roy. Soc. Edin.* **76B**: 215−22.

Shepherd, M.J. (1978). 'Some studies on the markets for farmed fish'. MSc thesis, University of Strathclyde, Glasgow.

Thorpe, J.P. (ed.) (1980) *Salmon Ranching*. Academic Press, London. 441 pp.

Yamaji, K. (1987) 'Protein in the Japanese diet'. In The Japan International Agricultural Council, Tokyo. *Eurofish report No. 248*. London: Agra Europe.

Appendix I Salmon farming — an economic case study

Atlantic salmon are typically reared in intensive marine farms on a two-year production cycle commencing in May or June. Juvenile stock may be purchased from freshwater hatcheries as 30–50 g smolts and introduced either into shore-based tanks using pumped seawater or, more commonly, into floating pens or marine cages. Large scale production units in Canada, Ireland, Norway and Scotland usually take in 100 000–200 000 smolts per year at an individual site. Capital investment is required in groups of floating pens, associated moorings, spare bag nets, boats, vehicles (including a fork-lift truck), a feed store and on-site housing for the manager. Assuming access roads and service points are available, total capital costs for a farm with a production capacity of 250 tonnes per annum are currently *c.* £200 000 in Scotland.

Table 1 indicates how this capital cost is shared between different elements with differing economic lives and hence depreciation schedules to give a total annual depreciation cost of £42 000. If such a farm stocks 120 000 smolts and achieves 85% overall survival to harvest, depending on genetic and environmental conditions, it is likely that about 21 000 fish will mature after 12 months as grilse

Table 1 Capital costs for 250-tonne salmon farms

Item	Capital cost in £000s	Estimated economic life in years	Annual depreciation in £000s
Pens and moorings	75.0	5	15.0
Nets	25.0	3	8.3
Boats	21.0	3	7.0
Vehicles	21.0	3	7.0
Accommodation	46.0	20	2.3
Miscellaneous	12.0	5	2.4
Total	200.0	4.8	42.0

at a mean weight of *c*. 1.5 kg and the balance will achieve a mean salmon weight of *c*. 3.2 kg after 24 months. In practice, many farmers prefer to harvest a proportion of their salmon as smaller 'pre-salmon' soon after the time of grilse harvest. Table 2 depicts the situation in which 55 000 fish are harvested after 12 months and 47 000 after 24 months representing 99.6 and 150.4 tonnes respectively. Thus 250 tonnes of fish are produced over an initial two-year production cycle at different unit prices, with larger fish achieving higher unit values compared with small fish such as grilse. Table 3 projects the annual operating costs over this two-year cycle, of which the dominant components are purchase costs for feed and smolts representing 36% and 26% respectively of the total. Including depreciation on capital, unit cost of production is estimated at £2.84/kg, but if it is assumed that total expenditure is financed from bank borrowings at an interest rate of 14% pa, interest charges further increase unit production costs to an estimated total value of £3.40/kg. This compares with a projected average delivered sales price of £3.98/kg to yield a positive balance of £146 000 over total expenditure, representing a pre-tax return on the initial £200 000 capital investment of 73% over the two-year period or 36.5% per annum (Table 4).

It is important to explore the sensitivity of any financial appraisal to possible changes in factor costs and market prices. A mere 15% reduction in average salmon sales price to £3.40/kg would mean a revenue loss of £145 000 over the cycle. At this point income and expenditure would exactly balance, hence £3.40 represents the breakeven selling price per kilo. By the same token, a stock loss of 10 538 salmon just prior to harvest would decrease sales revenue to breakeven point and it is clear that overall profitability is sensitive

Table 2 Harvest yield and sales revenue for a 250-tonne salmon farm over a two-year cycle

Year	Fish harvested	Mean weight in kg	Production in tonnes	Mean delivered price in £/kg	Net sales value in £000s
1	21 000 grilse	1.5	31.6	3.20	101.1
1	34 000 pre-salmon	2.0	68.0	3.65	248.2
2	47 000 salmon	3.2	150.4	4.30	646.7
Cycle 102 100		2.4	250	3.98	996

Table 3 Operating costs for a 250-tonne salmon farm over an initial two-year cycle

	Year 1 in £000s	Year 2 in £000s	Total costs in £000s	Percentage
1. Stock	162.0	—	162.0	25.9
2. Feed	120.0	105.0	225.0	35.9
3. Labour	38.0	38.0	76.0	12.1
4. Packaging and distribution	29.9	45.1	75.0	12.0
5. Insurance	24.0	24.0	48.0	7.7
6. Miscellaneous	20.0	20.0	40.0	6.4
Total operating costs	393.9	232.1	626.0	100.0

Notes
1. $120\,000 \times 35$ g smolts are purchased at £1.35 each, of which 85% survive to harvest.
2. Feed conversion rate is 2.0, hence 500 tonnes of salmon feed are purchased at an average price of £450/tonne.
3. Three workers plus manager represent an annual labour cost of £38 000.
4. Cost of boxes, ice and freight to market at 30p/kg fish.
5. Stock insurance premium is 4% of peak exposure.
6. Includes rents, fuel and electricity.

to relatively small changes in value and yield of the end product. A commonly used indicator of management skill is food conversion ratio, which has been claimed to average as low as 1.7:1 on certain farms compared with 2.0:1 in the model. This would equate to a saving of approximately £38 000 (including interest charges) over the cycle, yielding an increase in return on investment from 73% to 92% (i.e. 36.5% pa to 46% pa).

Sensitivity to changing input costs is well illustrated by considering the implications of a salmon farmer diversifying back into smolt production. Table 5 compares the production costs for smolts at two different levels of capacity, based on an initial 18-month cycle to produce 'S1' smolts. Capital and operating costs can be much increased if it is necessary to construct a dam and/or to install pumping equipment, but for the purpose of this analysis it is assumed that sufficient water is available on a gravity basis without the need for heavy capital expenditure other than in rearing tanks and staff accommodation. The latter is required whatever the level of output, hence capital costs per unit production reduce with increasing scale. Higher output also offers economies of scale due to more efficient use of labour, bulk purchase discounts off the costs of smolt feed and salmon eggs, etc. Estimates of total ex-

Table 4 Return on investment for 250-tonne salmon farm

1. *Capital cost* = £200 000. Depreciation at £42 000 pa costs £84 000 over the two-year cycle.
2. *Operating costs* over the two-year production cycle = £626 000. Including depreciation, this gives a unit production cost of 710 000/250 000 = £2.84/kg before finance charges.
3. *Interest on borrowed capital*   Total interest charges in years 1 and 2 are estimated (see footnote) to be £115 000 and £25 000 respectively, i.e. £140 000 over the two-year cycle.
4. Total and unit production costs over the two-year production cycle:

	Total costs in £	Unit costs in £/kg
Depreciation	84 000	0.34
Operating costs	626 000	2.50
Interest	140 000	0.56
Total	850 000	3.40

5. *Balance of income over expenditure*

	in £
Net sales value	996 000
Total expenditure	− 850 000

Profit = 146 000

Return on investment = 73% per cycle = 36.5% pa

Note   It is assumed that £362 000 is spent at the outset on fixed capital and smolts. At a current interest rate of 14% this would represent interest charges on borrowed capital over two years of £101 000. Since total expenditure over the two-year cycle is £826 000, there is a balance of £464 000 to be financed, but borrowings are reduced by the grilse and pre-salmon harvest at the end of year 1.

penditure including depreciation and finance charges increase from about 60 pence per smolt for a hatchery producing 500 000 smolts/cycle to about 89 pence per smolt for a hatchery producing 120 000 smolts/cycle. A 250-tonne capacity salmon farmer diversifying into smolt production to cover in-house requirements for 120 000 smolts per cycle would thus save 46 pence per smolt compared with an outside purchase price of £1.35, i.e. £55 000 plus £15 000 in reduced interest charges, which is equal to £71 000 over the $1\frac{1}{2}$-year cycle, which is reflected in the salmon farm's balance of income over expenditure. Because total capital costs of the integrated farm are increased by £130 000 in hatchery investment, the resulting return on total capital investment is reduced from 73% over the two-year initial salmon farming cycle to 66% by in-house smolt production. Construction of the larger hatchery producing 500 000 smolts at 60.4p would save £89 500 in smolt purchase costs plus £25 000 in

Table 5 The effect of scale on smolt production costs with a $1\frac{1}{2}$-year cycle

	Capacity of 120 000 × S1 smolts per cycle in £000s	Capacity of 500 000 × S1 smolts per cycle in £000s
Capital costs		
1. Hatchery, tanks, pipework accommodation	130.00	400.0
2. Depreciation over 18 months	10.0	30.0
Operating costs		
3. Purchase of salmon eggs	13.0	50.0
4. Cost of labour	33.0	57.0
5. Cost of feed	7.0	26.0
6. Miscellaneous	9.0	32.0
	62.0	165.0
Cost of borrowing at 14%		
On capital costs and stock over 18 months	30.0	95.0
On balance of expenditure	5.0	12.0
	35.0	107.0
Total production costs (including depreciation and finance)	107.0	302.0
Unit cost per fish produced	89.2p	60.4p

Notes
(a) Item 2 assumes an economic life of 15 years.
(b) Item 3 assumes a purchase of 300 000 eggs at £44 per thousand at small scale and 1 250 000 at £40 per thousand at large scale.
(c) Item 4 assumes manager plus one labourer at small scale and manager plus 3 labourers at large scale.
(d) Item 5 assumes 10 tonnes of feed at £700/tonne at small scale and 41 tonnes of feed at £634/tonne at large scale.
(e) Item 6 includes rents, diesel and electricity and assumes no pumping is required at either scale of production.

reduced interest charges, which is equal to *c.* £115 000 over the $1\frac{1}{2}$-year cycle. In this case capital costs of the integrated farm are increased by £400 000 in hatchery investment to a total of £600 000, whereas overall return on investment needs to take account of the revenue generated by 380 000 smolts available for third party sale.

It is preferable to view the options of smolt production and salmon on-growing as separate for the purpose of investment ap-

Table 6 Return on investment from smolt production at two levels of output

	Capacity of 120 000 smolts per cycle in £000s	Capacity of 500 000 smolts per cycle in £000s
Net sales value at £1.35	162.0	675.0
Total expenditure	107.0	302.0
Balance of income over expenditure	55.0	373.0
Initial capital cost	130.0	400.0
Return on investment per $1\frac{1}{2}$-year cycle	42.3%	93.3%
Return on investment per annum	28.2%	62.2%

praisal with all inputs and outputs (e.g. smolts) costed at true market rates. On this basis Table 6 shows how a high return on capital may be achieved with large-scale smolt production under the stated assumptions (e.g. favourable site conditions) due to economies of scale when compared with smaller hatcheries. Although return on investment in a small-scale, relatively high cost smolt unit producing 120 000 smolts would also appear somewhat limited compared with salmon farming, it should be noted that the break-even smolt price represents a 34% reduction in average sales price and viability of smolt production is thus less sensitive to price changes than salmon farming. In practice, it may be concluded that small-scale smolt production may offer some long-term security of supply by way of backward integration for the salmon farmer, but is unlikely to improve overall return on investment in the short term. The rationale may well be to obtain a reservoir of improved stock derived from on-farm selection with a known health status and performance potential.

Appendix II Useful information

AQUACULTURE JOURNALS AND RELATED PERIODICALS

1. *American Fish Farmer and World Aquaculture News*
2. *Aquaculture*
3. *Aqua-cultura* (in Spanish)
4. *Aquaculture Digest*
5. *Aquaculture and Fisheries Management*
6. *Aquaculture Ireland*
7. *Aquaculture Magazine*
8. *Bamidgeh* (Journal of Israeli Fish Breeders' Association)
9. *Bulletin of the Japanese Society of Scientific Fisheries* (Nippon Suisan Gakkaishi; English abstracts)
10. *California Department of Fish and Game (Fish Bulletin)*
11. *Canadian Journal of Fisheries and Aquatic Sciences*
12. *Commercial Fish Farmer*
13. *European Aquaculture Society* (quarterly newsletter)
14. *Fish Farmer*
15. *Fish Farming International*
16. *Freshwater and Aquaculture Contents Tables*
17. *Il Pesce* (Italian and English)
18. *Infofish Marketing Digest*
19. *Informationen für die Fischwirtschaft* (German)
20. *Journal of Aquaculture Engineering*
21. *Journal of Fish Biology*
22. *Journal of Fish Diseases*
23. *Journal of the World Aquaculture Society*
24. *NAGA* (ICLARM Newsletter)
25. *Progressive Fish-Culturist*
26. *Recent Advances in Aquaculture*
27. *Revista Latinoamericana de Acuicultura* (in Spanish)
28. *Salmon Farming* (Scottish Fisheries Research)
29. *SEAFDEC Asian Aquaculture*
30. *Transactions of the American Fisheries Society*
31. *Trout Cultivation*

Note The above list excludes certain scientific journals (e.g. *Comparative Biochemistry and Physiology*; *Marine Biology*) as well as less specialist publications (e.g. *Scientific American*; *New Scientist*) which frequently contain articles relating to aquaculture.

LIST OF SCIENTIFIC AND COMMON NAMES OF FARMED FISH

SCIENTIFIC	COMMON
Acanthopagrus schlegeli	black seabream
Anguilla anguilla	European eel
Anguilla japonica	Japanese eel
Anguilla rostrata	American eel
Aristichthys nobilis	bighead or spotted silver carp
Astronotus ocellatus	oscar
Carassius auratus	goldfish
Catla catla	catla
Catostomus commersoni	white sucker
Chanos chanos	milkfish
Cirrhina mrigala	mrigala
Coregonus spp.	whitefish
Crassostrea gigas	Pacific oyster
Crassostrea virginica	American oyster
Ctenopharyngodon idella	grass carp
Cyprinus carpio	common carp (includes the scale, line (zeil), mirror, leather and koi)
Dicentrarchus labrax	sea bass
Epinephelus akaara	Japanese grouper
Esox lucius	northern pike
Hypophthalmichthys molitrix	silver carp
Ictalurus catus	white catfish
Ictalurus furcatus	blue catfish
Ictalurus natalis	yellow bullhead
Ictalurus nebulosus	brown bullhead
Ictalurus punctatus	channel catfish
Ictiobus bubalus	smallmouth buffalo
Ictiobus cypinellus	bigmouth buffalo
Ictiobus niger	black buffalo
Labeo rohita	rohu
Lateolabrax japonicus	Japanese sea bass
Lepomis cyanellus	green sunfish
Lepomis macrochirus	bluegill
Lepomis microlophus	reader sunfish
Limanda yokohamae	right-eyed flounder
Liza ramada (Mugil capito)	grey mullet (thinlip)
Longirostrum (Caranx) delicatissimus	striped jack
Macrobrachium rosenbergii	freshwater prawn
Mercenaria campechiensis	southern quahog
Mercenaria mercenaria	northern quahog
Micropterus coosae	redeye bass
Micropterus dolomieui	smallmouth bass
Micropterus punctulatus	spotted bass

Micropterus salmoides	largemouth bass
Misgurnus anguillicaudatus	Oriental weatherfish
Morone saxatilis	striped bass
Mugil cephalus	grey mullet (flathead)
Mylopharyngodon piceus	black carp
Mytilus edulis	blue mussel
Notemigonus crysoleucas	golden shiner
Oncorhynchus gorbuscha	pink salmon
Oncorhynchus keta	chum salmon
Oncorhynchus kisutch	coho salmon
Oncorhynchus masou	masou salmon or gamame
Oncorhynchus nerka	sockeye salmon
Oncorhynchus tschawytscha	chinook salmon
Ophiocephalus spp.	snakehead
Oplegnathus fasciatus	striped knifejaw
Orconectes immunis	papershell crayfish
Orconectes rusticus	rusty crayfish
Oreochromis aureus	blue tilapia
Oreochromis mossambicus	Mozambique tilapia
Oreochromis niloticus	tilapia
Oryzias latipes	medaka
Ostrea edulis	European oyster
Ostrea lurida	Olympia oyster
Pacifastacus leniusculus	signal crayfish
Pagrus (Chrysophrys) major	red seabream
Paralichthys olivaceus	Japanese flounder
Penaeus monodon	jumbo/tiger prawn
Penaeus setiferus	white shrimp
Penaeus stylirostris	blue shrimp
Penaeus japonicus	kuruma prawn
Pimephales promelas	fathead minnow
Plecoglossus altivelis	ayu
Pleuronectes platessa	plaice
Pomoxis annularis	white crappie
Pomoxis nigromaculatus	black crappie
Procambarus acutus acutus	whiteriver crayfish
Procambarus clarki	red crawfish
Pylodictis olivaris	flathead catfish
Salmo aquabonita	golden trout
Salmo clarki	cutthroat trout
Salmo gairdneri	rainbow trout
Salmo gairdneri	steelhead trout
Salmo salar	Atlantic salmon
Salvelinus aureolus	Sunapee trout
Salvelinus fontinalis	brook trout
Salvelinus malma	dolly varden
Salvelinus namaycush	lake trout
Scophthalmus (Rhombus) maximus	turbot

Seriola quinqueradiata	yellowtail
Silurus silurus	wels catfish
Solea solea	Dover sole
Sparus aurata	gilthead bream
Stizostedion (Lucioperca) lucioperca	zander
Stizostedion canadense	sauger
Stizostedion vitreum vitreum	walleye
Takifugu rubripes	tiger puffer
Thunnus thynnus	bluefin tuna
Tilapia rendalli	tilapia
Tilapia zillii	redbelly tilapia

GLOSSARY

(See Tables 1.1 and 6.9 for the names of fish species and fish diseases respectively appearing in the text).

Adipose fin: soft fin in salmonids between dorsal and caudal fins; sometimes removed to mark fish because it does not regenerate.

Adjuvant: material added to in a vaccine to enhance the immunological response.

Adrenaline: fight or flight hormone produced in response to a sudden stress.

Alga (algae: plural): microscopic single-celled plant.

Aliquot: a single dose of a chemical or drug in solution.

Alkalinity: measure of ability to neutralise acidity.

Amino acid: constituent building block of proteins.

Ammonia: toxic excretory product of protein or nitrogenous metabolism in fish.

Anadromous: fish which lives most of its adult life in the sea but migrates to freshwater to spawn.

Anaemia: reduction in number of the oxygen-carrying red blood cells.

Annulus: growth ring present on scales, operculum and otoliths, used to age fish.

Anoxia/anoxic: little or no oxygen present.

Antibiotic: naturally occurring substance with antimicrobial activity used to treat some infectious diseases.

Antibody: protein produced by fish in an attempt to neutralise the effects of foreign material invading the body.

Anti-oxidant: chemical added to diets to prevent breakdown of fats.

Artemia: proper name for the brine shrimp, the crustacean used in feeding marine or cyprinid fish larvae.

Artificial selection: selection of parental fish in a programme of breeding designed to produce offspring or progency with specific characteristics.

Astaxanthin: principal natural pigment of salmonids.

Auricle: part of heart.

Auto-oxidation: breakdown of fats by oxidation at the carbon double bonds, usually followed by hydrolysis.

Automatic feeder: device powered by electric, clockwork, water, or air to feed fish at regular timed intervals.

Bacterium: ubiquitous usually single-celled microscopic organism with a rigid cell wall but no separate nucleus and no chlorophyll; about 1 micron (one thousandth of mm) in diameter.

Bait Fish: fish cultured for use as live bait by anglers.

Bends: gas bubble disease caused when blood becomes supersaturated with gases which then come out of solution in capillaries and damage tissues.

Biochemical oxygen demand (BOD): a measure of capacity of the organic materials in a solution (usually an effluent) to take up oxygen and become oxidised.

Biomass: total weight of organism(s) supported in an environment or habitat.

Biotechnology: use of living organisms in industrial production processes.

Borehole: shaft drilled down to provide water supplies from natural aquifers or groundwater.

Bridgestone cage: recently developed sea cage or pen which has a flexible float or collar; because of flexibility it can be used in exposed environments.

Brine shrimp: common name for *Artemia* used as live food for fish larvae.

Broodstock fish: parent fish cultivated to provide eggs or fry.

Caisson's disease: see **Bends**

Canthaxanthin: pigment often used in diets to colour fish flesh.

Capillary: smallest blood vessel where exchange of gases and nutrients occurs within organs and tissues.

Carotenoid: generic name for group of pigments used in diets to colour flesh, includes astaxanthin, canthaxanthin, carotene, etc.

Carrier: fish that does not show any signs of disease but carries the organism and may infect others.

Catadromous: fish that lives most of its adult life in freshwater but migrates to seawater to spawn.

Cataract: opacity in the eye caused by parasites, diet, or other stresses.

Chemotherapy: treatment with chemical or drug to remove parasites or other disease-carrying organisms.

Chromosome: thread-like structures found in all cells responsible for the carriage of genetic information.

Circular tank: one of the principal forms of fish-rearing facility; usually has outflow in the centre.

Commensal: organism which lives on, in, or within another; may offer some advantages but does not cause harm.

Competition index: a measure of the change in yield due to adopting polyculture of different species instead of monoculture with a single species.

Condition factor: weight/length relationship.

Conductivity: ability to conduct electricity, used as a measure of the ionic concentration of waters.

Conus arteriosus: elastic-walled blood vessel found immediately after the heart.

Copepod: crustacean often cultured or harvested from wild stocks for use in the feeding of cyprinid or marine fish.

Corticosteroid: hormone which reduces inflammation and controls ion balance; produced by the adrenal cortical tissue.

Culture-based fishery: the stocking of hatchery-reared juvenile fish into the natural environment for subsequent recapture when fully grown (see also **Ranching**).

Cuverian sinus: large vessels at back of gills which collect venous blood from the body and return it to the heart.

Decompression sickness: see **Bends**.

Degree days: composite unit of time and temperature; allows assessment of developmental times for processes at different temperatures e.g. 50 days is 5 days at 10°C or 10 days at 5°C.

Demand/pendulum feeder: type of feeder in which the fish are able to nudge a pendulum which operates a gate and releases food into the water.

Diadromous: an ability to migrate between fresh and seawater or vice versa.

Digestibility: extent to which a nutrient in the diet is digested and absorbed.

Diploid: having two sets of chromosomes, i.e. the number usually carried by all body cells.

Earth pond: elongated pond usually some 10 times longer than wide, made by simple excavation.

Economy of scale: economic advantage of large-scale production, mainly due to certain fixed costs being borne by a greater volume of product.

Effluent: outflow water from farm.

Elasticity of demand: a measure of the extent to which demand for a product varies with changing factors, such as price.

Elver: young stage of eel.

Enzyme: organic catalyst.

Epidemiology: study of the distribution and determinants of infection or disease.

Epidermis: surface or covering layer of cells on outside of boby.

Erythrocyte: red blood cell responsible for carriage of oxygen.

Essential amino acid (EAA): amino acid which must be present in the diet if a dietary deficiency is not to result, i.e. an amino acid that cannot be synthesised.

Essential fatty acid (EFA): fatty acid (component of fat) which must be present in the diet if a deficiency disease is not to result, i.e. a fatty acid which cannot be synthesised.

Euryhaline: ability to withstand exposure to both fresh and salt water environments (cf **Stenohaline**).

Eutrophication: enrichment of water with nutrients, often phospate and nitrate; this causes excessive plant and bacterial growth and hence creates an oxygen demand.

Extensive culture: low intensity aquaculture yielding only modest increase over natural productivity (e.g. usually involves no artificial feeding).

Eyed eggs/eyeing/eyeing up: stage of development of egg when the darkly pigmented retina (part of eye) appears.

Fatty acid: one of the constituent molecules of a fat.

Fecundity: numbers of eggs produced; sometimes expressed as number per fish, sometimes per kg of fish.

Feminisation: modification of gender of fish using a hormone, drug or genetic manipulation.

Fingerling: underyearling stage formed from a fry.

Fin nipping: biting of fins of cultured fish often caused by insufficient diet or too crowded conditions.

Fishmeal: dry and defatted protein concentrate derived from the trawling of fresh fish and used as the protein source in dry diets.

Fish pump: mechanical device which is used to transport fish around farms through wide-bore tubes; causes less physical damage than netting.

Fish kill: name for sudden mortality in a fish stock; usually caused by lack of oxygen or poisoning, never by disease.

Flashing: a sudden abnormal movement of fish in which they turn on their flanks, usually due to an external parasite of some kind.

Food conversion ratio (FCR): ratio of dry weight of food fed to wet weight of fish gain.

Fork length: length of fish measured from the tip of snout to point of division of tail or caudal fin.

Fry: young stage of fish formed from the egg.

Fungus: a type of plant which is often associated with fish diseases, usually as a secondary invader.

Furuncle: small area of infection in or just under the skin which develops into a raised abscess, notably in furunculosis disease.

Gamete: general name for sperm and eggs.

Genotype: genetic makeup of a fish; that portion which is inherited.

Geothermal water: groundwater which has been warmed by contact with heat from the earth's crust.

Gill filament/lamella/raker: parts of the gill apparatus.

Gizzard erosion (GE): abnormality of the gut or intestine caused by a dietary toxin.

Glucagon: pancreatic hormone involved in the control of sugar and protein levels in the blood.

Gonad: organ in which either sperm or eggs are produced; general name for either the testis or the ovary.

Gonadotropin: pituitary hormone which controls the production by the gonads (testis and ovary) of sperm and eggs and also the gonadal hormones.

Grading: husbandry process which involves the mechanical separation of different sized fish into groups or grades of similar weight; separation usually achieved on the basis of different girth size.

Grilse: Atlantic salmon which matures after only one year at sea.

Growth hormone/somatotropin: pituitary hormone which controls growth and repair of tissues.

Gynogenesis: genetic manipulation in which all the chromosomes in the progeny are derived from the mother.

HCG: human chorionic gonadotropin; commercially available, semi-purified hormone which is used to induce ovulation and spermiation, i.e. egg and sperm production.

Haemoglobin: oxygen-carrying red blood cell pigment.

Haemorrhage: loss of blood.

Hand-feeding: feeding of dry diet by husbandry staff, helpful in gauging health and vitality of stock.

Hapa: natural or synthetic structure of substrate on which fish eggs are deposited after artificial spawning of broodstock.

Haploid: having half the normal number of chromosomes; most gametes are haploid so that when fertilisation occurs the diploid condition is regained.

Hardness: total concentration of calcium and magnesium ions (usually carbonates) present in water.

Hectare: 2.47 acres.

Hemibranch: part of gill.

Heterosis: hybrid or interstrain vigour produced by crossing or cross-fertilising different species, strains or stocks.

Highly unsaturated fatty acid (HUFA): fatty acids with double carbon (unsaturated) bonds, especially those from vegetable sources.

Holobranch: part of gill.

Homeostasis: maintenance of constant internal conditions

Horizontal transmission: infection transmitted between different individuals of a stock or population.

Host (Parasitic): fish on, or in which, the parasite is found.

Human chorionic gonadotropin (HCG): see **HCG**.

Hybrid vigour: see **Heterosis**.

Hybridisation: crossing or fertilisation of one species by the sperm from another with the intention of producing progeny with the best charactistics of both parents.

Hypertonic/hyperosmotic: a solution or cell which has a higher ionic or solute concentration; this will attract or pull in water from a weaker solution.

Hypotonic/hypoosmotic: a solution or cell which has a lower ionic or solute concentration; this will give up water to a stronger solution.

Hypophysation: injection with an extract of the pitultary gland used to induce ovulation and spermiation.

Hypothalamus: part of brain which controls many internal body functions and the activity of the pitutary gland; produces releasing hormones.

Immunocompetence: the ability of fish to mount an immune response; often used to describe specific circulating cells within fish.

Insulin: pancreatic hormone which controls sugar balance in humans and fish, but in fish it is probably more involved in controlling protein metablism.

Intensive culture: high intensity aquaculture yielding far in excess of natural productivity levels; usually involves high stocking rates in purpose-built rearing facilities with use of controlled feeding (e.g. artificial diets).

Interferon: soluble protein produced as a response to specific viral infections but which imparts a general resistance to further attacks by viruses.

International unit (IU): measure of activity used when pure preparations are not available and hence dosage cannot be quantified by weighing; preparations are calibrated by comparison with international reference standards.

Intramuscular (injection): injection site usually in the dorsal musculature.

Intraperitoneal (injection): injection into the abdominal cavity.

Iodophor: chemical used for disinfection of eggs.

Isotonic/iso-osmotic: solution or cell with the same solute concentration.

Ion: inorganic material or element.

Jack: male Pacific salmon which matures earlier than normal.

11-Ketotestosterone: steroid hormone produced by the testis, along with testosterone responsible for the appearance of male secondary sex characters.
Kidney: organ which helps control ion and water balance.
Klinoptilite: naturally occurring inorganic material which is used to absorb some chemicals and gases, e.g. ammonia.
Kype: hooked jaw; a secondary sexual characteristic exhibited by many mature male salmonids.

LHRH/LHRHa/LHRH-A: releasing hormone which controls release of gonadotropin from the pituitary gland. Used as an alternative to hypophysation for the induction of spawning.
Lateral transmission: see **Horizonal transmission**.
Lesion: any pathological change in structure of an organ, tissue or cell.
Levée: earth embankment forming common wall between adjacent ponds.
Liver: major organ of the body, much involved in handling nutrients.

Macrophyte: large aquatic plant, rooted or floating.
Malachite green: chemical used to remove fungi and some protozoans from fish.
Mariculture: farming of organisms in the sea.
Masculinisation: modification of gender of fish by hormonal or genetic means.
Meiosis: cell division which results in the formation of gametes with a single set of chromosomes, i.e. haploid.
Metabolism: chemical processes occurring within any living organism.
Methaemoglobinaemia: under conditions of poor water quality the haemoglobin in catfish blood can combine with ammonium nitrite to cause brown blood disease due to resulting methaemoglobin.
Microgram: unit of weight; one thousandth of a milligram.
Micropyle: aperture in egg through which sperm enters for fertilisation.
Milligram: unit of weight; one thousandth of a gram.
Milt: common name for sperm or spermatozoa.
Monk: device used to control water levels in ponds.
Monosex culture: culture of either male or female fish only.
Monoculture: culture of a single species of fish.
Morula: embryo in process of cleavage or cell division.
Mould: superficial growth of fungus; also used for particular types of fungi.

Necrosis: area of dead cells or tissue in an otherise living or healthy organ.
Net protein utilisation (NPU): measure of the ability of an animal to digest and assimilate protein for growth.
Nitrofuran: synthetic chemical/drug with antimicrobial activity used for the treatment of some infectious diseases (cf *Sulphonamide*).

Oestradiol: hormone produced by the ovary.
Operculum: bony plate which covers the gills of bony fish.
Organophosphate: chemical used for the removal of ectoparasites.
Osmoregulation: regulation of water balance, sometimes includes regulation of ion balance.

Osmotic pressure: pressure generated by materials or ions in solution which results in an absorption or loss of water until any differences in concentrations are equalised.

Oviposition: release of eggs from the fish.

Ovulation: release of eggs from the ovary.

Ova: egg

Pancreas: organ which produces insulin and glucagon and some digestive enzymes.

Parasite: organism which lives on or in another organism (called the host) generally to its detriment and from which the parasite takes food.

Parr: freshwater stage of salmon with distinctive vertical markings on the flanks.

Pathogenic: disease-causing.

Phagocyte: cell found in the blood and tissues which provides an important defence mechanism by engulfing bacteria and other foreign materials.

Phenotype: physical appearance or characteristics of an organism (cf genotype).

Photosynthesis: synthesis by green plants of organic compounds from water and carbon dioxide using energy from sunlight.

Phytoplankton: microscopic plant life usually found at surface of water.

pH: measure of acidity and alkalinity on a scale from 1 (acid) to 14 (alkaline); 7 is neutral.

Pimozide: drug used to induce spawning artificially; works by stimulating gonadotropin release.

Pituitary extract: aqueous, alcoholic or acetone dissolved extract of pituitary gland used for hypophysation or the artificial induction of spawning.

Pituitary gland: endocrine or hormone-producing gland found on the underside of the brain just behind the eyes.

Poikilothermic: inability to control internal body temperature, hence changes passively with ambient conditions ('cold blooded').

Pollution: contamination of environment with foul, waste or toxic materials.

Polyculture: farming of several fish species in the same tank or pond, each dependent on a different source of feed or trophic level.

Polysaccharide: carbohydrate formed by polymerisation of many sugar (hexose or pentose) molecules, e.g. starch, cellulose, glycogen.

Polyunsaturated fatty acid (PUFA): a fat having many double (unsaturated) carbon bonds; also known as 'HUFA'.

Pop-eye/exophthalmos: protrusion of the eyes, caused by gases coming out of solution in the blood vessels behind the eye.

Precocious maturation/precocity: maturation at an earlier age than normal.

Primary pathogen: organism responsible for initiating disease process.

Productivity: capacity of a body of water to support animal and plant growth under natural conditions.

Progestagen: group of steroid hormones produced by the ovary.

Protein sparing: nutrient (often a fat or carbohydrate) whose presence in a diet allows the protein to be used for growth and not for energy formation.

Pyloric caecae: blind-ending tubular outgrowths of the posterior end of the stomach; thought to be involved with fat digestion.

Quaternary ammonium: chemicals used in the treatment of external bacterial or parasitic disease.

Ranching: culture strategy in which fish are released into the wild to complete their growth; subsequently recaptured for harvest (see also **Culture-based fishery**).

Rancidity: breakdown/oxidation of fats leading to the development of objectionable odours and flavours.

Raceway: elongated fish rearing facility usually fabricated from concrete; water enters at one end and leaves from the other.

Red tide: sudden population explosion of red algae (single-celled plants) which release neurotoxins (nerve toxins) which are sometimes lethal to fish.

Releasing hormone: peptide hormones which control all aspects of the pituitary gland.

Rheotactic: behavioural response to direction of water flow or current.

Rotifer: microscopic ciliated animals used as a food for certain fish larvae.

S1: smolt or sea-going stage of salmon which is transferred to seawater for on-growing after one year in freshwater.

S2: smolt or sea-going stage of salmon which is transferred to seawater for on-going after two years in freshwater.

Saturation value: maximum solubility of a chemical or gas in water; values increase with increased pressure and with reduced temperature and salinity.

Scoliosis: abnormal development or bending of the spine caused by diet or disease.

Secondary lamella: respiratory portion of the gills.

Secondary pathogen: organism whose infection of a host is secondary to other factors in the disease causing process.

Secondary sexual character: sex-specific characters which develop during maturation and are used for attraction during courtship behaviour.

Seed: term sometimes used for eggs and (more rarely) larvae and fry.

Selection: see **Artificial selection**.

Septicaemia: severe microbial infection which involves the presence of the infectious agent circulating in the blood.

Sex chromosomes: pair of chromosomes which determine gender; in salmonids the XX genotype produces a female and XY produces a male.

Sex reversal: modification of sex using genetic or hormonal manipulations.

Shocking: physical agitation of eggs which turns them white if infertile.

Smolt: stage of salmon development which is able to migrate to the sea (see **S1**, **S2**).

Somatostatin: hypothalamic releasing hormone which inhibits release of growth hormone.

Sparger: device for oxygenation or aeration of water.

Spawning can: container in which mature catfish are placed to spawn.

Spermiation: sperm or milt production.

Spore: infectious stage of fungi, often with a resistant covering to withstand adverse conditions (some bacteria also form spores).

Stenohaline: inability to tolerate wide variations of osmotic pressure, cf **euryhaline**.

Sterile: not able to produce gametes (eggs or sperm); may or may not still produce gonadal hormones.

Steroid: hormone produced either by the gonads or adrenal glands.

Stockard's solution: clearing solution used to visualise stages of embryonic development.

Stocking density: numbers or biomass of fish expressed either per unit volume of tank or pond capacity or per unit volume in unit time of water inflow.

Sulphonamide: synthetic chemical/drug with antimicrobial activity used for the treatment of some infectious diseases (cf **Nitrofuran**).

Supersaturation: greater than normal solubility of a gas or other chemical (e.g. oxygen in water), often due to prior exposure of mixture to high pressure or low temperatures or both.

Suspended solid: solids maintained in suspension which make water cloudy or opaque; includes organic wastes, inorganic materials and dead animal and plant life.

Swim up: post-hatch stage of development when fry rise to the surface to feed, particularly used to refer to salmonid fry.

Synahorin: commerically available gonadotropin preparation; used in the artificial induction of spawning.

Synergism: the action of drugs or hormones which produce a combined effect greater than the sum of each acting alone.

Taint/off-flavour: muddy flavour sometimes imparted to fish grown in earth ponds; thought to be due to algae.

Testosterone: steroid hormone produced by the testis; along with ketotestosterone it is responsible for development of male secondary sex characters.

Thyroid gland: endocrine gland which consists of scattered follicles along the ventral side of the fish just behind the mouth.

Thyroxine: hormone produced by the thyroid gland; necessary for many body functions.

Trash fish: commercial fish catch which is not used as a food for man but is either fed directly to farmed fish or used in the production of fish meal.

Triploid: having three sets of chromosomes; the homozygotic sex (i.e. XXX) is sterile.

Vaccination: inoculation of the host with a vaccine (usually made from dead or attenuated virus or bacteria) which provokes an immune response in order to protect the host from subsequent disease due to the organism in question.

Ventricle: part of heart responsible for pumping the blood round the body.

Vertical incubator: container for incubating eggs where water enters at the bottom and flows upwards through the eggs.

Vertical transmission: infection transmitted from parent fish to progeny via infected eggs.

Virus: minute intracellular organisms, only visible under an electron microscope, which cause various fish diseases.

XX chromosomes: homozygous (or like) pair of sex chromosomes; produces in salmonids a female.

XY chromosomes: heterozygous (or unlike) pair of sex chromosomes; produces in salmonids a male.

Zooplankton: microscopic animal life often found at the surface of water and used to feed fish larvae.

Zoug/zouger jar: container for incubating eggs where water enters at the bottom and gently turns the eggs over maintaining them suspended in the waterflow; especially used for incubating cyprinid eggs.

CONVERSION FACTORS

TO CONVERT	MULTIPLY BY
Acres to hectares	0.4047
Centimetres to inches	0.3937
Cubic inches to cubic centimetres	16.39
Cubic centimetres to cubic inches	0.06102
Cubic feet to cubic metres	0.02832
Cubic metres to cubic feet	35.31
Cubic yards to cubic metres	0.7646
Cubic metres to cubic yards	1.308
Cubic inches to litres	0.01639
Feet to metres	0.3438
Gallons to litres	4.546
Grains to grams	0.0648
Grams to grains	15.43
Grams to ounces	0.03527
Grams to pounds	0.002205
Hectares to acres	2.471
Inches to centimetres	2.54
Kilometers to miles	0.6124
Kilograms to pounds	2.205
Kilograms to tons	0.0009842
Litres to cubic inches	61.03
Litres to gallons	0.22
Litres to US gallons	0.26
Metres to feet	3.281
Metres to yards	1.094
Miles to kilometres	1.609
Ounces to grams	28.35
Pounds to grams	453.6
Square inches to square centimetres	6.452
Square centimetres to square inches	0.155
Square metres to square feet	10.76
Square feet to square metres	0.0929
Square yards to square metres	0.8361
Square metres to square yards	1.196
Square miles to square kilometres	2.599
Square kilometres to square miles	0.3861
Tons to kilograms	1016
Yards to metres	0.9144

CONVERSION TABLES

Metric conversion

Inches		Millimetres		Miles		Kilometres		Feet		Metres		Sq Feet		Sq metres		Yards		Metres		Sq yards		Sq metres
0.039	1	25.400		0.621	1	1.609		3.281	1	0.305		10.764	1	0.093		1.094	1	0.914		1.196	1	0.836
0.079	2	50.800		1.243	2	3.219		6.562	2	0.610		21.528	2	0.186		2.187	2	1.829		2.392	2	1.672
0.118	3	76.200		1.864	3	4.828		9.843	3	0.914		32.292	3	0.279		3.281	3	2.743		3.588	3	2.508
0.157	4	101.600		2.485	4	6.437		13.123	4	1.219		43.056	4	0.372		4.374	4	3.658		4.784	4	3.345
0.197	5	127.000		3.107	5	8.047		16.404	5	1.524		53.819	5	0.465		5.468	5	4.572		5.980	5	4.181
0.236	6	152.400		3.728	6	9.656		19.685	6	1.829		64.583	6	0.557								
0.276	7	177.800		4.350	7	11.265		22.966	7	2.134		75.347	7	0.650								
0.315	8	203.200		4.971	8	12.875		26.247	8	2.438		86.111	8	0.743								
0.354	9	228.600		5.592	9	14.484		29.528	9	2.743		96.875	9	0.836								

Cu feet		Cu metres		Cu yards		Cu metres		Pints		Litres		Gallons*		Litres		Ounces		Grams		Pounds		Kilograms
35.315	1	0.028		1.308	1	0.765		1.761	1	0.568		0.220	1	4.546		0.035	1	28.350		2.205	1	0.454
70.629	2	0.057		2.616	2	1.529		3.521	2	1.136		0.440	2	9.092		0.071	2	56.699		4.409	2	0.907
105.943	3	0.085		3.924	3	2.294		5.282	3	1.704		0.660	3	13.638		0.106	3	85.049		6.614	3	1.361
141.258	4	0.113		5.232	4	3.058		7.043	4	2.272		0.880	4	18.184		0.141	4	113.398		8.819	4	1.814
176.572	5	0.142		6.540	5	3.823		8.804	5	2.840		1.101	5	22.730		0.176	5	141.748		11.023	5	2.268
211.887	6	0.170		7.848	6	4.587		10.564	6	3.408		1.321	6	27.276		0.212	6	170.097		13.226	6	2.722
247.201	7	0.198		9.156	7	5.352		12.325	7	3.976		1.541	7	31.822		0.247	7	198.447		15.432	7	3.175
282.516	8	0.227		10.464	8	6.117		14.086	8	4.544		1.761	8	36.368		0.282	8	226.796		17.637	8	3.629
317.830	9	0.255		11.772	9	6.881		15.847	9	5.112		1.981	9	40.914		0.317	9	255.146		19.842	9	4.082

*Imperial gallons.

Temperature conversion

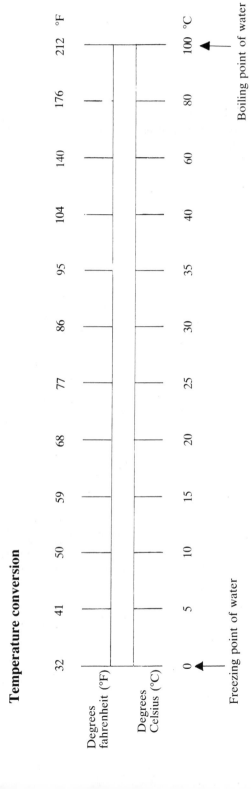

Index

See also Tables 1.1 (p. 8), 4.1 (p. 113) and 8.2 (p. 272−3) and Appendix II (p. 385−7) for lists of common and proper names of fish species and Table 6.9 (p. 237) for list of fish diseases.